14th European Workshop on Modern Developments and Applications in Microbeam Analysis (EMAS 2015 Workshop)

IOP Conference Series: Materials Science and Engineering
Volume 109

Portoroz, Slovenia
3 – 7 May 2015

Editors:

Xavier Llovet
Michael B. Matthews
Miran Ceh
Enrico Langer
Kristina Zagar

ISBN: 978-1-5108-2085-2
ISSN: 1757-8981

Printed from e-media with permission by:

Curran Associates, Inc.
57 Morehouse Lane
Red Hook, NY 12571

Some format issues inherent in the e-media version may also appear in this print version.

Printed by Curran Associates, Inc. (2016)

Institute of Physics
Dirac House, Temple Back
Bristol BS1 6BE UK

Phone: 44 1 17 929 7481
Fax: 44 1 17 920 0979

techtracking@iop.org

Additional copies of this publication are available from:

Curran Associates, Inc.
57 Morehouse Lane
Red Hook, NY 12571 USA
Phone: 845-758-0400
Fax: 845-758-2634
Email: curran@proceedings.com
Web: www.proceedings.com

Table of contents

Volume 109

14th European Workshop on Modern Developments and Applications in Microbeam Analysis (EMAS 2015 Workshop)
3–7 May 2015, Portorož, Slovenia

Accepted papers received: 22 December 2015
Published online: 9 February 2016

IOP Conf. Series: Materials Science and Engineering **109** (2016) 011001 doi:10.1088/1757-899X/109/1/011001

European Microbeam Analysis Society's 14th European Workshop on Modern Developments and Applications in Microbeam Analysis (EMAS 2015), Portorož, Slovenia, 3-7 May 2015

Xavier Llovet[1], **Michael B. Matthews**[2], **Miran Čeh**[3], **Enrico Langer**[4] **and Kristina Žagar**[3]

[1] Universitat de Barcelona, Scientific and Technological Centres, Lluís Solé i Sabarís 1-3, 08028 Barcelona, Spain

[2] AWE Plc., Aldermaston, Reading RG7 4PR, Great Britain

[3] Jožef Stefan Institute, Department for Nanostructured Materials, Centre for Electron Microscopy, Jamova cesta 39, 1000 Ljubljana, Slovenia

[4] Dresden University of Technology, Institute of Semiconductors and Microsystems, Nöthnitzer Strasse 64, 01187 Dresden, Germany

Preface

This volume of the IOP Conference Series: Materials Science and Engineering contains papers from the 14th Workshop of the European Microbeam Analysis Society (EMAS) on Modern Developments and Applications in Microbeam Analysis which took place from the 3rd to the 7th of May 2015 in the Grand Hotel Bernardin, Portorož, Slovenia.

The primary aim of this series of workshops is to assess the state-of-the-art and reliability of microbeam analysis techniques. The workshops also provide a forum where students and young scientists starting out on a career in microbeam analysis can meet and discuss with the established experts. The workshops have a unique format comprising invited plenary lectures by internationally recognized experts, poster presentations by the participants and round table discussions on the key topics led by specialists in the field.This workshop was organized in collaboration with the Jožef Stefan Institute and SDM – Slovene Society for Microscopy. The technical programme included the following topics: electron probe microanalysis, STEM and EELS, materials applications, cathodoluminescence and electron backscatter diffraction (EBSD), and their applications.

As at previous workshops there was also a special oral session for young scientists. The best presentation by a young scientist was awarded with an invitation to attend the 2016 Microscopy and Microanalysis meeting at Columbus, Ohio. The prize went to Shirin Kaboli, of the Department of Metals and Materials Engineering of McGill University (Montréal, Canada), for her talk entitled "*Electron channelling contrast reconstruction with electron backscattered diffraction*".

The continuing relevance of the EMAS workshops and the high regard in which they are held internationally can be seen from the fact that 71 posters from 16 countries were on display at the meeting and that the participants came from as far away as Japan, Canada, USA, and Australia. A selection of participants with posters was invited to give a short oral presentation of their work in three dedicated sessions. The prize for the best poster was an invitation to participate in the 24th Australian Conference on Microscopy and Microanalysis (ACMM 24) in Melbourne, Australia.

IOP Conf. Series: Materials Science and Engineering **109** (2016) 011001 doi:10.1088/1757-899X/109/1/011001

The prize was awarded to Aurélien Moy of the University of Montpellier (France) for his poster entitled: "*Standardless quantification of heavy metals by electron probe microanalysis*".

This proceedings volume contains the full texts of 9 of the invited plenary lectures and of 12 papers on related topics originating from the posters presented at the workshop. All the papers have been subjected to peer review by a least two referees.

November 2015

Participants of the 14th European Workshop on Modern Developments and Applications in Microbeam Analysis (EMAS 2015), held at Portorož, Slovenia, on 3-7 May 2015.

Acknowledgements

On behalf of the European Microbeam Analysis Society I would like to thank all the invited speakers, session chairs and members of the discussion panels for making the meeting such a great success. Special thanks go to Miran Čeh and Luc Van't dack who directed the organisation of the workshop giving freely of their time and talents. As was the case for previous workshops, the EMAS board *in corpore* was responsible for the scientific programme. The Workshop also included a commercial exhibition where many leading instrument suppliers were represented. Several companies that exhibited provided financial support, either by sponsoring an event or by advertising.

Below, in alphabetical order, is a list of exhibiting companies and sponsors of the workshop.

- Ametek GmbH, Edax Business Unit
- Bruker Nano GmbH
- Cameca SA
- European Institute for Transuranium Elements (Germany)
- IfG - Institute for Scientific Instruments
- Jeol (Europe) SAS
- Oxford Instruments NanoAnalysis Ltd.
- SCAN, d.o.o.
- Target-Messtechnik
- Thermo Fisher Scientific BV

Michael B. Matthews
EMAS President

IOP Conf. Series: Materials Science and Engineering **109** (2016) 011002 doi:10.1088/1757-899X/109/1/011002

Peer review statement

All papers published in this volume of *IOP Conference Series: Materials Science and Engineering* have been peer reviewed through processes administered by the proceedings Editors. Reviews were conducted by expert referees to the professional and scientific standards expected of a proceedings journal published by IOP Publishing.

Measurements of the quantitative lateral analytical resolution at sputtered gold-layers with the FEG-EPMA JEOL JXA-8530F

D Berger and J Nissen

Technische Universität Berlin, Zentraleinrichtung Elektronenmikroskopie (ZELMI), Strasse des 17. Juni 135, 10623 Berlin, Germany

E-mail: dirk.berger@tu-berlin.de, joerg.nissen@tu-berlin.de

Abstract. In this paper we present measurements of the quantitative lateral analytical resolution of the field emission (FEG) electron microprobe JEOL JXA-8530F. In particular, the minimum thicknesses of gold layers that are suitable for an accurate quantitative analysis are determined for several measurement parameters. Both, low voltage analyses at acceleration voltages from 15 kV to 5 kV and low overvoltage analyses at 13 and 15 kV are regarded. For both methods a minimum analytical resolution well below 170 nm is obtained while the X-ray signal is 5 times lower for the low overvoltage method. Special care is taken for the preparation of the test specimen. A Ag-layer reduces fluorescence effects and the cleaved surfaces are polished by a gentle focussed ion beam. The porosity of the gold layers is regarded by a separate measurement of the density (17.6 g/cm^3). The results coincide well with Monte Carlo simulations of the source volume of generated X-rays if the reduced density of the gold layers and experimental determined values of the beam diameter are used. Additionally the results are used to investigate methods to estimate the width of the source volume of emitted X-rays by the superposition of the beam diameter and theoretical predictions of the width of the electron interaction volume. The linear sum of both gives a quite good agreement.

1. Introduction

Many attempts have already been made to determine and increase the lateral analytical resolution of electron probe microanalysers, this includes both the scanning electron microscopes (SEM) with energy-dispersive X-ray spectrometers (EDS) and the microprobes (EPMA) with wavelength-dispersive X-ray spectrometers (WDS). To review the obtained results, the following two criteria have to be regarded.

Firstly, it is necessary to distinguish between the qualitative and the quantitative lateral analytical resolution. While the first describes the size of the smallest particles or features that are recognized, the latter refers to the ability of an accurate quantification. In the latter case the particles or features have to be equal to or larger compared to the source volume of the emitted X-rays taking absorption and fluorescence effects into account. As pointed out by several authors it is additionally necessary to keep in mind that the size of this source volume is influenced not only by the interaction volume of the electrons inside the sample but also by the beam diameter of the electron probe [1-5]. The left part of figure 1 illustrates the interaction volume of electrons if a point-shaped probe is used. The right part of figure 1 shows the spread of the interaction volume by an increased probe diameter. The source volume of emitted X-rays changes accordingly. Actual Monte Carlo (MC) simulation programmes are able to consider the finite beam diameter [6]. Merlet and Llovet [1] proposed to consider this effect in theoretical treatments by the mathematical convolution of the beam diameter d_0 and the classical range of X-ray generation R_x [19] reduced by depth of the maximum of the X-ray depth distribution Z_m. Then the minimum analytical diameter R_S is given by:

$$R_S = \sqrt{4 \cdot (R_x - Z_m)^2 + d_0^{\,2}} \tag{1}$$

in which R_x is calculated by the density ρ, the atomic number Z and atomic weight A, the primary electron energy E_0 and the critical edge energy E_i:

$$R_X = \frac{33A}{\rho \cdot Z}\left(E_0^{1.7} - E_i^{1.7}\right) \tag{2}$$

Both, MC simulations and theoretical predictions still have to be proven by measurements.

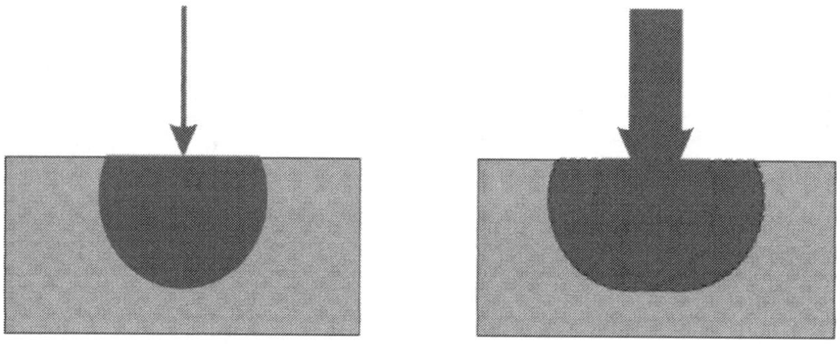

Figure 1. Simplified representation of the interaction volume of the electrons (blue) and its increase (red) with beam diameter (green).

Secondly, there are two measurement strategies to increase the lateral resolution, the low overvoltage and the low voltage analysis [7, 8]. The low overvoltage method implies the use of an electron energy close to the critical ionisation energy of the chosen X-ray line. In most cases this will be energies of several keV for which the lens aberrations are small and hence the diameter of the electron probe is small and negligible in comparison to the interaction volume. The advantage of this method is that the electrons having enough energy to ionise the considered shell have undergone few scattering processes only. Therefore, the interaction volume of these electrons is small compared to the overall interaction volume for complete stopping of electrons. Then the source volumes of generated and emitted X-rays are small as well.

For the low overvoltage analyses, few values for the lateral analytical resolution have been published only and even then solely for the quantitative analysis. McSwiggen [7] estimated from MC simulations values of 200 nm and 150 nm for Fe-Kα and Ni-Kα respectively in a Fe50Ni50 compound. Berger and Nissen [4] have measured a resolution of 200 nm for Au-Lα at an Au-Ag-interface.

The low voltage analysis implies that the interaction volume of the electrons inside the sample is reduced by decreasing the electron beam energy well below 7 keV. As mentioned, the beam diameter has to be taken into account. It has been proven that the low voltage method is unusable for thermionic electron sources since the diameter of their electron beam often exceeds the interaction volume of the electrons [24]. Therefore, a field emission electron source is required for the low voltage analyses.

For the low voltage analyses many authors have published values for the obtained analytical resolution. The results for the qualitative lateral analytical resolution (table 1) range between 12 nm for W at 3 keV and 390 nm for Fe at 6 keV. Some authors use the recognition of small particles or layers as a measure for the qualitative resolution while others analyze the increase of the X-ray intensity across interfaces with an intensity criterion of 16% - 64% or 10% - 90%. In recent years the examination has been extended to specific problems that occur for the quantification in the low voltage method [3] especially due to the application of L-lines [9]. The few published values of the quantitative resolution (table 2) range between 115 nm for Si and 300 nm for Ti.

Table 1. Published values for the lateral resolution for qualitative X-ray microanalysis with the low voltage method.

Element	Matrix	electron energy [keV]	qualitative lateral analytical resolution [nm]	source
Au	Ni-Au-Cu interface	5	70	[10]
Ni-layer semiconductor-layer		3 3	31 25	[11]
W-layer		3	12	[12]
Cu_3Sn Nb, C Ag, Cu	 steel Sn	6 7 8	100 100 100	[13]
Ti Fe		6 6	290 390	[1]
cementite	steel	?	100	[14]
olivine		7	250	[15]
orthopyroxene		5	350	[16]
Au Si	Si-Au interface	5 5	26 45	[4]

Table 2. Published values for the lateral resolution for quantitative X-ray microanalysis with the low voltage method.

Element	Electron energy [keV]	Quantitative lateral analytical resolution [nm]	source
Cu_3Sn	6	300	[13]
Ag_3Sn		250	[17]
Pd_2Ga	5	750	[18]
Titanite MgO	7 7	300 300	[15]
Cr	5	250 (Simulation)	[3]
Ni Au Si	5 5 5	200 120 115	[4]

Unfortunately, these published numbers do not give a satisfactory overview on the optimum quantitative resolution of a FEG-EPMA. Therefore, we started a systematic experimental determination of the quantitative lateral analytical resolution for different elements and different parameters, from which the first results for gold layers are presented in this paper. The resolution is determined by a standard-based quantitative point analysis in the layer. If the layer thickness is larger compared to the width of the source volume of emitted X-rays, then the sum of quantification is within (100.0 ± 0.5) %. In the next step the size of the layer is decreased until the sum is smaller than 100 %.

It is important to take absorption effects of the generated X-rays into account. Figure 2 shows Monte Carlo simulations (Casino v2.48 [6]) of the radial X-ray distribution with and without absorption. In the case of a primary electron energy of 5 keV the intensity of emitted X-rays is reduced at a maximum by 3 % with respect to the generated X-rays. For 15 keV the corresponding reduction is about 7 %. Therefore it is supposed that the difference of the widths of the source volumes of generated and emitted X-rays is negligible in our measurements.

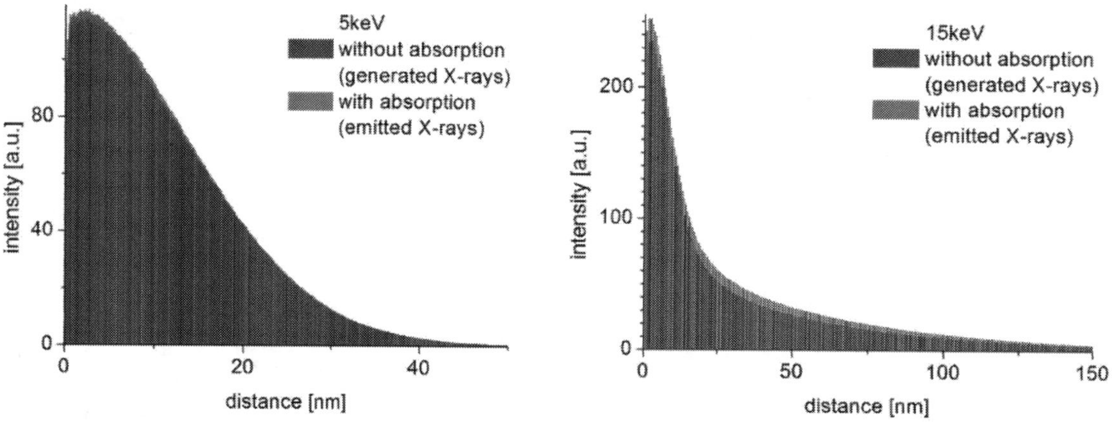

Figure 2. Monte Carlo simulations (Casino v2.48 [6]) of the radial X-ray distribution of Au-Mα with (red) and without (blue) absorption for an Au specimen excited with 5 keV (left) and 15 keV (right) electrons.

The focus of our investigations lies on the achievable resolution for an accurate quantification. All questions concerning signal intensity, precision and sensitivity are not regarded as a start. The experimental results are then compared to MC simulations and to predictions of theoretical considerations based on the ionisation range [1, 19, 20].

2. Experimental and results

2.1. Deposition and preparation of the gold layers

For the experimental determination of the quantitative analytical lateral resolution under different measurement conditions, test specimens were generated by sputtering gold layers with different thicknesses (1,000, 450, 335, 240, 200, 170 and 140 nm) on SIGRADUR glassy carbon substrates. The cross-sectional preparation of the specimen was done first by cleaving and then by polishing with a focussed Ga-ion beam (FEI Helios Nano Lab600 FIB). A gentle ion beam (30 kV, 90 pA) is used to avoid the deposition of Ga into the Au layer. The cross-section polishing process requires a protection layer consisting of Ga, Pt and C which is apparent on top of the specimen. To avoid fluorescence effects in the protection layer, all specimens were first covered with an intermediate Ag layer. Since the absorption coefficients of the considered Au-lines are nearly the same in Au and Ag and much smaller in C (substrate) it is supposed that X-ray absorption effects in the adjacent layers can be neglected in our case.

Figure 3 shows high resolution secondary electron (SE) images of the cross-section of the 6 test specimens with the thinnest Au layer thicknesses. To take the obvious porosity of the sputtered layers into account for the quantification, we use a 1,000 nm sputtered Au layer with the same porosity as a reference standard. Backscattered electron (BSE) images were made as well to verify that there is no significant diffusion of Ag into the Au grains. Since the absorption of the Au-lines in Au and Ag is nearly the same, the resolution measurements are not affected by the Ag condensation on the walls of the Au pores.

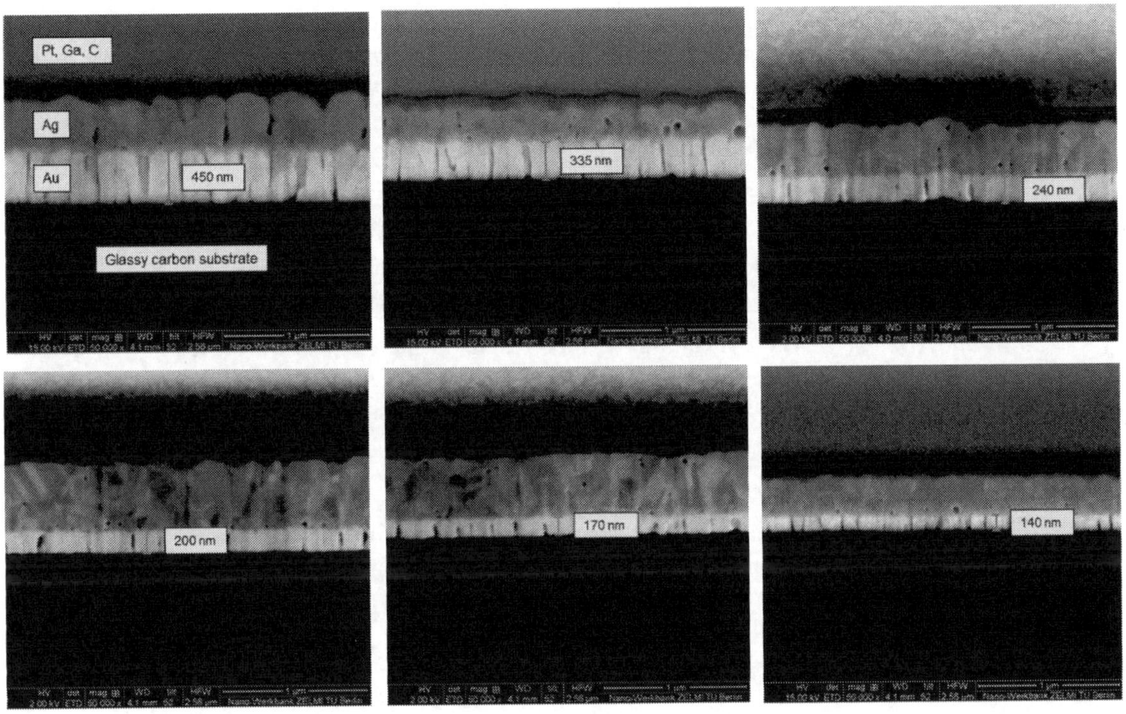

Figure 3. Cross-section images of the test specimens with different gold layers from 450 to 140 nm thickness.

For MC simulations the porosity is considered by a reduced density compared to that of bulk Au. The density of the layers is measured with the weighing method [21]. A Si wafer with size of 1 cm^2 is covered with an Au sputter layer whose thickness is measured to be 800 nm. The increase of the mass is determined with a precise balance and then the density is calculated to be (17.6 ± 1.1) g/cm^3.

2.2. Measurements of beam diameter

As already mentioned and as illustrated in figure 1, the diameter of the electron probe has a significant influence on the lateral resolution of the quantification. Since all measurements are being made with the FEG-EPMA JEOL JXA-8530F, the beam diameters for different beam energies and beam currents are well known since they have been accurately measured. The results obtained from the derivative (DR) method specified in the standard ISO/TS 24597 [4] coincide well with those obtained by Fourier transform [3]. We use the beam diameter values from [4] that are derived from the average about 30,000 edge profiles of SE-images of gold insulars. To consider nearly all electrons for the quantitative analytical resolution the edge profiles are analysed using the 1% - 99% criterion. The uncertainty of the beam diameters is expected to be less than 5 %. The relevant beam diameters for the measuring parameters of this paper are summarized in table 3. The beam currents are chosen from a practical point of view to get a sufficient WDS-signal. As a rule thumb it can be said that the beam diameter is reduced by 30 % if the current is reduced by a factor of 10. In most cases the resulting improvement of the analytical resolution is negligible since the electron range is much larger. But the reduced intensity prevents an accurate quantification.

Table 3. Measured beam diameter of the JEOL JXA-8530F [4].

electron energy / beam current	beam diameter d_0 [nm]
3 keV / 1 nA	52
5 keV / 1 nA	38
8 keV / 1 nA	31
15 keV / 5 nA	29

2.3 Quantitative measurements

2.3.1. Low voltage method. Table 4 summarizes the results of the element quantification of the gold layers with the low voltage method using the Au-Mα-WDS-signal (2.12 keV, PET-crystal) at 15, 8 and 5 keV electron energy. Measurements were made in the "Spot Mode" using probe currents according to table 3 with 30 s measurement time for the peak and 10 s for the background. All measurements of table 4 and table 5 are repeated 5 to 12 times.

Using the 1,000 nm Au layer, no signal from the surrounding carbon and silver is detected even at 15 keV, therefore, it is applied as reference standard for the quantification (indicated by 100.00 in tables 4 and 5). A quantification result of (100.0 ± 0.5) % indicates that the source volume of the emitted X-rays is completely inside the Au-layer. For 15 keV solely the 1 μm layer can be resolved quantitatively, whereas for 8 keV a minimum layer thickness of 335 nm and for 5 keV a minimum layer thickness of even 170 nm is suitable for reliable quantitative analysis. Measurements with 3 keV were not successful due to the low intensity of the WDS-signal at 1 nA. Moreover, the accurate positioning of the electron beam is impossible due to the poor quality of SE-images (beam diameter about 50 nm at 3keV).

2.3.2. Low overvoltage method. The results of the low overvoltage quantitative measurements of the gold layers are summarized in table 5. Here, the Au-Lα X-ray line (9.71 keV, LiF-crystal) was used at 15 and 13 keV. All other parameters are the same. For 15 keV an accurate quantification is possible down to 335 nm layer thickness.

To optimize the lateral resolution in the low overvoltage condition, the electron energy was reduced to 13 keV, 1.08 keV more than the critical ionisation energy of Au Lα at 11.92 keV (overvoltage: 1.09). Indeed, now the 170 nm layer could be resolved quantitatively, comparable to the low voltage method. However, the X-ray intensity of Au-Lα reduces by a factor of 5, if the primary electron energy is decreased from 15 keV to 13 keV.

Table 4. Results of the quantitative measurements of the gold layers in wt% (low voltage method). The quantitative analytical lateral resolution is measured to be between 140 and 170 nm for 5 keV and between 240 and 335 nm for 8 keV.

Au layer thickness [nm]	15 keV (Au-Mα)	8 keV (Au-Mα)	5 keV (Au-Mα)
1,000 (Reference)	100.00	100.00	100.00
450	92.04	99.93	99.89
335	/	99.81	99.82
240	/	88.08	99.48
200	/	/	99.70
170	/	/	99.71
140	/	/	91.04

Table 5. Results of the quantitative measurements of the gold layers in wt% (low overvoltage method). The quantitative analytical lateral resolution is measured to be between 140 and 170 nm for an overvoltage of 1.09.

Au layer thickness [nm]	15 keV (Au-Lα)	13 keV (Au-Lα)
1,000 (Reference)	100.00	100.00
450	99.95	99.88
335	99.76	99.83
240	93.34	99.86
200	/	100.70
170	/	100.18
140	/	97.93

3. Discussion

3.1. MC simulations

3.1.1. MC simulations of the low voltage method. To estimate the source volume of emitted X-rays inside the gold, MC simulations of the interaction volume of the electrons inside the specimen were accomplished with 15, 8, 5 and 3 keV electron energy (figure 4), using the programme Casino (v2.48) [6]. The MC simulations of the electron paths in the sample are based on the reduced density of sputtered gold which was measured to be about 17.6 g/cm^3. The uncertainty of about 6 % of the density measurements adds to the uncertainty of the MC simulations. All MC simulations incorporate the measured beam diameters. Therefore, the simulated interaction volume is expected to be a suitable estimation of the source volume of emitted X-rays and to the quantitative lateral analytical resolution. From MC simulations with and without absorption (figure 2) it has been proven that the absorption of the Au-lines in Au is negligible in the context of this work.

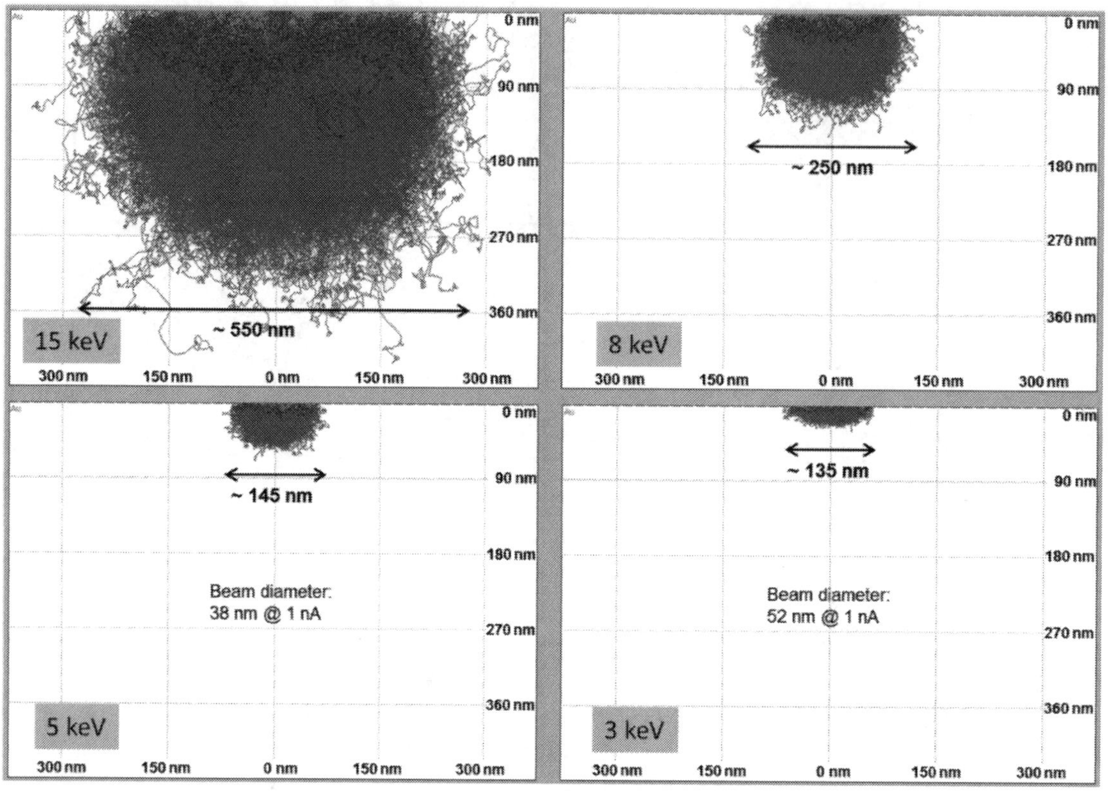

Figure 4. Monte Carlo simulations (Casino v2.48 [6]) of the electron scattering volume in gold with 15, 8, 5 and 3 keV electron energy for the low voltage method, the diagrams show electron paths (backscattered electrons in red, all others in blue).

Following the simulations, the lateral diameter of the interaction volume for 99 % of electrons decreases from about 550 nm at 15 keV to about 145 nm at 5 keV. A significant further decrease is not expected when the electron energy is further decreased to 3 keV, due to the increase of the beam diameter from 38 nm at 5 keV to 52 nm at 3 keV. At 5 keV, a decrease of the beam current by a factor of 10 would decrease the beam diameter by about 30 % only. Therefore, the source volume would not change much but the x-ray intensity would decrease at the same time to an impractical value.

The simulations are in good agreement with the experimental data as can be seen from comparison with table 4.

3.1.2. MC simulations of the source volume of generated X-ray lines in the low overvoltage method. The MC simulations in section 3.1.1 consider a complete stopping of the electrons inside the specimen. Therefore, these simulations are not suitable for the low overvoltage method for which the electron range of electrons with a certain energy loss matters only. This effect is considered by the MC programme PHI-RO-Z MONTE CARLO SIMULATION (v2.28) [22] that simulates the source volume of generated X-rays directly with respect to the energy loss of electron inside the sample. For our samples, the source volume of emitted X-rays is expected to be quite similar to the source volume of generated X-rays. This assumption is proved by the MC simulations of the radial X-ray distribution of emitted X-rays with and without absorption in figure 2. For our specimen and our measuring conditions the X-rays emission with absorption is at most 7 % lower.

Figure 5 shows the results for Au-Mα (2.12 keV) and Au-Lα (9.71 keV) X-rays. The electron energy is 15 keV (left) and 13 keV (right) respectively. The lateral diameter of the source volume for 99 % of X-rays decreases from about 500 nm for the Au Mα-line at 15 keV to about 140 nm for the Au Lα-line at 13 keV which is only 1.08 keV over the critical ionisation energy of Au-Lα (11.92 keV), resulting in an overvoltage of 1.09.

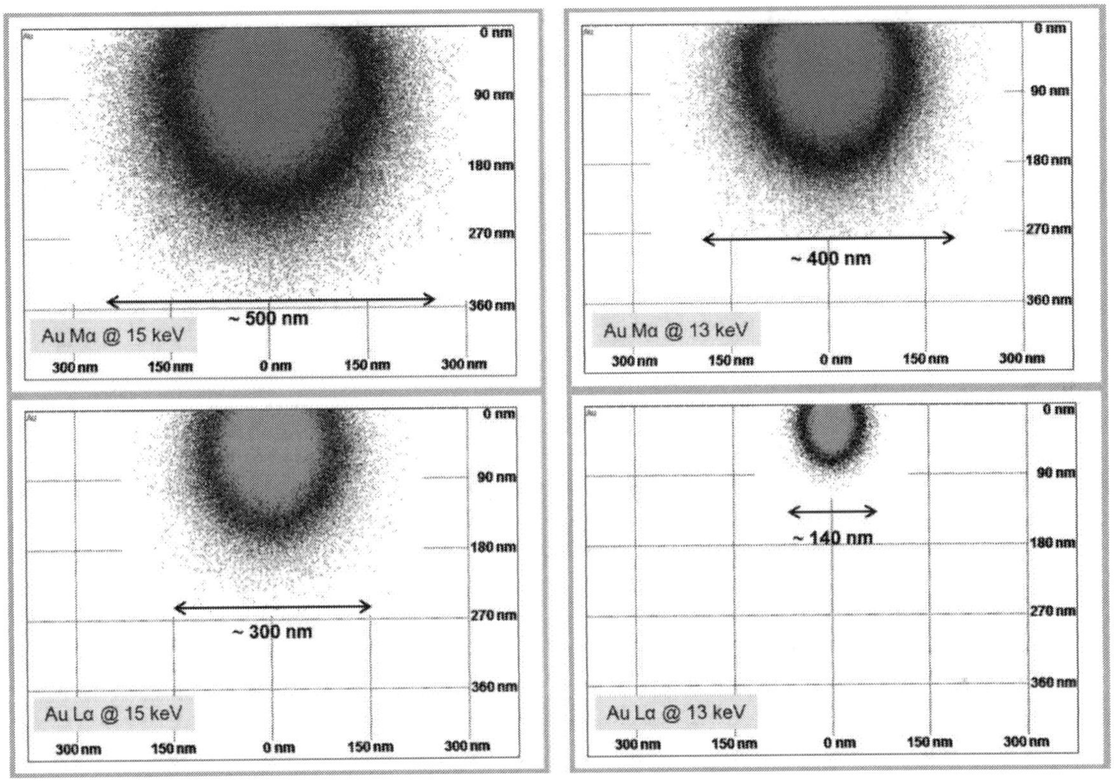

Figure 5. Monte Carlo simulations (Phi-Ro-Z Monte Carlo Simulation v2.28) of the source volume of generated Au-Mα and Au-Lα X-ray lines at 15 and 13 keV electron energy.

Unfortunately, the beam diameter of 29 nm is not included in these simulations. But even without it, the simulations coincide very well with the measurements as well. Not only low voltage but also low overvoltage result in a significant enhancement of the lateral resolution.

3.2. Comparison to X-ray range calculations

Instead of simulating the individual electron paths inside the specimen, the effective ionisation range of electrons R can be calculated [see paper by Fournelle *et al.* in this proceedings]. The most advanced formulae by Merlet and Llovet [1] consider the depth of the maximum of the X-ray depth distribution Z_m and the beam diameter d_0 [1], see equation (1).

The value of Z_m might be derived from MC simulations but are calculated by a formula from Merlet [23] here. The results from eq. (1) relevant to our measurements are summarized in table 6. The values of the range of X-ray generation R_x have to be multiplied by a factor of 2 to compare to the resolution measurements from section 2.3. The factor of 2 is chosen because the range of X-ray generation R_x describes the depth of the ionisation volume inside the sample and the width is expected to be the doubled value in Au, see figure 4.

Table 6. Calculated and measured values for the width of the ionisation volume and the lateral analytical resolution.

Case	Method	Z_m from [23] [nm]	2 R_X [nm] Eq. (2)	R_S [nm] Eq. (1)	Exp. Result [nm]	d_W [nm] Eq. (3)	d_i [nm] Eq. (4)
Au-Mα 5 keV	low voltage	6.9	108.4	101.8	140-170	92.2	146
Au-Mα 8 keV	low voltage	15.8	284.6	255	240-335	173.3	315
Au-Mα 15 keV	conventional	41.0	897.6	816	450-1,000	477.8	920
Au-Lα 15 keV	low overvoltage	7.6	302.0	287	240-335	180.0	324
Au-Lα 13 keV	low overvoltage	1.0	100.4	101.6	140-170	79.2	125

Obviously, eqs. (1) and (2) calculate rather too small values for the high resolution conditions Au-Mα 5 keV and Au-Lα 13 keV. The discrepancy could be caused by wrong Z_m values or due to fact that equation (2) has been derived for higher energies. Therefore, we will extend our measurements to layers consisting of further elements. The results will help to find the reasons for the discrepancies and to improve the formulae.

Willich and Bethke [24] proposed to calculate the lateral analytical resolution d_W by the sum of the range of X-ray generation R_x and the diameter width d_0 of the electron beam:

$$d_W = R_x + d_0 \tag{3}$$

Unfortunately the obtained values are not in good agreement with the measured resolution probably because eq. (3) assumes that the width of the source volume of emitted X-rays is equal to the depth. But from the MC simulations in figure 4 it can be seen that the width of the electron scattering volume is about two times that of its depth. Therefore another attempt to calculate the lateral analytical resolution d_i is proposed that linearly adds the width d_0 of the electron beam and the estimated width 2 R_x of the source volume of emitted X-rays:

$$d_i = 2 \cdot R_x + d_0 \tag{4}$$

From table 6 it can be seen that the values of (4) coincide well with the measurements, because this empirical formula considers better the shape of the interaction volume. Barkshire *et al.* [10] have observed and described this effect as well, finding a proportionality factor of 1.7 for Au, which obviously fits the measurements a little bit worse. However, eq. (4) will not be suitable for other elements than Au since the shape of the interaction volume depends on the density and probably on the primary energy. As soon as more measurements of the quantitative lateral analytical resolution of other elements are available, equation (4) has to be further developed.

4. Conclusions

In this paper it is experimentally proved that Au layers with thicknesses down to 170 nm are suitable for an accurate quantification in the EPMA. Therefore, the quantitative lateral analytical resolution is determined to be 170 nm for Au. This resolution is obtained with the low voltage method using 5 kV and the Au Mα-line as well as with the low overvoltage method using 13 kV and the Au Lα-line. The low voltage method requires an EPMA with a field emission source since the resolution is rather limited by the beam diameter than by the interaction volume of the electrons. A further decrease of the resolution for the low voltage method seems to be possible if a 3 kV-probe with a beam current well above 1 nA and a diameter well below 50 nm is available. For the low overvoltage methods no further improvements are to be expected since the resolution is independent from the probe properties and a further decrease of the overvoltage ratio decreases the X-ray intensity too strong.

For both methods the results of the quantitative lateral analytical resolution are in good agreement with MC simulations of the source volume of emitted X-rays. For the low voltage method it is mandatory to include the beam diameter in the simulation. Moreover, the achievable resolution can be estimated from measurements of the beam diameter and theoretical predictions of the range of X-ray generation. At least for Au, the linear sum of beam diameter and the doubled range of X-ray generation coincides well with the measured resolutions both in the low voltage and the low overvoltage method.

In the next step, the measurements of the quantitative lateral analytical resolution will be extended to further elements. Then experimental investigations of the quantitative depth-resolution in the EPMA will be of interest.

Acknowledgements

Financial support by the Europäische Fonds für Regionale Entwicklung (EFRE) within the project "Nano-Analytiklabor" is gratefully acknowledged. We thank Xavier Llovet for fruitful discussion and the support in applying eq. (1).

References

[1] Merlet C and Llovet X 2012 *IOP Conf. Ser.: Mater. Sci. Engng.* **32** 012016
[2] Berger D and Nissen J 2012 *JEOL User-Meeting* (Göttingen, Germany)
[3] Pinard P T and Richter S 2014 *IOP Conf. Ser.: Mater. Sci. Engng.* **55** 012016
[4] Berger D and Nissen J 2014 *IOP Conf. Ser.: Mater. Sci. Engng.* **55** 012002
[5] Hombourger C and Outrequin M 2013 *Microscopy Today* **21** 10
[6] Drouin D, Couture A R, Joly D, Tastet X, Aimez V and Gauvin R 2007 *Scanning* **29** 92
[7] McSwiggen P 2014 *IOP Conf. Ser.: Mater. Sci. Engng.* **55** 012009
[8] Newbury D 2002 *J. Res. Nat. Inst. Stand.* **107** 605
[9] Statham P and Holland J 2014 *IOP Conf. Ser.: Mater. Sci. Engng.* **55** 012017
[10] Barkshire I, Karduck P, Rehbach W P and Richter S 2000 *Mikrochim. Acta* **132** 113
[11] Sakurada T, Hashimoto S, Tsuchiya Y, Tachibana S, Suzuki M and Shimizu K 2005 *J. Surf. Anal.* **12** 118
[12] Hashimoto S, Sakurada T and Suzuki M 2008 *J. Surf. Anal.* **14** 428
[13] Mori N 2010 *JEOL News* **45** 42
[14] Bin W, Zhenyu L, Xiaoguang Z, Guodong W and Misra R D K 2013 *Mater. Sci. Engng. A* **575** 189
[15] Armstrong J T, McSwiggen P and Nielsen C 2013 *Microscopy and Analysis* **27** 17
[16] Saunders K, Buse B, Kilbrun M R, Kearns S and Blundy J 2014 *Chem. Geol.* **364** 20
[17] McSwiggen P, Mori N, Takakura M and Nielsen C 2012 *Microsc. Microanal.* **17** Supp. S2 624
[18] Müller D, Schwerin J, Gille P and Fehr K T 2014 *IOP Conf. Ser.: Mater. Sci. Engng.* **55** 01201
[19] Castaing R 1960 *Adv. Electron. El. Phys.* **13** 317
[20] Rinaldi R and Llovet X 2015 *Microsc. Microanal.* **21** 1053

[21] Berger D and Galbert F 2011 *Conf. Proceedings MC2011* (Kiel, Germany) ISBN 978-3-00-033910-3 IM7.P208
[22] Joy D 1995 *Monte Carlo Modeling for Electron Microscopy and Microanalysis* (Oxford: Oxford University Press)
[23] Merlet C 1994 *Mikrochim. Acta* **114/115** 363
[24] Willich P and Bethke R 1996 *Mikrochim Acta* **Suppl. 13** 631

A complementary approach to estimate the internal pressure of fission gas bubbles by SEM-SIMS-EPMA in irradiated nuclear fuels

C Cagna[1], I Zacharie-Aubrun[1], P Bienvenu[1], L Barrallier[2], B Michel[1] and J Noirot[1]

[1] CEA Cadarache, DEN, DEC, 13108 Saint-Paul-lez-Durance, France
[2] Arts et Métiers ParisTech, Laboratory MSMP, 2 Cours des Arts et Métiers, 13617 Aix-en-Provence, France

E-mail: celine.cagna@cea.fr

Abstract. The behaviour of gases produced by fission is of great importance for nuclear fuel in operation. Within this context, a decade ago, a general method for the characterisation of the fission gas including gas bubbles in an irradiated UO_2 nuclear fuel was developed and applied to determine the bubbles internal pressure. The method consists in the determination of the pressure, over a large population of bubbles, using three techniques: SEM, EPMA and SIMS. In this paper, a complementary approach using the information given by the same techniques is performed on an isolated bubble under the surface and is aiming for a better accuracy compared to the more general measurement of gas content. SEM and EPMA enable the detection of a bubble filled with xenon under the surface. SIMS enables the detection of the gas filling the bubble. The quantification is achieved using the EPMA data as reference at positions where no or nearly no bubbles are detected.

1. Introduction

The behaviour of rare gases produced by fission is of great importance for nuclear fuel in operation. It controls the release in the free volume and hence the internal pressure of the fuel rod, the fuel swelling and also part of the behaviour of the fuel during off-normal situations or accidental events. During the operation of a Pressurized Water Reactor (PWR), the fuel is subjected to different loads - mainly mechanical and thermal - that modify its behaviour when combined with the effects of irradiation. The irradiation generates fission gases (Xe and Kr) in the fuel pellets, which are generally found in bubble form, from nanometre to micrometre diameters. Fission gases behaviour in oxides fuels has been the focus of attention for decades because of their impact on the overall behaviour of the fuel rod. During accidents such as RIA (Reactivity Initiated Accident) or LOCA (Loss of Coolant Accident), increasing temperature in the pellets leads to grain boundary fracturing, release of intergranular gases and rod over-pressurisation. The internal pressure of the bubbles can be high and thus generate internal stress in the fuel. Evaluating this pressure is a key matter especially for nuclear safety aspects.

A method has been developed in CEA to improve the description of the fission gas behaviour [1-3]. This method consists in a collection of complementary experimental results. These measurements are supporting the fission gas model MARGARET [4].

It has been shown that the characterisation of irradiated samples with SIMS (secondary ion mass spectrometry) is possible, and that xenon can be measured on irradiated UO_2 fuel [1-3]. Xenon filling the bubbles is detected as peaks measured during a depth profile [3-5]. It has been demonstrated that the total xenon inventory can be quantitatively measured from a large number of detected peaks and the pressure inside the bubbles deduced [6, 7]. The statistical analysis of the gas quantity inside bubbles can be compared to the statistical characterisation of porosity obtained with SEM (scanning electron microscopy) in order to get a mean atomic volume of xenon inside the bubbles.

In this paper, a complementary methodology is proposed to evaluate the internal pressure of xenon in fission gas bubbles in an irradiated nuclear fuel pellet using three microanalysis techniques: SEM/SIMS/EPMA (electron probe microanalyser). In order to get more refined results and a comprehensive gas inventory of the nuclear fuel, it is no longer based on statistical method from many bubbles but on a single bubble analysis. This work was initiated [8] but the main difficulty was to isolate one bubble.

This breakthrough in the description of fission gas inventory will enrich the data base used for code validation and is expected to lead to significant improvement in the understanding of fission gas behaviour.

2. Experimental

2.1. Sample

The sample used for this study was a large grain UO_2, with an initial ^{235}U enrichment of 4.89 %. It was taken from a French PWR fuel rod with zircaloy cladding, irradiated for 2 cycles and ramped at 470 $W.cm^{-1}$ during 12 h. The local burn-up at the sampling position was calculated to be 38.8 GWd/t_M. The sample was embedded in a metallic alloy with a low melting point to ensure good electrical conductivity and polished to get a plane surface.

2.2. Equipment

The same sample was examined by shielded SEM, EPMA and SIMS for radioactive material characterisation.

SEM was performed with a shielded XL 30 model (PHILIPS) with a Centaurus KE developments BSE (backscattered electron) and SE (secondary electron) detectors and a SIS (Soft Imaging Solutions) ADDA image acquisition device to obtain 4096 x 4096 pixels images covering large areas with high definition.

EPMA was performed with a shielded CAMEBAX model (CAMECA). Xenon was analysed using the Lα–line of the characteristic X-ray with a PET crystal. Xenon radial distribution and mapping were measured at an electron acceleration potential of 30 kV and an incident beam current of 250 nA. The acquisition time was 20 s. Quantitative analysis of xenon was carried out using an UO_2 (irradiated at low temperature to 25 GWd/t_U) standard, multi-characterized [9].

SIMS was performed with a shielded IMS 6f (CAMECA). The ^{132}Xe depth profile was measured with a 20 nA oxygen primary beam focussed on a 40 μm diameter area. ^{132}Xe signal was measured during 0.2 s periods. On irradiated samples, the xenon signal measured with SIMS must be separated from the molecular interferences produced by fission products, especially from $^{100}Mo^{16}O_2^+$. The interferences were avoided thanks to an energy filter (offset + 50 V): the ions energy depends on the location at which the atoms or molecules are ionized. The xenon ions are created over the sample surface while most of the other ions are created at the surface, allowing us to neglect the local environment effect on the measurement of the gas out of the bubble.

2.3. Procedure

2.3.1. Selection of the region of interest. The first step was to localize an area where bubbles were large and where the bubble density was low enough to isolate a single bubble under the surface. A global view of the sample surface and an image of the pellet radius are shown in figure 1. In this paper, the positions of the examination points are given relatively to the radius of the pellet: 1R is the periphery of the pellet, 0R is its centre.

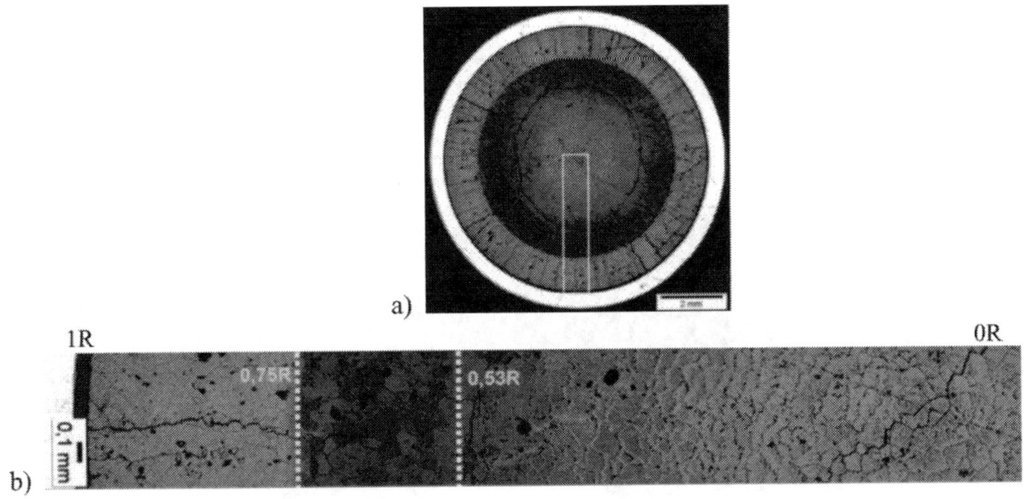

Figure 1. Macrograph of the polished cross-section sample (a) and detail along a radius (b).

The global methodology [4] for the determination of the bubbles internal pressure was carried out on this sample. Xenon was measured with SIMS by making depth profiles regularly spaced along the same radius that was analysed by EPMA. The xenon signal was measured as a function of time while a 80 μm diameter area was sputtered with a 60 nA O_2^+ primary current during 5,000 s (figure 1).

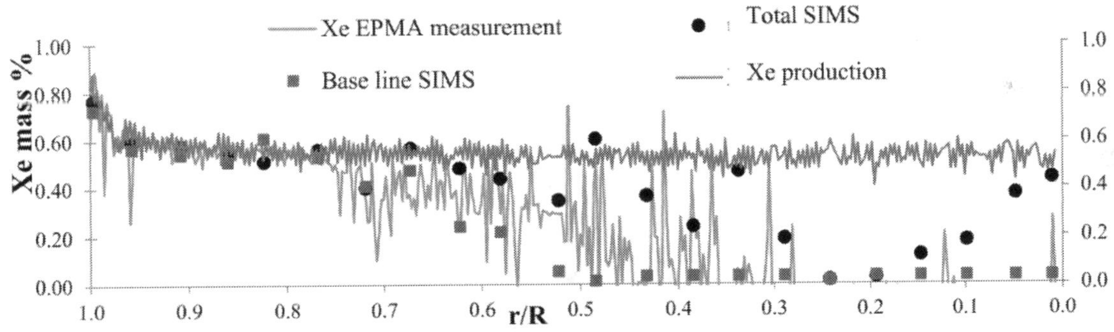

Figure 2. Comparison between the EPMA xenon measurement and the SIMS baseline (R: pellet radius, r: radial position).

The three techniques, SEM, EPMA and SIMS provide a consistent and complementary description of the gas inventory. Between 0.8R and 0.9R, in the bulk of the grains, xenon is either dissolved or precipitated in too small bubbles to be evidenced with SIMS or EPMA; a relative lack of xenon around the grain boundaries and scarce bubbles are observed. At the centre, the total xenon concentration is lower than the production.

In the central area, mean pressure was deduced from the global methodology and estimated at 9.1 MPa and bubble diameters were around 1.2 μm.

To implement the new approach on a single bubble, the centre is a suitable area (see figure 3). The large grain size is interesting for our studies in particular compared to the beam diameter of the SIMS that is around 40 - 50 μm. Figure 3b shows grain boundaries with lenticular bubbles (green) and grains with spherical intragranular bubbles (red) with an average equivalent diameter between 1 and 2 µm. Their density (108 mm^{-3}) is well suited to allow counting a small number of bubbles. The area selected for the study, figure 3b, is situated inside a grain and consequently only concerns intragranular bubbles.

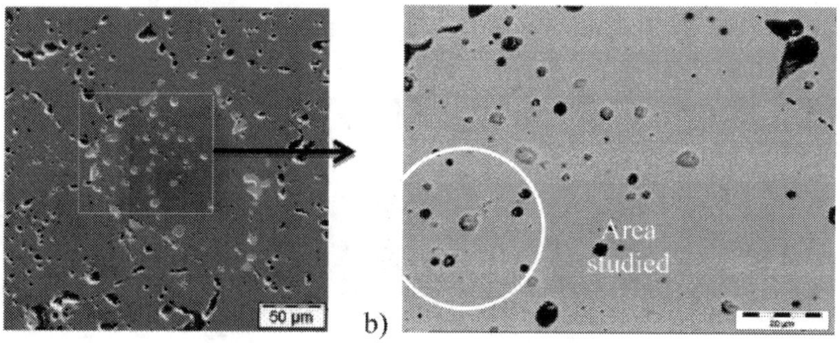

Figure 3. SEM images of a) the central zone, and b) grain selected for the study (BSE - 25 kV).

2.3.2. Localisation of bubbles under the surface. The localisation of bubbles just under the surface was carried out by both SEM and EPMA observations. Figures 4a and 4b present SEM images and xenon X-ray maps obtained with EPMA on a small part of the working area before and after ion sputtering with SIMS. With SEM, the investigation was performed by working with two detectors (SE and BSE) and at different electron acceleration voltages, from 10 to 30 kV. This method is discussed in the next section. In the following paragraphs, SEM images were obtained at 25 kV. In figure 4a, the green circles point out the presence of a bubble uncut by polishing and still filled with fission gases. BSE images show the presence of a cavity (green circle). Not detected by the SE image, this cavity is below the polished plane. The EPMA X-ray map of Xe shows that this cavity, though close to the surface, is still filled with Xe (bright spot).

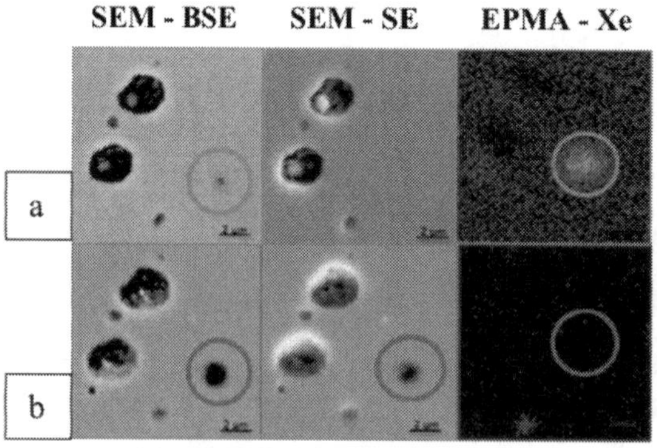

Figure 4. SEM SE, BSE images and EPMA X cartographies showing the evolution of one bubble before (a) and after (b) SIMS abrasion (enlarged images on a bubble).

2.3.3. Abrasion with SIMS. When a bubble is selected, a first depth profile with SIMS is performed. The diameter of the ion beam is controlled by the voltage and the current. The diameter of the spot, obtained with 15 kV and 20 nA, was around 50 µm. The acquisition was stopped when a peak with a high intensity was observed (hypothesis, one peak = one bubble). At this stage, we assumed that one bubble has been opened and that the new surface after sputtering was situated in the first half of the bubble above its median plan. Work is ongoing in order to improve the measurement of the bubble diameter (see section 4.2).

2.3.4. Control of bubble opening. The peak obtained with SIMS needs to be associated with a bubble. In order to check if it does correspond to the bubble selected previously, the sample was observed a second time with EPMA. Figure 4b, shows the same area as figure 4a, but after the short sputtering period. On BSE and SE images, most of the bubbles appear unchanged. However, the initially uncut bubble is now appearing more clearly on the BSE image, and is detectable with the SE image. With the X-ray map, the release of Xe from the bubble appears clearly.

2.3.5. Evaluation of bubble diameter. The determination of the bubble diameter was performed after SIMS abrasion. The sample was placed a second time in the SEM and it was measured by image analysis taking into account the black halo around the bubble considered as the maximum diameter. The accuracy of this method is discussed in the next section.

2.3.6. Synoptic of the overall methodology. Figure 5 is a schematic view of the single bubble methodology.

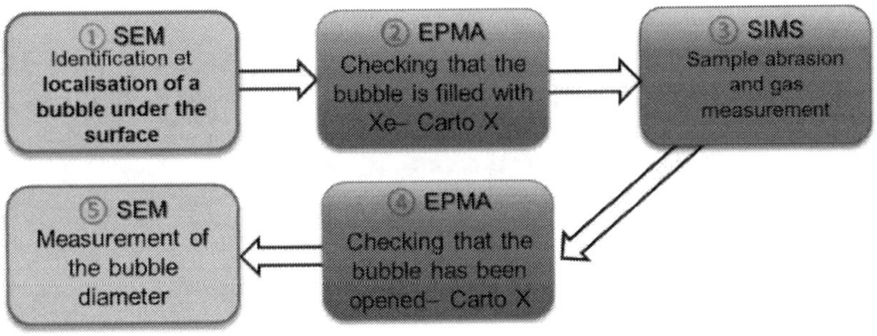

Figure 5. Procedure for the single bubble methodology using SEM, EPMA and SIMS.

3. Results

The procedure has been applied on the UO_2 sample and the results are described and explained up to the calculation of the bubble pressure.

3.1. SIMS results

Figure 6 presents the xenon signal measured with SIMS as a function of the sputtering time. The measurement has been performed with ^{132}Xe isotope. As soon as one large peak is observed, the acquisition is stopped because it is considered that the number of peak is related to the number of bubbles observed by SEM. Two components are seen:

- a baseline which is associated to ^{132}Xe solid solution or nanobubbles,
- 3 significant peaks coming from the bubbles.

Between 15 and 120 seconds, an average baseline can be determined. It is then subtracted from the whole profile. The intensity of the highest peak is assumed to be associated to xenon that was contained in the largest opened bubble.

3.2. Quantification of the xenon inventory

A quantitative evaluation of xenon concentration is possible using the experimental results obtained with SIMS and EPMA [1]. At the centre, the EPMA measurement shows almost no xenon in solution

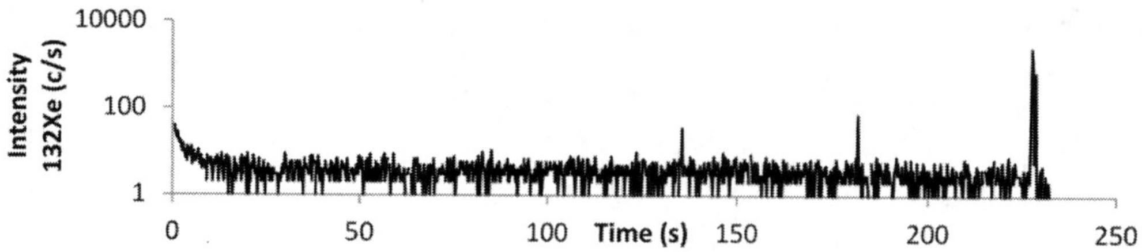

Figure 6. SIMS depth profile of Xe-132.

(figure 2). However, at the mid-radius position, the xenon is considered to be totally in solid solution because no bubble was detected by SIMS or observed by SEM. At this position, both EPMA and SIMS measurements are equal to the quantity produced during the irradiation and can be directly compared and correlated. For a given material, the sputtering yield depends on the microstructure (including grain crystalline orientation) and the impurities it contains. The sputtered surface is not large enough to average the effect of the crystalline orientation of the UO_2 grains but several SIMS acquisitions at the same position (mid-radius) and a work on the average data helps reducing the effect of grain orientation. Besides, the compositions and burn-up of the mid-radius and centre areas are relatively close. For example, figure 2 showing the radial Xe production profile deduced from Nd, exhibits no major change from the mid-radius to the centre. Thus the composition should not have a significant effect on the sputtering yield.

The mean volume of the SIMS craters is calculated and with the abrasion time, we can deduce a sputtering speed and a volume per count. The mass of UO_2 corresponding to this volume is obtained and thanks to the xenon concentration measured from EPMA, it is possible to calculate the quantity of xenon corresponding to one count.

The quantity of xenon n_{xe} (mol) is obtained through the following relation (1):

$$n_{Xe} = I_{(c/s)} f \tag{1}$$

with the factor $f = \dfrac{W_{\%xe} V_c \rho_{UO_2}}{I_m M_{Xe}}$ and $I_{(c/s)}$ the intensity of the peak corresponding to one bubble,

where: $W_{\%Xe}$: weight % of xenon (EPMA results), V_c: mean crater volume (m^{-3}), ρ_{UO2}: UO_2 density ($g.m^{-3}$), M_{Xe}: molar mass of xenon ($g.mol^{-1}$), and I_m: mean intensity in the mid-radius position ($c.s^{-1}$).

3.3. SEM results

SEM observations are useful to localize bubbles under the surface prior to SIMS analysis. Then, it allowed us to determine the diameter, *d*, and thus the volume, *V*, of the bubbles opened by SIMS with the hypothesis that the bubbles are spherical. The SEM images of the other bubbles, cut during the polishing process show that this hypothesis is reasonable for intragranular bubbles, even if a perfect sphericity is not observed (figure 4).

3.4. Pressure results

From the SIMS results and the bubble volume measured, it is then possible to obtain a molar volume of xenon for the bubble with the relation (2):

$$V_m = \frac{V}{n_{Xe}} \tag{2}$$

IOP Conf. Series: Materials Science and Engineering **109** (2016) 012002 doi:10.1088/1757-899X/109/1/012002

Xenon is not the only atom present in its gas form at room temperature [10]. Indeed, krypton has to be considered in the calculation of the pressure. A ratio is used between Xe and Kr obtained from neutronic calculations assuming the same behaviour for both gases. In order to calculate the bubble pressure, an equation of state is needed. Several equations can be found in the literature and behave differently depending on the domain of temperature and molar volume. Some experimental data are available for Xe. For example, Ronchi [11] has extrapolated confidently the high temperature and high density pressure–volume–temperature data of Xe using a Leonard–Jones (LJ) interatomic potential.
Comparisons have been carried out between those experimental results and equations of state.

The Soave equation (3) [12] is in good agreement with the experimental data for a molar volume larger than 40 $cm^{-3}.mol^{-1}$ which is the case of the measurements here. With the application of mixing rules, the equation can be extended to multicomponent calculations (Xe + Kr).

$$P_g = \frac{RT}{(V_m - b_M)} - \frac{a_M}{(V_m^2 + cb_M V_m + db_M^2)} \quad (3)$$

Mixing rules: $a_M = (\Sigma x_i a_i^{0.5})^2$, $b_M = \Sigma x_i b_i$ and x_i fraction of the element i (depending on the ratio xenon/krypton calculated).
With:

$$a_i = \frac{a_0 R^2 T_C^2}{P_C} \text{ and } a_0 = 0.42748(1 + f(1 - (\frac{T}{T_C})^{1/2}))^2$$

$$b_i = b_0 R \frac{T_C}{P_C}, \; b_0 = 0.0866;\; c = 1;\; d = 0;$$

For xenon: critical temperature, T_C = 289.70 K; critical pressure, P_C = 5.84 x 10^6 Pa and f = 0.492.
For krypton: critical temperature, T_C = 209.35 K; critical pressure, P_C = 5.50 x 10^6 Pa and f = 0.488.

The methodology has been applied on several bubbles. Table 1 gives their diameter, molar volume and pressures at room temperature calculated with (3).

Table 1. Pressure results for several bubbles.

	Bubble diameter (nm) (± 200)	Molar volume ($cm^3.mol^{-1}$)	Pressure at 300 K (MPa)
1	300	59.1	32.8
2	1,600	52.52	56.4
3	400	68.3	19.3
4	2,100	182	7.1
5	2,100	178	7.1
6	400	177	7.1
7	600	440	4.3
8	1,900	10,397	0.2
9	1,900	641	3.2

4. Discussion

4.1. Comparison with experimental results

The results can be compared with other experimental data obtained with different techniques on irradiated fuels. Nogita and Une [13] determined the Xe density in nanometric bubbles with EDS. Noirot [6] and Walker [7] used SIMS to measure the Xe quantity. Noirot's and Walker's values correspond to the high burn-up structure (HBS) from UO_2 and MOX fuels at the periphery and mid-radius for the first author and from a UO_2 fuel at the periphery for the second. All these published data therefore correspond to situations very far from the sample studied here.

The results obtained with the method presented in this paper are represented figure 7 and data from the literature are also reported. The graph gives a schematic correlation between bubble internal pressure and bubble radius for irradiated UO_2 fuels, in which the equilibrium pressure ($Peq = 2\gamma/r$; γ: surface energy; r: bubble radius) is given. The pressures were calculated at room temperature (300 K).

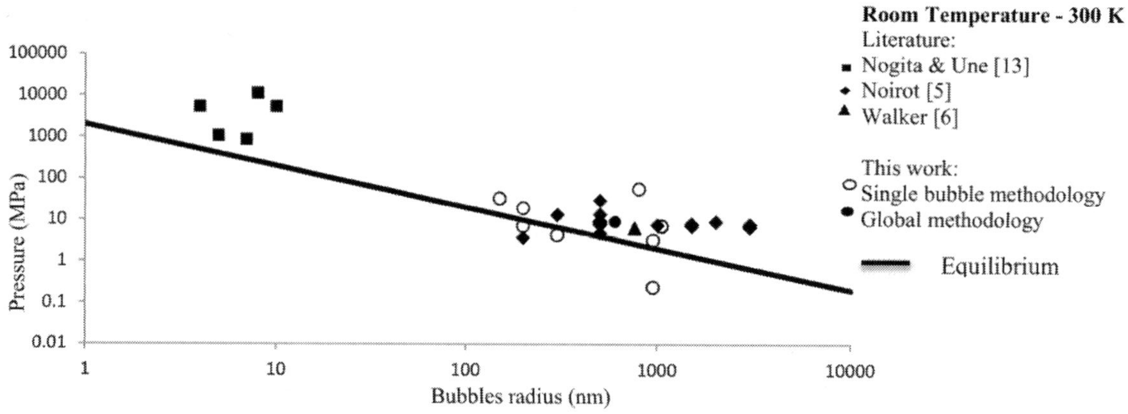

Figure 7. Experimental data of fission gas bubbles, pressure of xenon versus their radius.

The single bubble approach and the global methodology were all applied on the same sample. The mean pressure is close to the equilibrium (3 MPa) at room temperature. This means that in spite of the ramp and high temperatures, it was, like HBS bubbles at the periphery or nanobubbles, clearly over-pressurized in pile. The individualized bubble pressures are also around this equilibrium, at room temperature, but highly scattered, with pressures above equilibrium and pressures under equilibrium. Those results cannot be directly compared to the values from the literature. The areas studied are different but they globally seem to be in the same order of magnitude (around 10 MPa) and they tend to be over-pressurized or close to the equilibrium except for one value.

Besides, the pressure results do not conform to the evolution of the equilibrium law and figure 7 shows a large variation of the pressures from 50 MPa to less than 10 MPa. Indeed, the pressures are not inversely proportional to the radius. For a diameter of 400 nm (see table 1), the equilibrium pressure is 10 MPa and a bubble contains gas at a pressure of 22.9 MPa (over-pressurized), whereas the other shows its pressure decrease at 6 MPa (slightly under-pressurized). Without more work on the uncertainty, it is difficult to conclude if this observation is a result of a real variation in the nuclear fuel or if it is a consequence of the uncertainty of the method.

4.2. Accuracy

The errors introduced in the pressure calculation come from the measurement of the quantity of xenon and the measure of the bubble diameter.

The accuracy of the xenon concentration obtained with this procedure is a function of the EPMA xenon concentration accuracy and of SIMS experimental parameters which are difficult to measure. Their evaluation is investigated by performing tests on the different sources of uncertainty such as the sample and standard, instruments, effects of experimental conditions (current, counting rate, background noise, …) and data post-treatment.

The second source of error is of great importance. The diameter is obtained by the analysis of 2D images involving 3D effects and is raised to the cube in the calculation of the volume. Some hypotheses are taken into account and work is ongoing in order to validate or improve them.

With the single bubble approach, it is complicated to determine if the surface is situated just at the median plan of the bubble. Several measurements have been carried out with SEM images obtained with different high tensions, before and after SIMS abrasion. It seems that the bubbles open at their top with a burst of gas, the pressure being too strong. In the case of micrometric bubbles just opened after a SIMS abrasion, BSE detector at high tensions enables to perceive the bubbles diameter close to its maximum (figure 8).

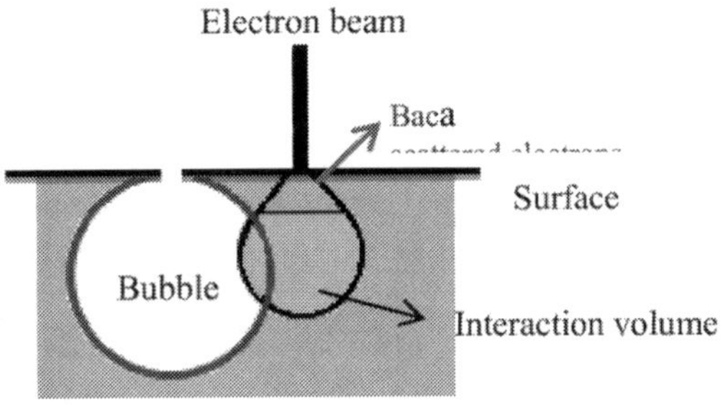

Figure 8. Localisation of the new surface after a SIMS abrasion - electron interactions.

The uncertainty associated with the determination of the diameter is estimated at ± 200 nm for micrometric bubbles. A new SEM replacing the device used for this study, with higher capacities and especially a FIB (focussed ion beam), will be used to better characterise the bubbles and reduce this major uncertainty.

Besides, during the first SIMS abrasion, 3 bubbles were opened (figure 6). For the post-treatment of the SIMS depth profiles, we assumed that one bubble corresponds to one pic and that the highest pic is associated to the largest bubble. This is tested by changing the order of attribution in order to identify aberrations (such as a pressure of 1 GPa in a bubble from the centre of the nuclear fuel).

The sources of error are identified and work is carried out in order to evaluate the total uncertainty.

4.3. Detection of bubbles under the surface

The detection of bubbles with SEM is the first step that enables us to get easily a view of the bubbles located under the surface. Figure 9 presents SEM images of the same area that were carried out with both detectors, BSE and SE and at different tensions, from 10 to 30 kV.

It appears that a bubble close enough to the surface becomes visible by increasing the tension and with a better contrast using BSE detector sensitive to electrons that come from deeper. For image analysis, in order to obtain the bubble size distribution from opened bubbles, it is then recommended to work at low voltage (10 kV). On the contrary, high voltage (25 - 30 kV) is necessary in order to visualize with good contrast bubbles that are around 0.25 to 0.5 µm deep in UO_2 samples (Monte Carlo simulation).

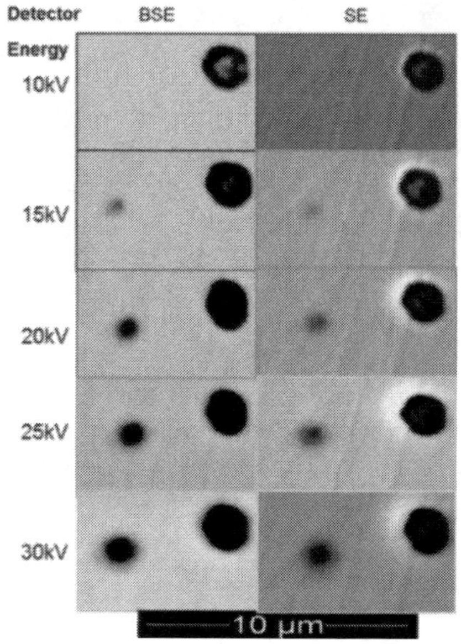

Figure 9. Appearance of a bubble detected with SEM using BSE and SE detectors at electron acceleration voltages from 10 to 30 kV.

5. Conclusion

We have described a new approach that combines three techniques in a high activity laboratory that enable the determination of individual fission gas bubble pressure in irradiated fuels. The method has been developed to get more refined results on a few bubbles, for the best case, one. It is complementary to the global methodology established a decade ago. The method has been applied to the central part of a ramped UO_2 irradiated fuel. This paper has shown that:

- The quantity of xenon coming from a single bubble can be measured, quantified and used to determine the pressure.
- In order to get closer to the case of a single bubble, some improvement will be made especially for the ion beam diameter. A compromise is needed between the diameter and the intensity of the beam (xenon ionisation rate).

The uncertainty coming from the measurement of the bubble diameter has been estimated, but the total uncertainty needs to be completely investigated. It will be improved with a FIB-SEM available soon in the laboratory, in order to get a 3D view of the microstructure. Several sources of errors were highlighted and work is carried out in order to obtain the total uncertainty.

Acknowledgments

The authors are indebted to Electricité de France and Areva for their support.

References

[1] Desgranges L and Pasquet B 2004 *Nucl. Instr. Meth. B* **215** 545-551
[2] Noirot J, Noirot L, Desgranges L, Lamontagne J, Blay T, Pasquet B and Muller E 2004 *Proc. of the International topical meeting on LWR fuel performance.* (Orlando, FL, USA) **1019** 329
[3] Lamontagne J, Desgranges L, Valot C, Noirot J, Blay T, Roure I and Pasquet B 2006 *Microchim. Acta* **87** 183-187
[4] Noirot L 2011 *Nucl. Engng. Des.* **241** 2099-2118
[5] Desgranges L, Valot C, Pasquet B, Lamontagne J, Blay T and Roure I 2008 *Nucl. Instr. Meth. B* **266** 147-154
[6] Noirot J, Desgranges L and Lamontagne J 2008 *J. Nucl. Mater.* **372** 318-339

[7] Walker C T, Bremier S, Portier S, Hasnaoui R and Goll W 2009 *J. Nucl. Mater.* **393** 212-223
[8] Lamontagne J, Noirot J, Desgranges L, Blay T, Pasquet B and Roure I 2004 *Microchim. Acta* **145** 91-94
[9] Boidron M, Tourasse M, Boussard F, Piron J P and Pasquet B 1991 *Proc. of the International topical meeting on LWR fuel performance.* (Avignon, France) 21-24
[10] Valin S 1999 PhD-thesis (Grenoble, France: INP)
[11] Ronchi C 1981 *J. Nucl. Mater.* **96** 314-328
[12] Soave G 1972 *Chem. Engng. Sci.* **27** 1197-1203
[13] Nogita K and Une K 1998 *Nucl. Instr. Meth. B* **11** 481-486

Low voltage EPMA: experiments on a new frontier in microanalysis – analytical lateral resolution

J Fournelle[1], H Cathey[2,4], P T Pinard[3] and S Richter[3]

[1] University of Wisconsin-Madison, Department of Geosciences, 1215 W. Dayton Street, Madison, WI 53706, USA
[2] Arizona State University, LeRoy Eyring Center for Solid State Science, 901 S. Palm Walk, Tempe, AZ 85287-1704, USA
[3] RWTH Aachen University, Central Facility for Electron Microscopy (GFE), Ahornstrasse 55, 52074 Aachen, Germany

E-mail: johnf@geology.wisc.edu

Abstract. Field emission (FE) electron gun sources provide new capabilities for high lateral resolution EPMA. The determination of analytical lateral resolution is not as straightforward as that for electron microscopy imaging. Results from two sets of experiments to determine the actual lateral resolution for accurate EPMA are presented for Kα X-ray lines of Si and Al and Lα of Fe at 5 and 7 keV in a silicate glass. These results are compared to theoretical predictions and Monte Carlo simulations of analytical lateral resolution. The experiments suggest little is gained in lateral resolution by dropping from 7 to 5 keV in EPMA of this silicate glass.

1. Introduction

Recent papers [1-5] provide good summaries on field emission electron probe microanalysis (FE-EPMA). The most obvious advantage of the FE electron probe, relative to the more conventional W- and LaB_6-probes, is that the FE source produces a higher electron flux from a smaller source point and thus a smaller *initial* probe size, which itself is a function of both accelerating voltage on the gun as well as beam current. There is a 'sweet spot' for minimal spot size in terms of operating conditions (current work and informal discussion suggests 6 - 8 keV and ~ 10 nA). Nothing is gained by going to lower keV, as electrons with lower kinetic energy cannot be as tightly focussed, resulting in a wider beam. As beam current decreases, so does beam diameter but quantitative analysis requires abundant counts for good statistics, so generally one does not go below 10 nA unless it is feasible to count longer at lower beam current.

Operating at low keV introduces a plethora of issues which must be addressed, particularly carbon contamination [6] and complications of using non-traditional X-ray lines [7]. The issue of sample damage resulting from the intense FE beam impact has received slight attention, and will be addressed in a later document.

[4] Now at Central Analytical Research Facility, Institute for Future Environments, Queensland University of Technology, 2 George St., Brisbane, QLD 4000, Australia

The ability to acquire sharper images of nm-scale features does not equate to the ability to determine an accurate chemical composition of the same nm-scale feature. When considering X-rays generated by the electron beam, it is essential to appreciate the fact that virtually all of the X-rays collected by the detectors are generated at some depth within the sample (figure 1). The incident high energy electrons scatter and produce an interaction volume which has less to do with beam 'spot' size and much more to do with both the gun high voltage (keV) and the material composition (atomic number and density) and secondarily with beam current (nA).

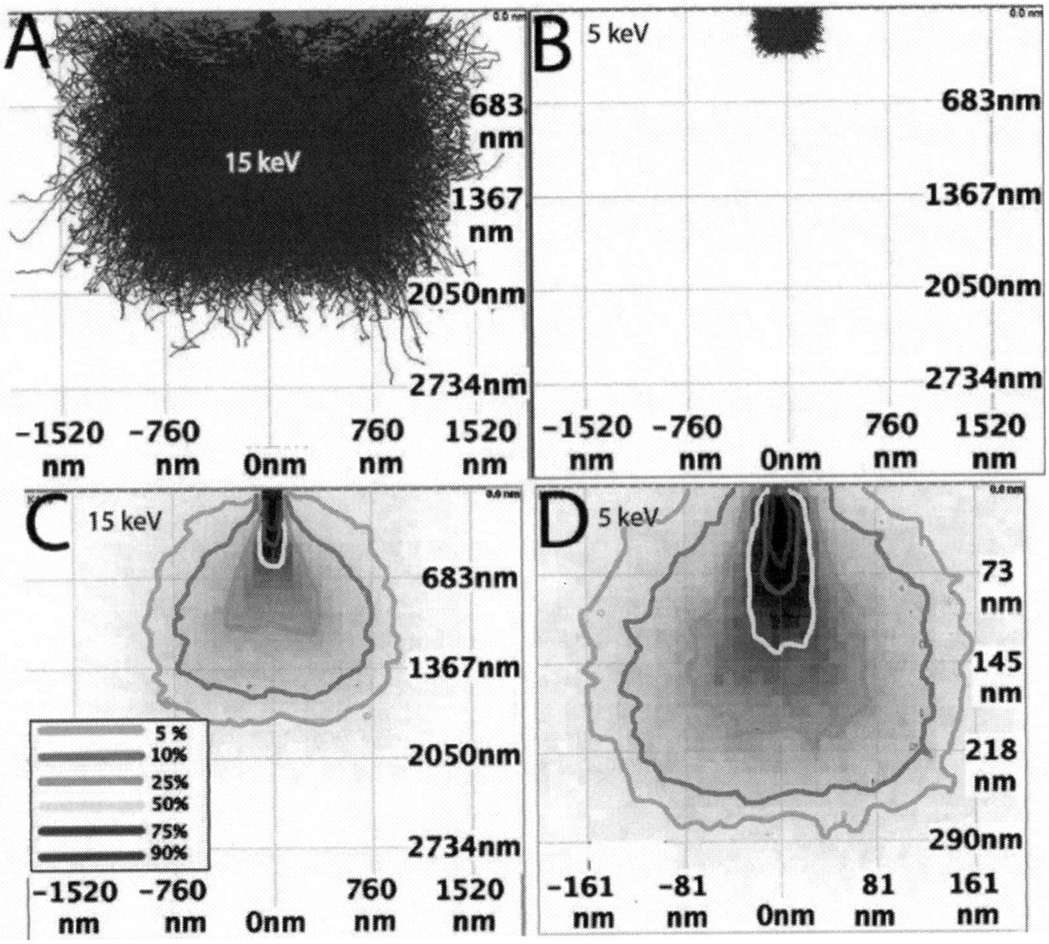

Figure 1. CASINO simulations of interaction volume in silicate glass K409, at 15 and 5 keV, with 10,000 trajectories, and 100 nm incident beam diameter. A-B-C at same scale, D magnified by a factor of 940 %. A, B indicate maximum extent of electron scattering. C, D show remaining energy of E_0 beam electrons in the glass. CASINO simulations indicate that higher energy electrons are concentrated in a more constrained cylindrical region (basis for the "low overvoltage approach" [1]), whereas lower energy electrons show more lateral scatter; for 15 keV beam electrons, this is for 4 keV and lower, whereas for 5 keV, this is for 1.2 keV and lower.

IOP Conf. Series: Materials Science and Engineering **109** (2016) 012003 doi:10.1088/1757-899X/109/1/012003

2. Background: X-ray analytical lateral resolution in low keV EPMA

The promise of low voltage FE EPMA is enhanced lateral resolution for quantitative analysis. With a field emission source, clearly electron (SE and BSE) image resolution is greatly improved. However, the problem of quantifying the lateral resolution of lower energy X-ray lines – including ones both traditionally and non-traditional used – assumes new and particular relevance.

2.1. Definitions of resolution

There is a potential for confusion regarding the word "resolution" in electron microanalysis, mainly because of the intimate association between electron microanalysis and electron microscopy imaging with secondary and backscattered electrons. The ability to "resolve" or distinguish two objects in an electron image is based on contrast measurements techniques.

Barkshire *et al.* [8] provided a good summary of the terminology of image resolution: (a) *point-to-point resolution* (an example being gold spheres on a graphite substrate) is defined as the minimum distance between two points over which the signal intensity drops by at least 75 % of the maximum intensity, and (b) *edge resolution* refers to the distance over which some percentage change in signal intensity occurs (e.g., 16 - 84%) in a scan over a sharp edge between two features.

Barkshire *et al.* [8] also give a definition of lateral resolution for EPMA, which is worth repeating:

> *... instead of simply being able to resolve the feature in an image, the entire X-ray information volume must be contained within the feature. Thus, for this purpose a more reasonable definition of lateral resolution is the diameter of a feature such that it will include some fraction of the total X-ray intensity, e.g., 99 %.*

2.2. X-ray range

Additionally, definition of X-ray range is important, as two concepts – X-ray range and X-ray lateral resolution – are sometimes used interchangeably, when they are not the same. X-ray range refers to an "average path length that electrons travel before slowing down to an energy equal to the excitation energy of the considered electron shell" [9]. As it encompasses all directions of electron travel within the sample, this solitary definition assumes that the average path length applies equally to vertical as well as lateral distances travelled. As Monte Carlo electron beam simulations show, this is true only for certain materials and clearly not all (e.g., in silicate minerals the beam penetrates relatively more deeply than wide, unlike in high Z metals). Moreover, as figure 1 shows, for a constant keV and material, the electron energy distribution is rather tightly constrained for higher energy X-ray lines, whereas for lower energy lines, the remnant electron energy is spread much wider, albeit deeper. Thus it is incumbent to appreciate that X-ray range values are involved with X-ray lateral resolution, but are not generally equivalent.

In his 1952 thesis, Castaing [10] proposed that X-ray range is determined by five factors according to the following equation:

$$R_x(nm) = \frac{33*A}{\rho*Z}(E_0^{1.7} - E_{Cr}^{1.7}) \quad (1)$$

where three of which are material dependent (atomic number, atomic weight, and density) the beam energy E_0; and the critical excitation energy E_{Cr} for a particular X-ray line of interest.

Since then, several other researchers have modified the equation: Anderson and Hasler [11], Reed [12], and Hovington *et al.* [13], but it retains the same overall formulation. In these formulations, the beam diameter is not a necessary factor for consideration.

2.3. X-ray lateral resolution

In the early 1960s when the newly invented electron probe microanalyser was being used to quantify elemental compositions of small objects, the question of exactly how small was the analytical volume came to the fore. Duncumb [14] proposed that to determine the analytical lateral resolution, the calculated X-ray range value needed to be multiplied by a factor of 1.6. A decade and half later Reed [15] proposed that the X-ray range actually needed to be multiplied "by a factor of about 3".

Recently, Merlet and Llovet [9] re-addressed this question in two important ways. They included a term for the beam (spot) diameter, and they accounted for materials where electron penetration and scattering differ in the lateral and depth dimensions. This formulation of minimum analytical diameter includes an X-ray range value that is modified first by the depth of maximum X-ray generation (the deeper the maximum, the narrower the resolution, as seen in figure 1) and then includes the beam diameter. Their formulation for lateral resolution (R_s = diameter) uses Castaing's X-ray range (2) where R_x is that X-ray range, Z_m is the depth where the maximum distribution (i.e., phi-rho-Z) occurs, and d_0 is the beam diameter:

$$R_S = \sqrt{4(R_x - Z_m)^2 + d_0^2} \tag{2}$$

Another approach to estimating X-ray lateral resolution is Monte Carlo simulation, i.e., CASINO [16] and PENEPMA [17]. These simulations utilize a variety of approximations of the physics involved in the electron scattering and X-ray generation and absorption with the sample. Microanalysts commonly utilize these simulations to estimate the interaction volume prior to actually EPMA operation, to set the beam energy for a particular small size feature of interest.

The goal of this study is to utilize a well suited experimental material to generate data sets to evaluate predicted X-ray lateral resolution, fore quantitative resolution for accurate EPMA analysis, and compare estimates from both X-ray range and Monte Carlo estimates, with actual experimental data.

3. Experimental evaluation of X-ray lateral resolution

3.1. NIST K409, a useful material for FE EPMA

Several decades ago, the U.S. National Bureau of Standards (today NIST) developed several different reference materials as EPMA standards, and two glasses were publicized and distributed: K411 and K412 [18].

The K409 material was "re-discovered" at UW-Madison by John Fournelle when he took charge of the EPMA laboratory at the time Everett Glover retired from the position in 1992. A material labelled "NBS K409" with a nominal composition containing Na-Fe-Si-Al-O (table 1) was among the multitude of reference materials in the UW-Madison EPMA lab collection. After noting the ubiquity of micrometre-sized "fuzzy balls" in the material once it was mounted and polished for EPMA, Fournelle consulted Dale Newbury at NIST about the material, who explained that it was a "failed experiment." Apparently the oxygen partial pressure was too high and an iron-rich phase separated out. The glass was measured by EPMA using PROBE FOR EPMA's time-dependent-intensity correction (extrapolating to time = 0 to correct for elemental migration under the beam).

NIST K409 is a very useful material for assessing lateral resolution using low voltage EPMA by virtue of its unusual texture and composition [19]. It consists of iron oxide crystals that are rather uniform in size and shape (~ 800 nm maximum diameter quasi-octagons) that are densely distributed in a glass matrix (figure 2). Multiple measurements (table 2) suggest that dispersed within the iron-oxide crystals there is ~ 10 - 15 wt% of glass nano-inclusions. PEMEPMA calculations suggest that secondary fluorescence cannot account for this level of glass component measured in the crystals. Figure 2b shows tiny sub-100 nm-size regions of different intensity which would be consistent with trapped silicate liquid during rapid crystallisation (image is collected with SE detector, but backscattered electrons are also incident upon the detector, as evident in the contrast difference between the interstitial glass (dark grey) and the lighter grey Fe-oxide crystals

EBSD measurements were performed on the iron oxide crystals and the best fit (with low mean angular deviation, MAD) is consistent with the crystals being magnetite (figure 3).

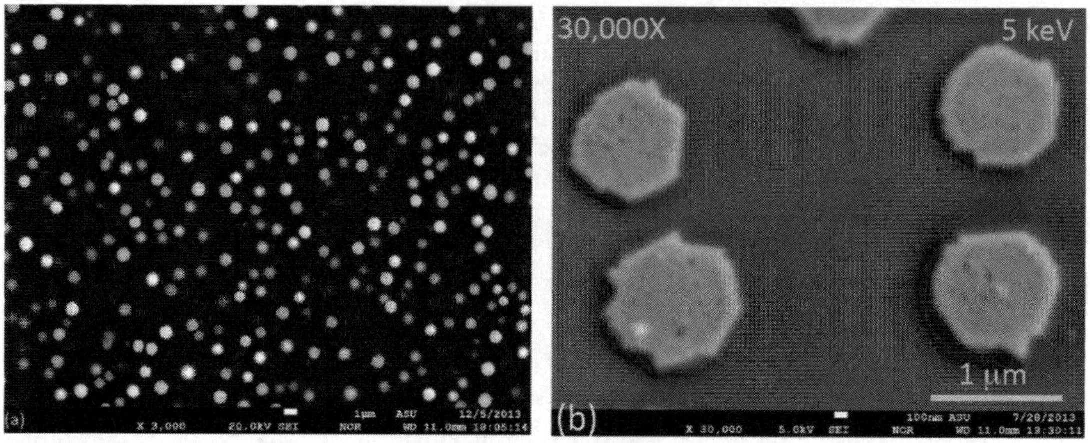

Figure 2. Two SEM images of NIST K409 glass. (a) At low magnification, the Fe-oxides appear circular, but at high magnification (b), they appear as equant crystals.

Table 1. Oxide weight% compositions for (1) original bulk material estimated by NBS (now NIST) for the glass synthesis, (2) EPMA defocussed beam (20 µm) analysis of the bulk material, and (3) EPMA measurement of the interstitial glass using a focussed beam and time dependent intensity (TDI) correction (PROBE FOR EPMA). (2) and (3) EPMA with the UW SX51, 15 keV, 10 nA.

	(1) Bulk - NBS/NIST nominal	(2) Bulk - EPMA defocussed	(3) Glass - EPMA TDI
SiO_2	55	56.6	59.9
Al_2O_3	15	15.1	17.0
FeO	20	15.7	6.4
Na_2O	10	9.8	10.8
MnO		0.2	
CaO		0.1	
Excess O		1.8	5.9
Total	100	99.3	100
(from)	on vial	SX51 2007	SX51 2012

3.2. Experiments with NIST K409: looking in the gaps

The Fe_3O_4 crystals are scattered throughout the NIST K409 glass with variable gaps separating them (figure 4). This experiment hypothesizes that at in regions furthest from any Fe-oxide crystals, the measured glass composition reflects the 'true' composition of the glass. Then at a series of smaller inter-crystal spacing, one would measure glass compositions that would show "contamination" by the adjacent Fe-oxide crystals. And at the point where the effect of the adjacent Fe-oxides were not observed, would then be an experimental determination of the analytical lateral resolution of the X-ray lines being evaluated: Si-Kα, Al-Kα and Fe-Lα. Na-Kα was attempted but unsuccessful (see below).

Table 2. Quantitative FE-EPMA of magnetite in NIST K409 using the Fe-Lα X-ray line. Results in (a) and (b) include direct measurement of oxygen on the RWTH-Aachen JEOL 8530F (O = excess measured oxygen). (c) EPMA using Arizona State University (ASU) JEOL 8530F with oxygen calculated by stoichiometry. Fe recast as Fe_3O_4. Although there are some discrepancies in Si and Al between the two datasets at 5 keV, each is internally consistent and shows significant measurement of Si and Al. Monte Carlo simulations by PENEPMA do not predict this amount of Si and Al by either primary or secondary fluorescence. A possible explanation for Si and Al differences would be variable amounts of nanoscale glass trapped within the Fe_3O_4 crystals.

					(a) 4 keV 10 nA (RWTH Aachen JEOL 8530F)				
Analysis#	SiO2	Al2O3	Fe3O4	MgO	CaO	Na2O	K20	excess O	Total
1	10.50	4.23	87.9	0.08		1.02		-1.67	102.1
2	11.07	3.60	87.2	0.08		1.06		2.12	100.9
3	9.86	3.88	85.4	0.12		0.93		0.00	100.2
4	11.14	4.29	85.3	0.17		1.06		-1.32	100.6
5	11.32	4.03	85.6	0.06		0.59		-1.64	99.9
Average	10.78	4.01	86.3	0.10		0.93		-1.35	100.7
Std Dev	0.60	0.28	1.2	0.04		0.20		0.81	0.8
					(b) 5 keV 10 nA (RWTH Aachen JEOL 8530F)				
Analysis#	SiO2	Al2O3	Fe3O4	MgO	CaO	Na2O	K20	excess O	Total
1	10.40	4.08	83.4	0.16	0.52	0.77	0.00	-0.06	99.2
2	9.23	2.16	88.8	0.12	0.08	0.64	-0.06	-0.93	100.0
3	9.58	2.71	87.4	0.09	-0.14	0.87	0.01	-0.70	99.8
4	10.17	3.01	86.9	0.18	-0.04	0.92	-0.05	-0.52	100.6
5	10.40	2.65	88.2	0.19	0.16	0.83	0.28	0.09	102.7
6	10.41	3.25	84.7	0.11	0.01	1.01	0.07	-0.17	99.2
7	9.82	3.29	86.7	0.16	0.34	0.95	-0.02	-0.99	100.2
8	10.70	2.79	86.0	0.13	0.30	0.75	0.06	-0.29	100.4
9	9.22	2.80	88.7	0.07	0.30	0.86	-0.15	-0.94	100.9
Average	9.99	2.97	86.7	0.13	0.17	0.84	0.00	-0.50	100.3
Std Dev	0.55	0.54	1.8	0.04	0.21	0.11	0.12	0.41	1.1
					(c) 5 keV 10nA (ASU JEOL 8530F)				
Analysis#	SiO2	Al2O3	Fe3O4	MgO	CaO	Na2O	K20	excess O	Total
1	7.36	2.11	89.3	0.02		0.72			99.5
2	7.86	2.19	86.1	0.19		0.36			96.7
3	7.20	2.27	85.6	0.20		0.65			95.9
4	6.48	2.00	83.5	0.21		0.70			92.9
5	7.06	2.22	89.5	0.31		0.61			99.7
6	7.32	2.35	89.1	-0.10		0.69			99.3
7	6.73	2.07	90.6	0.16		0.52			100.0
8	7.52	2.25	85.5	-0.20		0.63			95.7
9	6.99	1.74	91.5	0.02		0.65			100.9
10	6.86	1.94	88.9	-0.04		0.59			98.2
11	7.16	2.15	89.0	-0.02		0.66			98.9
Average	7.14	2.12	88.0	0.07		0.62			98.0
Std Dev	0.38	0.18	2.5	0.16		0.10			2.4

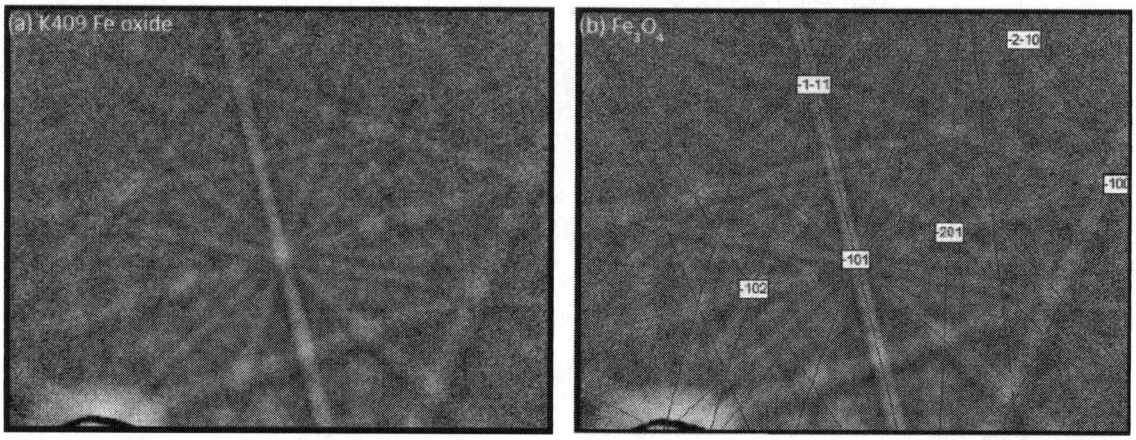

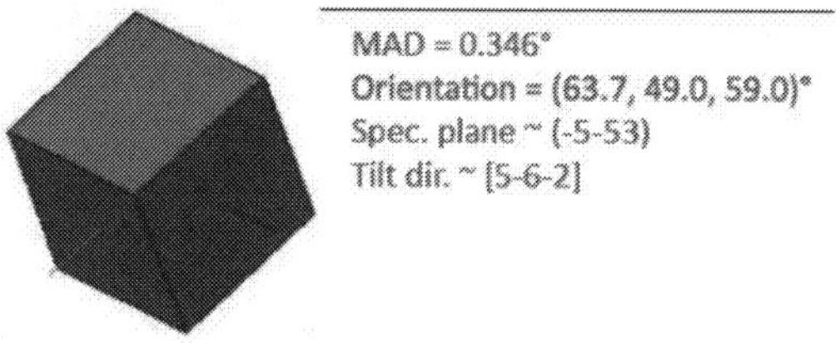

Figure 3. (a) Diffraction pattern from the K409 Fe-rich phase. (b) Superposition of Kikuchi lines for pure magnetite upon K409's lines gives a high level of confidence (low mean angular deviation, MAD) for identification. (HKL system with Channel 5 software, Hitachi S3400N VP-SEM, 20 keV, UW-Madison).

An additional test would be whether a given X-ray line intensity reached a plateau in the gap between two Fe-oxides, suggesting little contribution from the adjacent phases.

A key part of this project is to evaluate the various theoretical predictions for X-ray analytical resolution with the experimental results. In addition to the X-ray range calculation and predictions using the multiplicative factors for X-ray ranges of Duncumb and of Reed, CASINO and PENEPMA models were performed, all using the glass composition in table 1 and a density of 2.6 g/cm^3.

The experiments were performed at 5 and 7 keV with the UW-Madison SXFive FE probe (#944) with faraday cup current of 10nA. A focussed beam was used with a measured 80 - 100 nm diameter. Beam deflection was used with ~ 100 nm steps. The liquid nitrogen (LN)-anti-contamination plate was in operation, and generated some small vibrations (not believed to be of major significance here). The NIST K409 glass was mounted in epoxy, polished down to 0.3 μm diamond, and coated with 4 nm of Ir.

Two series of experiments were performed. The initial ones were a series of EPMA line profiles through the centres of the largest of these crystals – looking for the smallest gap that gives the same X-ray intensity as seen far from the crystals. This is an indication of the ability to resolve the intervening glass composition without direct contribution from the iron oxide crystals.

Upon further evaluation of the counting statistics of the first set of experiments, a second series were run, for long count times as well as duplicate spectrometers for Si, with a stationary beam at the midpoint between series of pairs of the crystals.

3.2.1. Line scans. Figure 4 shows line scans along pairs of magnetite crystals with different spacings. Counts were collected for 5 seconds per spot, at ~ 100 nm distance between analyses. Background counts were not subtracted; gross counts were plotted. Na loss was rapid and no evaluation of Na-Kα resolution was attempted. Any Si and Al 'grow in' (increase in counts) did not appear to affect the results as the Si, Al and Fe counts between sufficiently widely spaced crystals were similar in magnitude to the distant glass count levels for these elements.

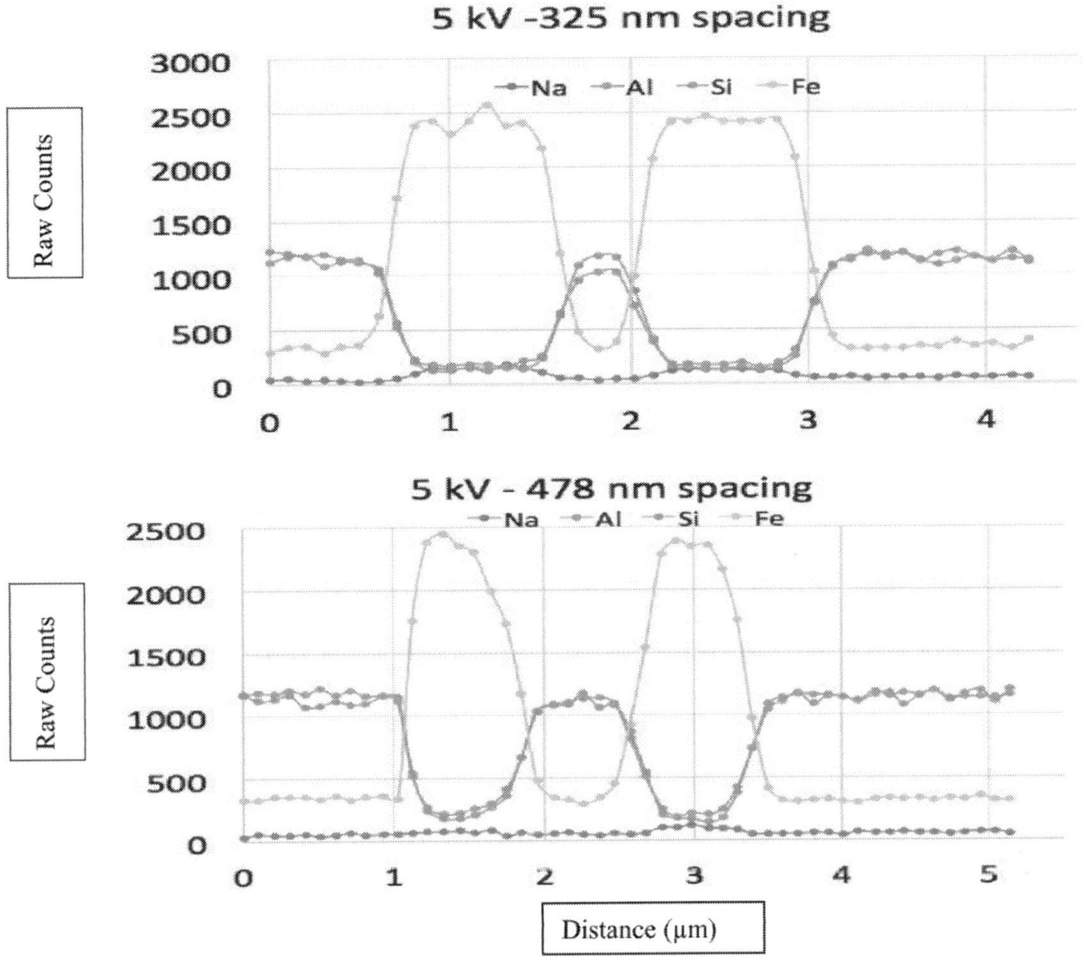

Figure 4. Examples of two of the traverses at 5 keV, through the centres of Fe_3O_4 crystals. Measurements with CAMECA SXFive FE at University of Wisconsin-Madison.

In the Figure 4 line scans, "spacing" refers to the measured distance between Fe-oxide crystals, measured from the closest edges using secondary electron images, and using the largest and equal sized crystals.

In the initial experiments, the numbers of traverses being evaluated were: 7 at 5 keV, and 10 at 7 keV. Shown below in figure 4 are two traverses at 5 keV, where one spacing shows some clear interaction with the adjacent Fe_3O_4 crystals as indicated by the reduced intensity of Al-Kα (orange dots) relative to that at distance. For the wider spacing, the Al-Kα matches that at distance.

The ratios in table 3 were determined by taking the maximum X-ray intensity for Al-Kα, Si-Kα and Fe-Lα on the glass in the gap between two Fe_3O_4 crystals, and dividing by the glass X-ray intensity 1 - 1.5 µm "far away" from the crystals.

An evaluation of the precision of the measurements based upon counting statistics shows that while a qualitative estimate can be made of analytical resolution based on these results, the error bars are large: taking the traverses at 7 keV as an example, for a 2 sigma error there is variability in the Al and Si-Kα measurements of ~7% (thus values from 0.93 - 1.08 are indistinguishable from 1.00); variability

Table 3. Results from 5 and 7 keV experiments, with ratios relative to glass composition in regions far from the Fe_3O_4 crystals.

5 keV	X-ray intensity relative to far away		
spacing nm	Al Kα	Si-Kα	Fe Lα
300	0.81	0.91	1.20
325	0.87	1.03	0.95
360	0.97	0.97	1.03
416	0.81	0.87	0.91
478	0.94	0.99	0.98
490	1.01	0.96	0.99
528	0.95	1.00	1.00

7 keV	X-ray intensity relative to far away		
spacing nm	Al-Kα	Si-Kα	Fe Lα
454	0.92	0.96	1.19
485	0.88	0.94	1.18
571	0.95	0.95	1.05
586	0.93	0.97	1.18
670	0.85	0.99	0.87
675	0.93	0.97	0.87
686	1.01	0.99	0.96
749	0.98	1.01	0.93
776	0.98	1.01	0.89
882	1.01	0.99	0.96

in Fe-Lα is worse, with errors reaching 20 %. These errors were calculated by taking, for each measurement, the average value of the gap measurements (i.e., one 5 second count), and then determining the 2 sigma value.

3.2.2. Point measurements. In order to reduce the counting statistical error in the measurements, an additional experiment was designed. In order to determine whether a point midway between two Fe_3O_4 crystals encompassed only the excitation volume of the glass and not the Fe-oxides, spot analyses with an on-peak counting time of 50 seconds were performed at each of X midpoints between crystals (20 spots for the 5 keV and 30 for the 7 keV experiments), representing a range of midpoint distances between crystals of ~ 300 - 900 nm. Similarly, counts were collected for 50 seconds in at several spots in glass that were distant from obvious Fe-oxides. As before, a beam current of 10 nA was used, with spot size of ~ 100 nm. However, here 3 spectrometers (2 LPETs + 1 TAP) counted simultaneously on Si-Kα and their counts were summed, yielding a 1 sigma counting statistical error of 0.4 % / 0.3 % (5 keV / 7 keV values). Al-Kα (on a LTAP) yielded a counting statistics error of 0.5 % / 0.4% and Fe-Lα (on a PC1 60 Å W/Si) had a 1 sigma statistical error of 1.3 % / 1.1%. Results are shown in figure 5.

These experiments provide additional data to consider the analytical lateral resolution in silicate glasses similar to the K409 composition. Clearly it lacks other cations commonly abundant in many natural glasses as well as in common silicate minerals (e.g., Mg, Ca); nevertheless these data can be used to evaluate the theoretical analytical lateral resolution for this composition.

With respect to the experiments at 7 keV at 800 nm and higher, the normalized value for Si-Kα counts is at or very close to 1.00, whereas from 600 - 800 nm values range from 0.95 to 1.00. The data for Al-Kα are roughly similar to those for Si-Kα, but with more scatter down to 0.90. Data for

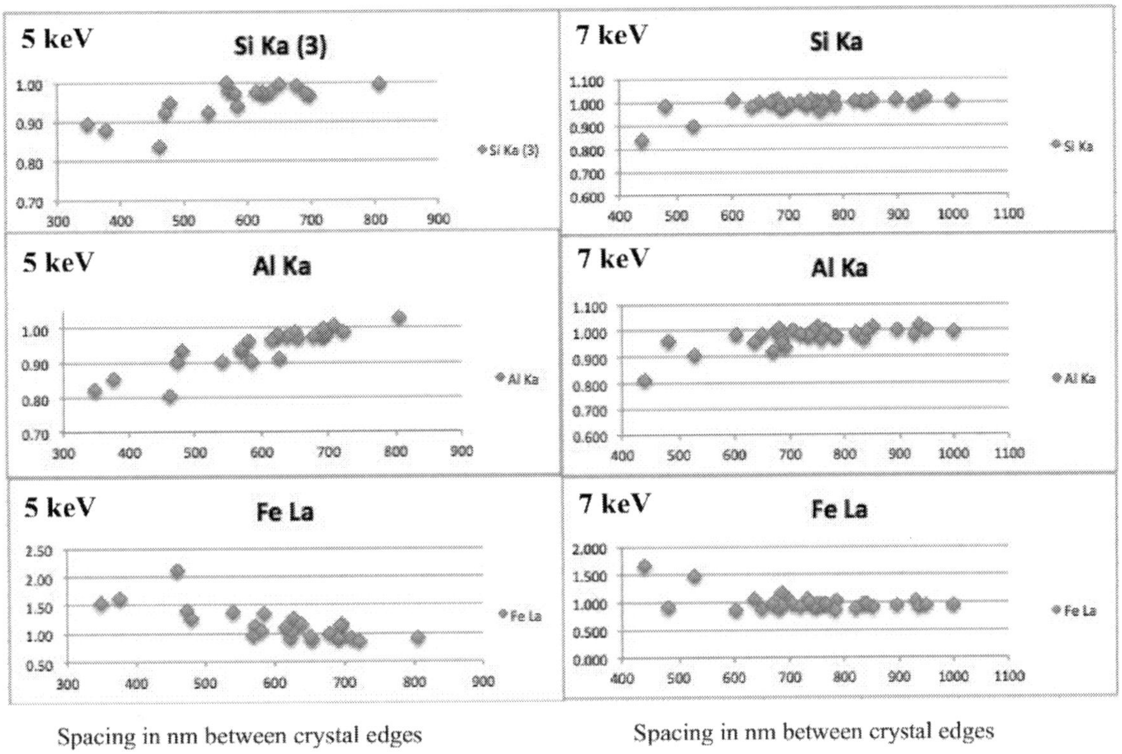

Figure 5. Plots of counts on NIST K409 glass normalized to glass composition far from Fe_3O_4 crystals. X-axis is nanometre spacing between Fe-oxide crystals.

Fe-Lα follow this general behaviour (glass regions representing shorter midpoint distances clearly contain Fe from the adjacent Fe-oxides). One would conclude from these data that the minimum analytical lateral resolution for these X-ray lines is somewhere between 600 - 700 nm. One would expect the lower energy Al-Kα and Fe-Lα lines (compared with Si-Kα) to have decreased (wider) lateral resolution, but the data do not permit such a refined determination (and the issue of statistics must play a role, particularly for the lower count rate Fe-Lα).

An obvious point of concern is whether Fe-oxides reside within the sample interaction volume just below the surface. Their presence was detected in some of the first experiments, by excursions of count intensities in some of the traverses. In the second series of experiments, the evidence would include elevated Fe compared with lowered Al and Si counts, and that is seen in some of the data around 700 nm.

At 5 keV (figure 5), the data for all of the X-ray lines – Si-Kα, Al-Kα and Fe-Lα – suggest a minimum lateral resolution of 575 - 650 nm.

4. Theoretical results using NIST K409

4.1. CASINO modelling

CASINO v2.42 [13] was run assuming a single phase (i.e., bulk K409 glass) at 5 and 7 keV, and the results for X-ray radial distribution were output and evaluated to determine the lateral distance (2 times the radial distance) at which 99.9 % of the X-ray intensity was contained. Figure 6a shows the result for Si-Kα at 5 keV from the CASINO simulation.

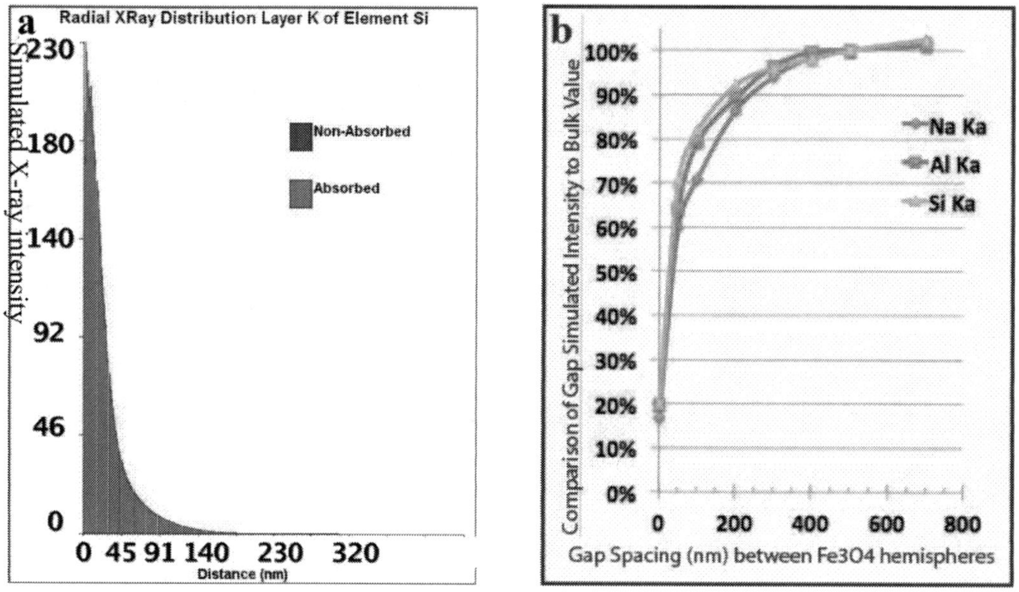

Figure 6. Monte Carlo simulations of low voltage EPMA of K409 glass. (a) Left, CASINO X-ray radial distribution for 5 keV simulation of Si-Kα in bulk glass K409 with beam diameter of 100 nm. The CASINO model shows negligible absorption of Si-Kα. (b) Right, PENEPMA 5 keV simulation of X-rays detected with beam halfway between two 800 nm Fe-oxide hemispheres, with the annular detector at a 40° take off angle.

4.2. PENEPMA modelling

PENEPMA Monte Carlo modelling allows a full simulation of a geometry of two 800 nm diameter Fe_3O_4 hemispheres at variable spacings in a matrix of K409 glass composition. Simulations were run for 18 - 24 hours. Figure 6b shows the results of the PENEPMA 5 keV set of simulations. The data are all from simulations using actual K409 glass composition (table 1, 3rd column). Nine simulations were done: 8 for spacings from 0 nm (touching) to 700 nm, and one for bulk material – which was the reference to which each of the various spacings was compared. Thus, at 200 nm spacings, the simulated X-ray intensities, relative to what they 'should' be, is around 87 % for Na, ~ 89 % for Al and ~ 91 % for Si. At 0 nm spacing, the spheres touch at a point and there is exposed glass, so the value is 15 – 20 %, not zero. (The 2014 version of PENEPMA provides a nice quick graphical representation of 3D X-ray generation within the target material, but is not shown here. The approach used here, while slower, includes the absorption at the given detector take-off angle, and is more directly comparable with the actual experimental data).

PENEPMA modelling here indicated that all 3 elements reach a value of 100 % of the bulk K409 composition only when the spacing is at 500 nm. Experiments were only done in no tighter than 100 nm increments, and optimally future work would include tighter spacing (e.g., 50 nm steps near the putative resolution point).

This PENEPMA modelling also indicates that there should be less than ~ 1 % secondary fluorescence (including continuum) involvement.

5. Experimental versus theoretical results using NIST K409

Table 4 compares the results of theoretical calculations with experimental results on analytical lateral resolution at 5 and 7 keV. Key observations include:

1. For silicate glass of K409 composition, all the calculated X-ray ranges using published formulations at 5 keV are smaller than the observed X-ray analytical lateral resolution, whereas for 7 keV, those of Castaing and Anderson-Hasler are close to the experimental values.
2. The formulations of Duncumb and of Reed do not match the experimentally observed lateral resolution (either over or underestimate).
3. Merlet and Llovet's [9] estimation is close for the 5 keV case (using the CASINO depth of maximum phi-rho-Z distribution) but overestimates lateral resolution in the 7 keV case.
4. PENEPMA calculations agree closely with experimental measurements 7 keV, but underestimate by ~ 20 % for 5 keV.
5. CASINO's "X-ray radial distribution" (multiplied by 2 to give an analytical resolution value) is actually close to the Castaing X-ray range value for both 5 and 7 keV, and is lower than the experimentally observed 5 keV lateral resolution, but matches that for 7 keV.

Table 4. Comparison of experiments at 5 and 7 keV with various estimations of analytical X-ray lateral resolution, for Si-Kα, Al-Kα and Fe-Lα in K409 glass, using various published formulations and results from Monte Carlo simulations. For Merlet-Llovet formula, (a) uses phi-rho-Z max from CASINO, (b) uses phi-rho-Z max from Merlet (pers. comm.). PENEPMA for Fe-Lα not included due to poor statistics.

X-ray ranges	5 keV Al-Kα	5 keV Si-Kα	5 keV Fe-Lα	7 keV Al-Kα	7 keV Si-Kα	7 keV Fe-Lα
Castaing 1952	351	321	405	664	625	730
Anderson-Hasler '66	322	311	352	602	590	632
Reed '66	273	257	314	491	475	531
Hovington et al '97	297	281	333	558	543	1065
Analytical resolution						
Duncumb (x1.6)	437	411	502	785	759	849
Reed (x3)	820	772	940	1472	1424	1593
Merlet-Llovet '12 (a)	578	512	619	989	909	1065
Merlet-Llovet '12 (b)	515	465	607	974	900	1091
PENEPMA (100 %)	500	500	*	700	700	*
CASINO (99.9 %)	360	360	436	674	674	755
K409 point experiments	575-650	575-650	575-650	600-700	600-700	600-700

6. Conclusions

The theoretical estimates of X-ray ranges indicate that there is a significant improvement in dropping from 7 to 5 keV. Similarly, the theoretical estimates for analytical resolution also indicate the same improvement in dropping from 7 to 5 keV. However, the K409 experiments show that there may be only a slight improvement to be gained by operating at 5 keV versus 7 keV. Any gain in a slightly smaller degree of electron scattering is offset by beam broadening (and perhaps issues of surface contamination).

With respect to Si-Kα and Al-Kα lines versus the lower energy Fe-Lα line, the theoretical analytical volume encompassing excitation of the Fe-Lα line would be larger than that for only the Si-Kα and Al-Kα lines, given that less energy is needed to generate the Fe-Lα line. All the Fe-Lα

X-ray lateral estimates in Table 4, based on theoretical calculations, show such a results. All these cited values (PENEPMA not included there), have one thing in common – they consider only one phase, the glass, and not the adjacent Fe-oxides.

This predicted wider Fe-Lα X-ray lateral resolution is not observed in the experiments. Absorption of the Fe-Lα X-ray energy may account for this observation, such that resolution of Fe-Lα ends up matching that of Si-Kα and Al-Kα. This behaviour is actually beneficial for silicate analysis where one would like the different elemental X-ray intensities contributing to the compositional analysis by EPMA to represent the same volume.

Further work needs to be done to better understand the behaviour of the low energy Fe-Lα X-ray line here. The SE image (figure 2) may indicate some polishing-induced relief between the glass and the Fe-oxides, and it would be helpful to check to verify this. If relief is present, then any impact on X-ray intensity could be modelled in a more complex Monte Carlo geometry. And longer PENEPMA models could be run to achieve the required statistical significance for modelling the Fe-Lα intensity therein, where the multi-phase (glass plus two Fe-oxides) assemblage could be simulated.

Acknowledgements
The following researchers have been generous with their comments and advice concerning low voltage EPMA over the past several years: Xavier Llovet, Phil Gopon, Michel Outrequin, Chrystel Hombourger, Dieter Rhede, John Armstrong, Peter McSwiggen, Stuart Kearns, and Chi Ma. Two anonymous reviewers are thanked for their comments. JF acknowledges NSF-MRI award EAR-1337156 to UW-Madison for the SXFive FE electron probe.

References

[1] McSwiggen P 2014 *IOP Conf. Ser.: Mater. Sci. Engng.* **55** 012009
[2] Hombourger C and Outrequin M 2013 *Microscopy Today* **21**(3) 10
[3] Pinard P T and Richter S 2014 *IOP Conf. Ser.: Mater. Sci. Engng.* **55** 012016
[4] Rinaldi R and Llovet X 2015 *Microsc. Microanal.* **21** 1053
[5] Berger D and Nissen J 2014 *IOP. Conf. Ser.:Mater. Sci. Engng.* **55** 012001
[6] Buse B and Kearns S 2015 *Microsc. Microanal.* **21** 594
[7] Gopon P, Fournelle J, Sobol P E and Llovet X 2013 *Microsc. Microanal.* **19** 1798
[8] Barkshire I, Karduck P, Rehbach W P and Richter S 2000 *Mikrochim. Acta* **132** 112
[9] Merlet C and Llovet X 2012 *IOP Conf. Ser.: Mater. Sci. Engng.* **32** 012016
[10] Castaing R 1952 PhD-thesis. *Publication ONERA* no. 55
[11] Anderson C A and Hasler M E 1966 *Proc. 4th IXCOM.* (Castaing R, Deschamps P and Philibert J; Eds.) (Paris: Hermann) p 310
[12] Reed S J B 1966 *Proc. 4th IXCOM.* (Castaing R, Deschamps P and Philibert J; Eds.) (Paris: Hermann) p 339
[13] Hovington P, Drouin D, Gauvin R and Joy D C 1997 *Microsc. Microanal. 3 suppl* 2 885
[14] Duncumb P 1960 in: *X-ray Microscopy and X-ray Microanalysis. Proc. 2nd Intl Symposium.* (Engstrom A, Cosslett V and Pattee H; Eds.) (Amsterdam: Elsevier) 364
[15] Reed S J B 1975 *Electron Microprobe Analysis* (Cambridge: Cambridge University Press)
[16] Hovington P, Drouin D and Gauvin R 1997 *Scanning* **19** 1-14
[17] Salvat F, Fernandez Varea J M and Sempau J 2013 *PENELOPE - A code system for Monte Carlo Simulation of Electron and Photon Transport.* (Issy-les-Moulineaux, France: OECD/Nuclear Energy Agency)
[18] Marinenko R B 1982 Standard Reference Materials: Preparation and Characterization of K-411 and K-412 Mineral Glasses for Microanalysis SRM 470. *NBS Special Publication* 260-274.
[19] Cathey H, Gopon P and Fournelle J 2013 *AGU Fall Meeting Abstracts* 2793

IOP Conf. Series: Materials Science and Engineering **109** (2016) 012004 doi:10.1088/1757-899X/109/1/012004

The role of intergranular chromium carbides on intergranular oxidation of nickel based alloys in pressurized water reactors primary water

F O M Gaslain[1], H T Le[1], C Duhamel[1], C Guerre[2] and P Laghoutaris[3]

[1] MINES ParisTech, PSL - Research University, MAT - Centre des Matériaux, CNRS UMR 7633, P.O. Box 87, 91003 Evry cedex, France
[2] CEA, DEN, DPC, SCCME, Laboratoire d'Etude de la Corrosion Aqueuse, 91191 Gif-sur-Yvette, France
[3] CEA, DEN, DPC, SCCME, Laboratoire d'Etude de la Corrosion Non Aqueuse, 91191 Gif-sur-Yvette, France

E-mail: fabrice.gaslain@mines-paristech.fr

Abstract. Alloy 600 is used in pressurized water reactors (PWRs) but is susceptible to primary water stress corrosion cracking (PWSCC). Intergranular chromium carbides have been found beneficial to reduce PWSCC. Focussed ion beam coupled with scanning electron microscopy (FIB/SEM) 3D tomography has been used to reconstruct the morphology of grain boundary oxide penetrations and their interaction with intergranular Cr carbides in Alloy 600 subjected to a PWR environment. In presence of intergranular Cr carbides, the intergranular oxide penetrations are less deep but larger than without carbide. However, the intergranular oxide volumes normalized by the grain boundary length for both samples are similar, which suggest that intergranular oxidation growth rate is not affected by carbides. Analytical transmission electron microscopy (TEM) shows that the intergranular oxide consists mainly in a spinel-type oxide containing nickel and chromium, except in the vicinity of Cr carbides where Cr_2O_3 was evidenced. The formation of chromium oxide may explain the lower intergranular oxide depth observed in grain boundaries containing Cr carbides.

1. Introduction

Currently, there are plans to extend the lifetime of pressurized water reactors (PWRs) of French nuclear power plants up to 60 years. Among the materials used in PWRs, nickel based alloys such as Alloy 600 (Ni-15Cr-10Fe) are used as structural materials in the primary circuit. Alloy 600 provides a good overall corrosion resistance but is susceptible to primary water stress corrosion cracking (PWSCC) [1, 2]. As a consequence, PWSCC of Alloy 600 and parent weld metals (Alloys 182/82) is a significant cause of failure in the primary circuit of PWRs [3]. This mode of damage leads to a brittle-type intergranular cracking of the material under the combined action of the environment and a tensile stress state. The SCC phenomenon is composed of several stages: development of oxide penetrations at the grain boundaries, fracture of the oxide leading to crack initiation, and then propagation of an intergranular crack [4]. However, SCC is a complex process and even though it has been studied extensively over the last thirty-five years, the cracking process is still under question [5-11]. To prevent PWSCC of Alloy 600, the presence of chromium carbides at the grain boundaries have shown to be beneficial [12-16]. Depending on the final thermal treatment, two kinds of chromium carbides can be formed: $Cr_{23}C_6$ and/or Cr_7C_3.

The aim of this work is to study the effect of intergranular chromium carbides on intergranular oxidation, which is a key step in the PWSCC mechanism. Indeed, intergranular Cr carbides are assumed to delay oxidation along the grain boundaries [17]. For this purpose, two A600 samples with model microstructures have been analysed after exposure to simulated nominal primary water in order to determine the influence of intergranular Cr carbides. The intergranular oxidation kinetics as well as the nature of the intergranular oxides was studied. To date, most of the studies describing such oxides are either based on cross-section observations (SEM, TEM), or volumetric techniques with either atomic resolution (atom probe tomography) or macroscopic approaches (SIMS or Auger depth profiling). Recently, Lozano-Perez and co-workers have shown that 3D tomography using a dual column scanning electron microscope associated with a focussed ion beam (FIB/SEM) is an appropriate technique for investigating such oxides [18, 19]. In this work, the oxidation depth and path along given grain boundaries were characterized by 3D FIB/SEM tomography. Cubic micrometre volumes of the two model microstructure materials were reconstructed with nanometric resolution to reveal the oxide shapes and their interactions with intergranular carbides. Additional characterisations by analytical transmission electron microscopy (TEM) were performed to investigate the chemical composition and structure of the intergranular oxide by coupling information from energy filtered TEM (EFTEM) analyses, electron diffraction patterns and high resolution images. Using this original analytical approach, we will give a better description of the role of Cr carbides on intergranular oxidation in nickel based alloys.

2. Materials and experimental procedures

2.1. Materials and oxidation test

Two specimens (thickness: 1.5 mm) with model microstructures have been prepared from commercial grade Alloy 600 (Ni-16Cr-10Fe), by applying two different thermal treatments under vacuum [20]. The two samples are named, *SA* for "Solution Annealed" and *SA+TT* for "Solution Annealed followed by Thermal Treatment". All the experimental conditions are detailed in table 1.

Table 1. Overview of the thermal and oxidation treatments applied to the A600 samples.

	SA	*SA+TT*
Thermal treatment	1,050 °C for 1 h + water quenching	1,050 °C for 1 h + water quenching + tempering at 700 °C for 16 h
Aim	Microstructures without carbides	Microstructures with a dense intergranular carbide precipitation
SEM image	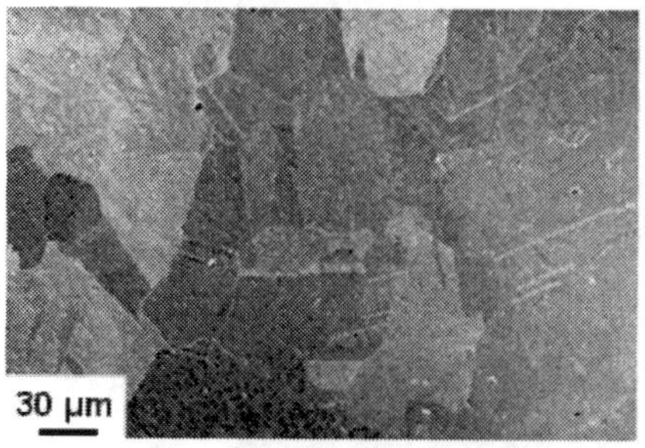	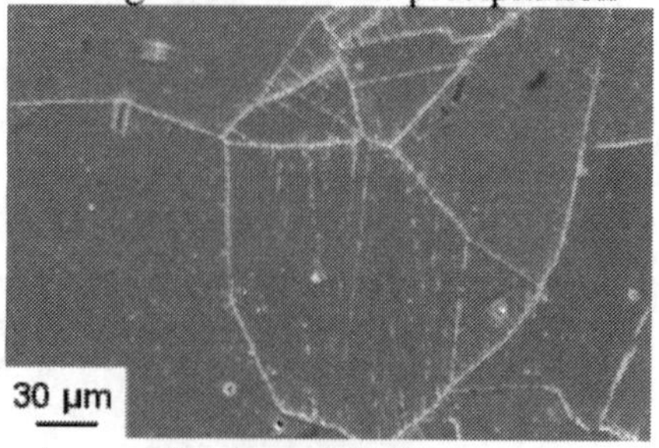
Oxidation test conditions	1,400 h in simulated PWR water conditions at 325°C (P = 120 bar, $[H_2]$ = 30 mL (TPN)/kg H_2O (200 mbar à 325°C), $[H_3BO_4]$ = 1,000 ppm, [LiOH] = 2 ppm)	

The *SA* microstructure exhibits no chromium carbides whereas SEM observations of the *SA+TT* one reveals a high density of intergranular chromium carbides. The stoichiometry of these carbides ($Cr_{23}C_6$ or Cr_7C_3) was not identified at this stage. Oxidation tests were then performed at the CEA Saclay (France) in stainless steel autoclaves under simulated pressurized water reactor (PWR) primary water conditions at 325 °C for 1,400h.

2.2. 3D FIB/SEM tomography

The oxidation depth and path along given grain boundaries of samples *SA* and *SA+TT* were characterized using a FEI Helios 660 dual column scanning electron microscope associated with a focussed ion beam (SEM/FIB). Before sectioning, a Pt protective coating was deposited above the area of interest. Then three trenches were excavated at the front and at the sides of the volume of interest to clear up the cross-section view imaged by SEM and to avoid any material re-deposition. The samples were orientated such that the focussed Ga^+ ion beam was at a perpendicular incidence to the surface (x-z plane). SEM secondary electron (SE) images of the cross-section were taken after each sequential 2D slice (in the x-y plane) at a tilt angle of 52° (figure 1). FIB sectioning parameters and SEM image acquisitions conditions are summarized in table 2.

Figure 1. SEM image showing an example of the volume isolated for the reconstruction after coating the exposed surface with Pt and milling two sides and one front trenches.

Table 2. FIB parameters and SEM image acquisition conditions.

	SA	*SA+TT*
Ga^+ FIB parameters	HV 30 kV, 2.5 nA	HV 30 kV, 2.5 nA
Slice thickness (nm)	20	10
Number of slices	640	800
SEM image acquisition parameters	HV 5 kV, 0.1 nA, Non-immersion mode, Everhart-Thornley detector	HV 5 kV, 0.2 nA, Immersion mode, TLD detector in SE mode
Voxel size or '3D pixel' size (nm^3)	13x13x20	10x10x10

The 3D reconstructions were carried out using FEI software Avizo Fire 8.1. For both stacks of 2D SE images, stretching caused by the 38° tilt was corrected during image acquisition. Those images were then aligned and corrected for drift using cross-correlation of static features on the sample original surface following the method described in [21]. Then an area containing the region of interest, with its topside coinciding with the sample top surface, was extracted from each image and arranged as a 3D dataset.

The different regions of interest appearing in the 2D images (e.g., protecting Pt layer, oxides, carbides and metal matrix) were automatically segmented using intensity thresholding tools. When the contrasts between oxides, carbides and/or grains were too small, the automated segmentation methods failed and manual corrections were done until the result was satisfactory. One of these images extracted from the sample *SA+TT* stack of images is shown in figure 2 as an example.

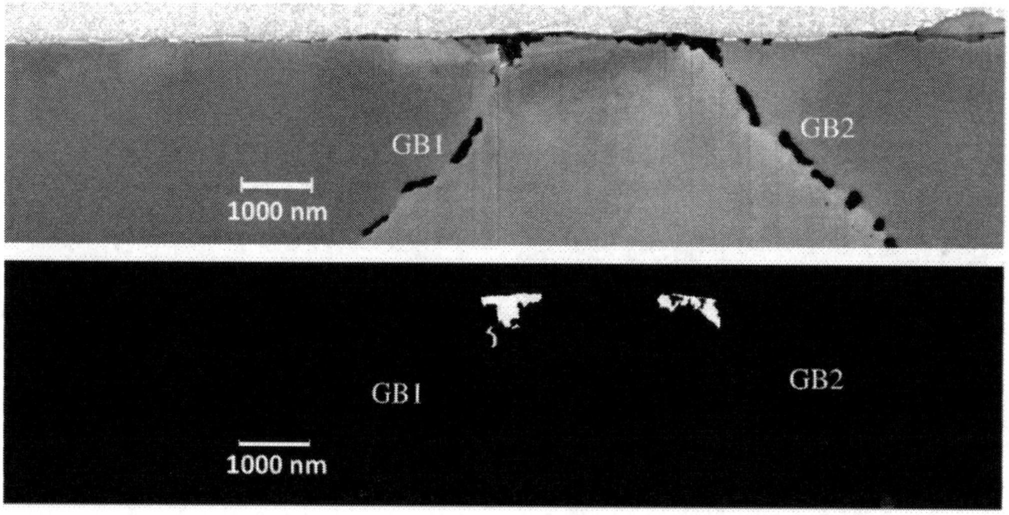

Figure 2. a) Example of a SEM image showing the cross-sectional view from sample *SA+TT*. b) Binary mask of the oxide penetration (white) and the Cr carbides (grey) after segmentation.

2.3. Analytical TEM

Additional characterisations by analytical transmission electron microscopy (TEM) were performed on a FIB prepared thin foil. The thin foil was carefully micro-sampled then thinned down following the "lift-out" method described in [22, 23]. TEM observations and analyses were carried out at 200 kV using a FEI Tecnai F20 ST field emission gun microscope equipped with a Gatan Imaging Filter (GIF) device.

Analytical TEM was used to investigate the chemical composition and structure of the intergranular oxide by coupling information on composition from energy filtered TEM (EFTEM) analyses and on crystallography from electron diffraction patterns.

3. Results and discussions

3.1. 3D SEM/FIB tomography

For both samples, a 3D reconstruction model of the intergranular oxidation was created (figure 3) and three grain boundaries were analysed. The dimensions of the reconstructed volumes were 18.0 x 11.4 μm^2 for sample *SA* and 14.0 x 15.4 μm^2 for sample *SA+TT*. For each grain boundary, the intergranular oxide depths were automatically measured for all the segmented images using a Python script. The average intergranular oxidation depths are presented in table 3 and the intergranular oxide penetrations distribution are shown in figure 4.

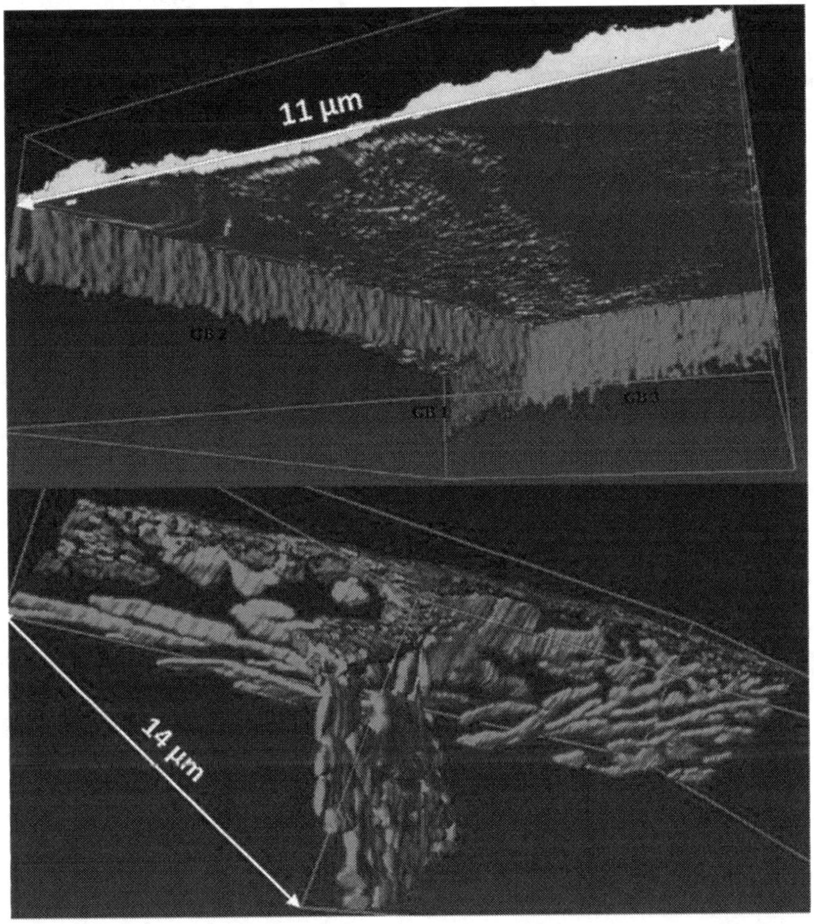

Figure 3. Reconstructed 3D volumes of the oxide penetrations represented in red of both sample *SA* (top) and sample *SA+TT* (bottom). The surface oxide present on sample *SA* is represented in yellow and the Cr carbides present on sample *SA+TT* are represented in green.

Table 3. Average intergranular oxidation depth for each grain boundary (GB).

	SA		*SA+TT*	
	Average intergranular oxidation depth (nm)	Standard deviation (nm)	Average intergranular oxidation depth (nm)	Standard deviation (nm)
GB 1	2168	180	661	183
GB 2	1113	93	430	98
GB 3	1824	174	378	136

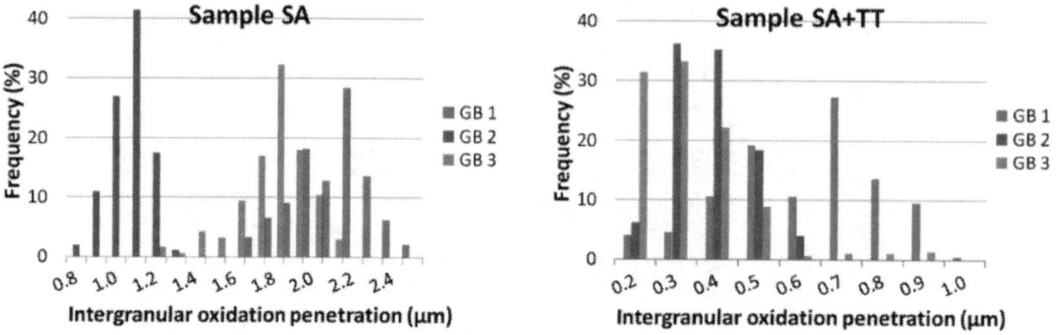

Figure 4. Intergranular oxidation depth distributions of each grain boundary for samples *SA* and *SA+TT*.

For sample *SA*, oxide depths along each grain boundary are relatively even and continuous but the average lengths are relatively different from one grain boundary to another. Oxidation depth distributions vary from 0.9 to 2.6 µm. The deepest oxide penetration for this sample is found at the triple junction.

In contrast, for sample *SA+TT*, oxide depths are much shallower, uneven and wider along grain boundaries than the previous sample, but the average lengths seem to be more homogeneous from one grain boundary to another. In this sample, the oxidation depth distribution varies from 0.2 to 1.0 µm. Again, the deepest oxide penetration for this sample is found at the triple junction. The deepest oxide penetration is found along GB1 and can be explained by a less dense carbide distribution along this grain boundary. 3D reconstruction also highlights two features when an oxide reaches a Cr carbide: either the oxide is stopped by the carbide or the oxide is surrounding the carbide forming a cap on it.

Comparison of both samples highlights that intergranular oxidation goes much deeper along grain boundaries in Alloy 600 when Cr carbides are not present, since they probably act as obstacles for the oxidation to get through. We have also noticed that for sample *SA+TT*, intergranular oxide penetrations are much wider than in sample *SA* (table 4) by a factor of 3. This suggests that the oxidation penetrations are inhibited by Cr carbides, but oxide will nevertheless expand in the adjoining grains as shown by the oxide width reported in table 4.

Table 4. Average depths and surface widths of intergranular oxides.

	SA	*SA+TT*
Average intergranular oxide depths (nm)	1679	447
Standard deviation (nm)	445	179
Average surface width of intergranular oxides (nm)	247	719
Standard deviation (nm)	111	327

In order to have a global vision, the intergranular oxide volumes have been evaluated for both samples. The measurements were performed using the "Label analysis" tool from the Avizo software and normalized by the total length of the grain boundary to make the results comparable. The results are summarized in table 5. Despite evident heterogeneities from one intergranular oxide penetration to another one, the total volumes of oxide penetration normalized by the total GB length are similar for each sample (i.e., 0.160 $\mu m^3/\mu m$ for sample *SA* against 0.140 $\mu m^3/\mu m$ for sample *SA+TT*, respectively). These results are interesting but need to be moderated since we have extracted values from a relatively small volume compared to the total sample volume.

Table 5. Intergranular oxide volumes.

	SA			*SA+TT*		
	Oxide volume (μm^3)	GB length (μm)	Normalized oxide volume ($\mu m^3/\mu m$)	Oxide volume (μm^3)	GB length (μm)	Normalized oxide volume ($\mu m^3/\mu m$)
GB 1	1.419	9.3	0.153	0.513	7.0	0.073
GB 2	1.000	14.0	0.071	0.412	6.0	0.068
GB 3	2.276	6.0	0.455	1.522	4.5	0.338
Total	4.695	29.3	0.160	2.447	17.5	0.140

3.2. Analytical TEM studies

The FIB thin foil was taken close to a triple junction visible at the surface of sample *SA* and has the particularity to reveal a cross-section of two oxidized grain boundaries with two oxide penetration features. As can be seen from bright field TEM micrographs of the thin foil (figure 5), there is a relatively deep oxide penetration along a grain boundary exempt of carbide (oxide penetration 1) and a second one that is stopped by an intergranular carbide close to the surface (oxide penetration 2). In addition, these micrographs reveal that the oxide found at the surface is made from large discontinuous crystallites of about 1 μm wide and 0.3 μm high and from a thin continuous inner layer of 0.1 - 0.2 μm thickness.

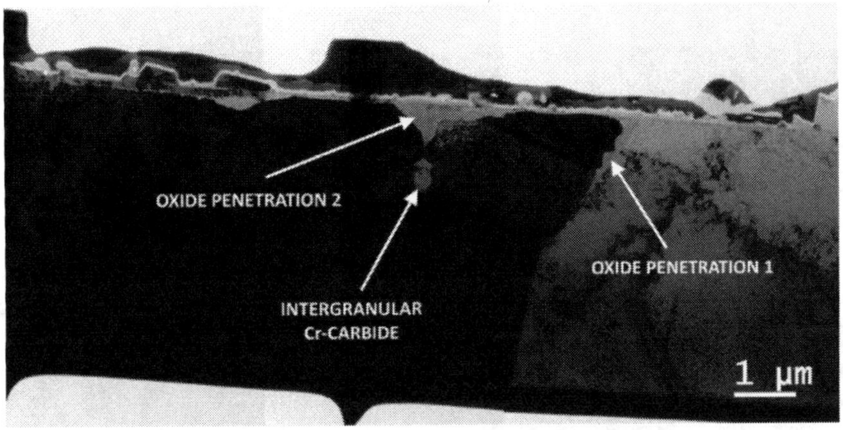

Figure 5. Bright field TEM micrograph overview of the thin foil sampled from sample *SA*.

EFTEM analyses reveal that the large crystallites are iron rich oxides and that the inner layer is composed of chromium rich oxides. Diffraction patterns also confirm EFTEM observations: outer crystallites are identified as nickel ferrite $NiFe_2O_4$ whereas the inner oxides are of spinel-type and close to the nickel chromite $NiCr_2O_4$ composition.

The oxide penetration 1, along a GB which does not contain any carbide (figure 6), is representative of oxide penetrations found in sample *SA*. It is composed of about 2 µm long chromium and nickel rich oxide. It is important to notice that the central part and the penetration tip are richer in chromium. Rings on the diffraction patterns acquired on the oxide penetration show that it is nanocrystalline. The distances match with a spinel-type oxide.

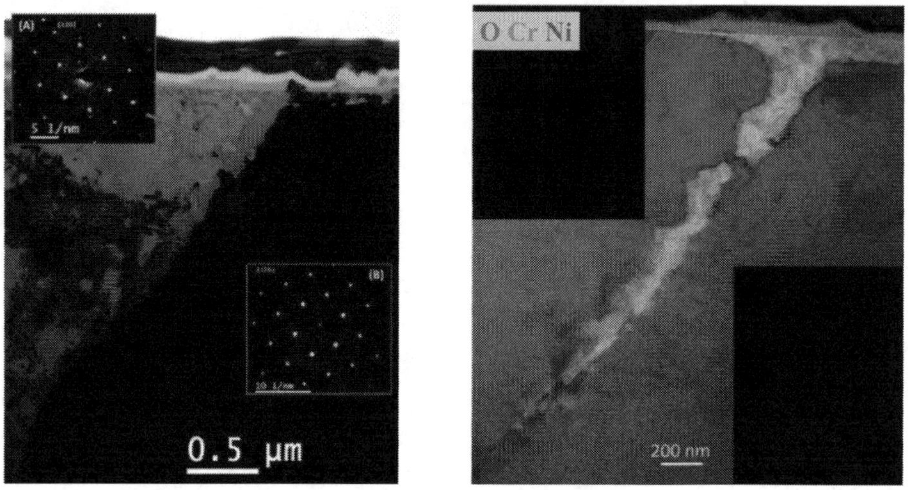

Figure 6. Left: Bright field TEM micrograph of oxide penetration 1 with insets showing the diffraction patterns of the intergranular oxide (A) and of the A600 matrix (zone axis: <110>) on one side of the oxide (B). Right: Corresponding EFTEM RGB composition map of respectively oxygen, chromium and nickel elemental maps.

The oxide penetration 2, which in this case is ended by a carbide (figure 7) like those found in sample *SA+TT*, is composed of a much broader chromium and nickel rich oxide penetration stopped by a chromium carbide particle. The oxide forms a cap on top of the carbide and seems to consume the surrounding matrix as it could be observed by FIB/SEM tomography.

4. Conclusions

Intergranular oxide penetrations in model Alloy 600 samples oxidized under simulated PWR primary water conditions have been characterized by FIB/SEM 3D tomography and analytical TEM. The effect of intergranular Cr carbides on the intergranular oxide depth has been studied. It was found that intergranular Cr carbides reduce intergranular oxide depths but enlarge them compared to intergranular oxide depths without carbide. However, the intergranular oxide volumes normalized by the GB length for both samples are similar, which suggest that intergranular oxidation growth rate is not affected by carbides. The FIB/SEM 3D tomography has allowed the reconstruction of volumes containing several oxide penetrations along grain boundaries with a remarkable nanometre spatial resolution. It also produced results of statistical significance, where intergranular oxidation depth

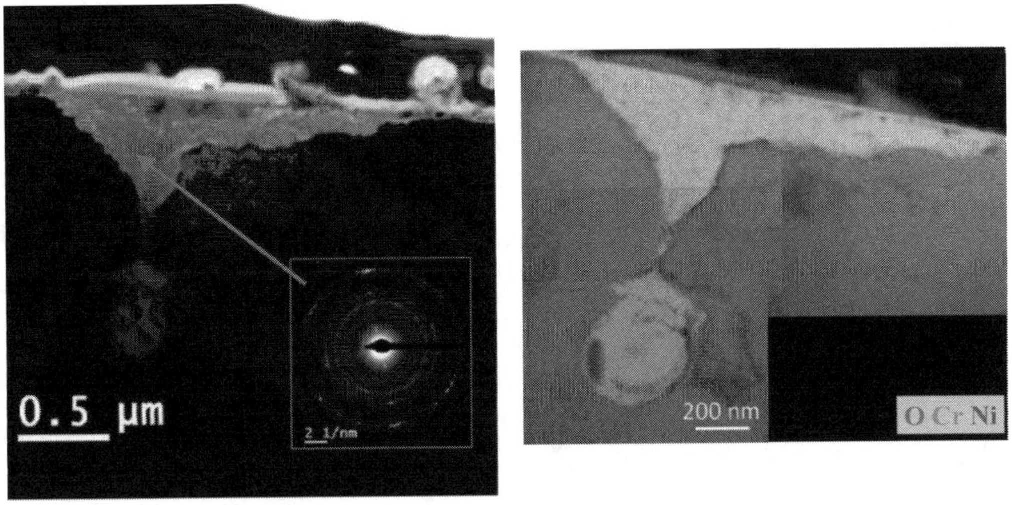

Figure 7. Left: Bright field TEM micrograph of oxide penetration 2 with an inset showing the oxide diffraction pattern. Right: Corresponding EFTEM RGB composition map of respectively oxygen, chromium and nickel elemental maps.

distribution histograms are used for characterisation. TEM observations have given a complete insight of the oxide penetration compositions and were identified as Cr and Ni rich spinel type oxides with the end tip richer in Cr.

Acknowledgement
This work was carried out within the MATMECA consortium and supported by the ANR under contract number ANR-10-EQUIPEX-37. It has benefited from the facilities of the Laboratory MSSMat (UMR CNRS 8579), CentraleSupélec, France.

References

[1] Combrade P, Scott P M, Foucault M, Andrieu E and Marcus P 2005 Oxidation of Ni base alloys in PWR water: oxide layers and associated damage to the base metal. in: *12th International Conference on Environmental Degradation of Materials in Nuclear Power Systems Water Reactors* (Salt Lake City, USA) 883-890

[2] Bruemmer S M and Was G S 1994 Microstructural and microchemical mechanisms controlling intergranular stress corrosion cracking in light-water-reactor systems. *J. Nucl. Mater.* **216** 348-363

[3] Turluer G, Cattiaux G, Monnot B, Emond D, Reuchet J and Chartier P 2003 *The French regulatory experience and views on nickel-base alloy PWSCC prevention and treatment.* (Fontenay aux Roses, France: Institut de Radioprotection et de Surete Nucleaire, IRSN)

[4] Wehbi M 2014 *Modélisation de l'amorçage de la corrosion sous contrainte dans les alliages base nickel 182 et 82 en milieu primaire des réacteurs à eau sous pression.* Thesis (Paris, France: Ecole Nationale Supérieure des Mines de Paris)

[5] Perusin S, Viguier B, Monceau D, Ressier L and Andrieu E 2004 Injection of vacancies at metal grain boundaries during the oxidation of nickel. *Acta Mater.* **52** 5375-5380

[6] Delabrouille F, Viguier B, Legras L and Andrieu E 2005 Effect of the chromium content on the corrosion of nickel based alloys in primary water of pressurised nuclear reactors. *Mater. High Temp.* **22** 287-292

[7] Panter J, Viguier B, Cloué J-M, Foucault M, Combrade P and Andrieu E 2006 Influence of oxide films on primary water stress corrosion cracking initiation of alloy 600. *J. Nucl. Mater.* **348** 213-221

[8] Lecoester F, Chêne J and Noel D 1999 Hydrogen embrittlement of the Ni-base Alloy 600 correlated with hydrogen transport by dislocations. *Mater. Sci. Engng. A* **262** 173-183

[9] Scott P M 1999 An overview of internal oxidation as a possible explanation of intergranular stress corrosion cracking of alloy 600 in PWRs. in: *Proc. 9th Int. Symp. on Environmental Degradation of Materials in Nuclear Power Systems-Water Reactors* (Newport Beach) 3-14

[10] Magnin T, Foct F and De Bouvier O 1999 Hydrogen effects on PWRSCC mechanisms in monocrystalline and polycrystalline alloy 600. in: *Proc. 9th Int. Symp. on Environmental Degradation of Materials in Nuclear Power Systems-Water Reactors* (Newport Beach) vol 27

[11] Chêne J and Brass A M 2004 Role of temperature and strain rate on the hydrogen-induced intergranular rupture in alloy 600. *Metall. Mater. Trans. A* **35** 457-464

[12] Kergaravat J-F 1996 *Influence de la microstructure sur la sensibilité à la corrosion sous contrainte des Alliages 600 et 690 en milieu primaire des réacteurs à eau sous pression.* Thesis (Grenoble, France: Institut National Polytechnique de Grenoble)

[13] Thomas L E and Bruemmer S M 2000 High-Resolution Characterization of Intergranular Attack and Stress Corrosion Cracking of Alloy 600 in High-Temperature Primary Water. *Corrosion* **56** 572-587

[14] Panter J 2002 *Étude de la corrosion sous contrainte des alliages 690 et 600 en milieu primaire de R.E.P.: influence des procédés de fabrication des tubes de générateur de vapeur sur la phase d'amorçage.* Thesis (Toulouse, France: Institut National Polytechnique)

[15] Laghoutaris P 2009 *Corrosion sous contrainte de l'Alliage 600 en milieu primaire des Réacteurs à Eau Pressurisée : apport à la compréhension des mécanismes.* Thesis (Evry, France: Mines Paristech, PEMAT- Centre des matériaux PM Fourt)

[16] Dugdale H, Armstrong D E J, Tarleton E, Roberts S G and Lozano-Perez S 2013 How oxidized grain boundaries fail. *Acta Mater.* **61** 4707-4713

[17] Sennour M, Laghoutaris P, Guerre C and Molins R 2009 Advanced TEM characterization of stress corrosion cracking of Alloy 600 in pressurized water reactor primary water environment. *J. Nucl. Mater.* **393** 254-266

[18] Lozano-Perez S, Kruska K, Iyengar I, Terachi T and Yamada T 2012 Understanding surface oxidation in stainless steels through 3D FIB sequential sectioning. *J. Phys. Conf. Ser.* **371** 012086

[19] Lozano-Perez S, Kruska K, Iyengar I, Terachi T and Yamada T 2012 The role of cold work and applied stress on surface oxidation of 304 stainless steel. *Corros. Sci.* **56** 78-85

[20] Guerre C, Laghoutaris P, Chêne J, Marchetti L, Molins R, Duhamel C and Sennour M 2012 Stress Corrosion Cracking of Alloy 600 in PWR Primary Water: Influence of Chromium, Hydrogen and Oxygen Diffusion. in: *15th Int. Conf. Environmental Degradation of Materials in Nuclear Power Systems-Water Reactors*. (Busby J T, Ilevbare G and Andresen P L; Eds.) (New York, NY: John Wiley & Sons, Inc.) 1477-1488

[21] Schaffer B, Grogger W and Kothleitner G 2004 Automated spatial drift correction for EFTEM image series. *Ultramicroscopy* **102** 27-36

[22] Giannuzzi L A and Stevie F A 1999 A review of focused ion beam milling techniques for TEM specimen preparation. *Micron* **30** 197-204

[23] Giannuzzi L A, Drown J L, Brown S R, Irwin R B and Stevie F A 1998 Applications of the FIB lift-out technique for TEM specimen preparation. *Microsc. Res. Tech.* **41** 285-290

Electron probe microanalysis of Ni-silicides at low voltage: difficulties and possibilities

E. Heikinheimo[1], P.T. Pinard[2], S. Richter[2], X. Llovet[3] and S.Louhenkilpi[1]

[1] Aalto University, Department of Materials Science and Engineering, P.O. Box 16200, 00076 Aalto, Espoo, Finland
[2] RWTH Aachen University, Central Facility for Electron Microscopy (GFE), Ahornstrasse 55, 52074 Aachen, Germany
[3] Universitat de Barcelona, Scientific and Technological Centres (CCiTUB), Lluís Solé i Sabarís 1-3, 08028 Barcelona, Spain

E-mail: erkki.heikinheimo@aalto.fi

Abstract. Interest in the use of EPMA at low voltage has grown considerably in recent years, mainly because of the availability of electron-beam instruments equipped with field-emission guns. However, EPMA at low voltage is marred by both experimental and analytical problems which may affect the accuracy of quantitative results. In the case of the analysis of transition elements, both the emission and absorption of X-rays are still poorly understood when they originate from electron transitions involving the partially filled 3d-shell. This is the case for the most intense Lα (L_3-M_5 transition) and Lβ (L_2-M_4 transition) lines. In this communication, we point out anomalies which appear to afflict the accuracy of EPMA of Ni-silicides using the Ni-Lα X-ray line and we discuss possible solutions.

1. Introduction

The incorporation of the field-emission (FE) gun into electron-beam instruments has stimulated the use of these instruments at low voltage and has opened up new opportunities for the characterisation of complex materials at a sub-micrometre scale. However, electron probe microanalysis (EPMA) at low voltage is marred by both experimental and analytical difficulties which may affect the accuracy of quantitative results [1-4]. On the one hand, at low energies, a surface layer several nanometres in thickness may represent a non-negligible fraction of the volume from which X-rays escape and therefore the influence of carbon contamination, surface oxidation, or the quality of the sample polish becomes crucial [1]. On the other hand, for most elements, conventional K-lines cannot be excited and therefore we have to rely on the use of L- and M-lines, which are less intense and more sensitive to chemical effects, especially below 1 keV photon emission energy.

Using conventional matrix correction procedures, quantitative analysis of the L-lines of transition elements is particularly problematic when the electron transitions involve the partially filled 3d-shell ($M_{4,5}$-subshells). This is the case for the most intense Lα (L_3-M_5 transition) and Lβ (L_2-M_4 transition) lines. Fialin showed that both the peak shape due to satellite emission and self-absorption of the Zn-Lα line emitted from Zn minerals largely depend on the electronic structure of these materials [5]. Pouchou illustrated the difficulties found when using the Ni-Lα line for the analysis of Ni aluminides [2]. The limitation of the use of the Lα- and Lβ-lines for the quantitative analysis of transition elements was recently discussed in a round-robin study [6], where participants analysed three steel alloy samples at low voltage. The results showed an underestimation of Cr and an overestimation of Fe and Ni, with relative deviations from the expected values which amounted up to 30 - 40 %. Jonnard *et al.* [7] reported the results of another round-robin study of a metallic glass sample. The results obtained at low voltage using Lα-lines showed an overestimation of the Ni and Co content by 13 % rel. and 11 % rel., respectively.

Gopon *et al.* [8] showed that the use of the less intense Ll (L_3-M_1 transition) line, which does not involve transitions from the unfilled 3d-shells, may be a suitable alternative for the analysis of Fe-silicides at low voltage. Statham and Holland [9] also used the Ll-lines to analyse a high Cr-Ni steel. These authors concluded that it may be a route to improve quantitative analysis, although Cr-concentrations were found to be 9 % below expected values. Pinard *et al.* [10] used a similar strategy to re-analyse the steel alloy samples from Llovet *et al.*'s round-robin study [6] as well as other steel samples. Although good agreement was obtained for the major elements using the Ll-lines, the authors observed large discrepancies for the minor elements such as Mn (0.7 - 1.6 wt%) and Co (0.3 - 1.5 wt%). They were attributed to the low intensity of the Ll-lines and the presence of several interferences. It is therefore of great importance to further understand the quantification problem using the Lα-lines and to identify whether these difficulties are also found in other types of materials.

In this study, we have carried out systematic EPMA measurements on Ni-silicides in order to investigate the anomalies which appear to afflict the analysis of such materials using Lα-lines and we discuss possible solutions. The Ni-silicide system was selected for its simplicity as it consists of several binary compounds with a different composition and crystallographic structure. Furthermore, Ni-silicides play an important role in semiconductor technology, where methods for an accurate analysis of sub-micrometre Ni-silicide phases are needed.

The rest of the paper is structured as follows. Section 2 provides information on the experimental methods. In Section 3 we briefly discuss the theoretical calculations performed. We present and discuss the results of the different measurements and calculations performed in Section 4, and Section 5 concludes the paper.

2. Experimental method

The Ni-silicides were prepared by the diffusion-couple technique. The starting metals were bulk samples of Ni and Si (99.95+ purity). Diffusion couples were prepared by hot pressing the polished Ni and Si pieces together. The diffusion experiments were carried out by annealing the couples at 850 °C for 16 hours in a reducing atmosphere. Cross-sections of the diffusion couples, perpendicular to the interface, were cut, embedded in conductive resin, ground and finally polished.

The compounds Ni_5Si_2, Ni_2Si, Ni_3Si_2, and NiSi were identified by their different atomic number contrast in the backscattered electron image (figure 1) and quantified by conventional EPMA at 15 kV using the Kα (K-L_3 transition) line for both Ni and Si, pure Ni and Si metals standards, and the matrix correction algorithm XPP [11], as implemented in the microprobe software (table 1). The deviation (percent error) of the measured concentrations with respect to the nominal values is 1.9 % for Si and 0.4 % for Ni.

X-ray measurements were performed on three different instruments, namely a LEO 1450 scanning electron microscope (SEM) equipped with an INCA wavelength-dispersive X-ray spectrometer (WDS) at the Aalto University (Finland), a JEOL JXA-8230 electron microprobe at the University of Barcelona (Spain) and a JEOL JXA-8530F electron microprobe at the RWTH Aachen University (Germany). The aim of using different instruments (as well as operators, procedures, samples, etc.) was mainly to assess the reproducibility of the different measurements. The advantage of the FE microprobe (JXA-8530F) was the more reliable analysis of small NiSi-phases close to pure silicon. Furthermore, in the SEM the take-off angle is 30°, a much lower value than that of the two electron microprobes (40°). The use of the SEM-WDS reinforces confidence in the data as regards the influence of X-ray absorption effects on the analysis.

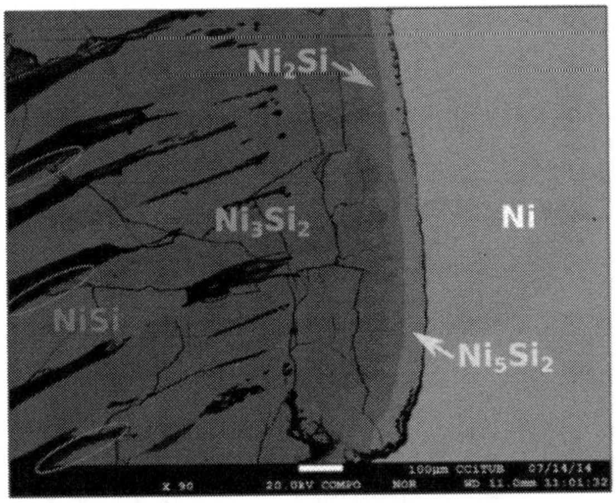

Figure 1. Backscattered electron image of the Ni-Si diffusion couple cross-section showing the different Ni-silicide phases.

Table 1: Nominal composition and results of EPMA analysis at 15 keV using the Ni-Kα and Si-Kα lines for the studied Ni-silicide phases. The EPMA analyses are the average of 10 measurements, with the corresponding standard deviation (at 1σ level) in parenthesis. Pure metal standards were used.

	Nominal concentration (wt%)		EPMA analysis (wt%)		
Phase	Ni	Si	Ni	Si	Total
Ni_5Si_2	83.9	16.1	84.2 (0.2)	15.6 (0.04)	99.8
Ni_2Si	80.7	19.3	80.9 (0.2)	19.1 (0.05)	99.9
Ni_3Si_2	75.8	24.2	76.1 (0.2)	23.8 (0.09)	99.9
NiSi	67.6	32.3	67.9 (0.3)	31.7 (0.3)	99.6

Four series of measurements were performed on the four Ni-silicides. In all cases, when *k*-ratios were measured pure Ni and Si standards were used, and measurements were made at 10 different positions, with typical counting times of 30 s. The dispersing crystals were TAP crystals. Firstly, conventional EPMA measurements using the Ni-Lα and Si-Kα lines, pure Ni and Si metals standards, and the matrix correction procedure XPP, as implemented in the JEOL microprobe software, were performed at 6 keV. The beam energy of 6 keV was selected to reach a compromise between X-ray production efficiency and spatial resolution (a lower beam energy would not improve the spatial resolution because of the larger beam size) [3]. Secondly, Ni-Lα net intensities and *k*-ratios, extracted from peak-height intensity measurements, were collected for incident beam energies in the range 2 - 30 keV on all Ni-silicides and pure Ni. Although a different beam current was used at each accelerating voltage, it was measured before and after each measurement to diminish errors due to instrument instabilities. In the case of the SEM-WDS as well as the FE-microprobe, a peak search was also performed for each accelerating voltage and silicide. Thirdly, Ni-Ll *k*-ratios were acquired using the same procedure using a TAPH crystal (only on the JEOL JXA-8230 instrument). Finally, X-ray emission spectra around the positions of the Ni-Lα and -Lβ peaks at 2 and 30 keV were acquired using a Johann spectrometer with a 300 μm diameter exit slit in an attempt to ensure the highest spectral resolution. Measuring time was 7500 s per spectrum, with currents of 500 nA at 30 keV and

250 nA at 2 keV. The beam was defocussed to 5 μm. In order to minimize carbon contamination and its potential influence on peak shape changes [12], the sample was cleaned with plasma cleaner for 10 min prior to its introduction in the microprobe chamber.

3. Calculations

k-ratios for the different X-ray lines and studied compounds were calculated using the Monte Carlo (MC) simulation programmes PENEPMA [13] and MONACO [14], and the analytical model XPP [11], as implemented in the software DTSA-II [15].

Based on the general-purpose MC simulation package PENELOPE [16], PENEPMA simulates the transport of both electrons and photons using interaction models which combine results from first-principles calculations, semi-empirical models and evaluated databases (see Ref. [16] for details). In the case of photon absorption, which is non-negligible for soft X-rays, PENELOPE uses total cross-sections for photoelectric absorption from the Evaluated Photon Data Library (EPDL) [17]. By definition, the commonly used mass absorption coefficients (MAC) are directly proportional to the photoelectric cross-sections. PENELOPE simulates the emission of almost all known K-, L-, M- and N-lines, which include the Ll-line.

MONACO is a MC simulation programme of electron transport based on elastic and inelastic single scattering events calculated by using the Mott cross-sections and the modified Gryzinski cross-sections, respectively [13]. The mean energy loss is calculated by means of the Bethe formula and corrected by the energy losses of the single inelastic scattering events. It neglects any solid state structure. X-ray absorption is analytically computed from the conventional Beer-Lambert exponential relation using MACs from a selection of MACs from Henke [18] and Heinrich [19]. MONACO allows the simulation of the most commonly used X-ray lines, namely the Kα-, Kβ-, Lα-, Lβ-, Mα- and Mβ-lines.

DTSA-II is a multi-platform software package for quantitative EPMA where several correction algorithms and sets of fundamental parameters, including MACs, can be used. The implementation of the XPP model from Pouchou and Pichoir [11] was used for all analytical calculations. For the quantification of the Ni-Lα results, the MACs as defined in [14] were used. They consist of an experimentally determined MAC for Ni-Lα in Ni and MACs from Heinrich [19] for the rest of single-element materials. For the quantification of Ni-Ll results, the MACs from Chantler *et al.* [20] were used.

Furthermore, DTSA-II was modified to allow the use of different MACs to correct for the absorption in both the unknown and standard samples. Notice that most EPMA correction programmes assume the same MAC for both corrections. Finally, using the libraries and databases within DTSA-II, an in-house programme was developed to derive empirical MACs from the EPMA measurements of the studied materials. This programme follows the method described by Pouchou and Pichoir [21] and is implemented in the software package XMAC [2], which was also used for comparison purposes. For a given X-ray line and sample, our programme considers the MAC of each absorber as a free parameter in the XPP model. The input values are experimental X-ray intensities measured at different accelerating voltages. The programme yields the MAC values which give the best fit between the experimental and calculated X-ray intensities.

4. Results and discussion

4.1. k-ratios versus energy (Lα)

The results of the analysis of the different Ni-silicide phases at 6 keV using the Ni-Lα line are tabulated in table 2. The deviation (percent error) of the obtained Ni concentrations with respect to the nominal values adds up to about 20 % - 30 % rel. This obviously throws serious doubts upon such results. Notice that only the Ni concentrations are overestimated, the Si values satisfactorily match the nominal concentrations as well as the measurements obtained at 15 keV (see table 1).

Table 2. Analysis of three of the silicide phases at 6 keV using the Ni-Lα and Si-Kα lines, Ni and Si pure metal standards and the matrix correction algorithm XPP, as implemented in the JEOL microprobe software.

	Nominal concentration (wt%)		EPMA results (wt%)		
Phase	Ni	Si	Ni	Si	Total
Ni_5Si_2	83.9	16.1	102.1	15.8	117.9
Ni_2Si	80.7	19.3	105.3	19.1	124.5
Ni_3Si_2	75.8	24.2	100.7	24.6	125.3

Figure 2 shows the experimental and calculated Ni-Lα *k*-ratios as a function of incident electron energy for the four Ni-silicides. There is good agreement between the experimental *k*-ratios measured on the three different instruments, which increases confidence in the data. The results obtained from the two MC programmes agree well with each other. The simulated results slightly differ from those obtained by XPP, especially at high energies. However, all calculated *k*-ratios differ greatly from the experimental values. The largely increasing differences with increasing beam energy suggest a problem with the absorption correction and seem to point to incorrect MAC values used in the calculations.

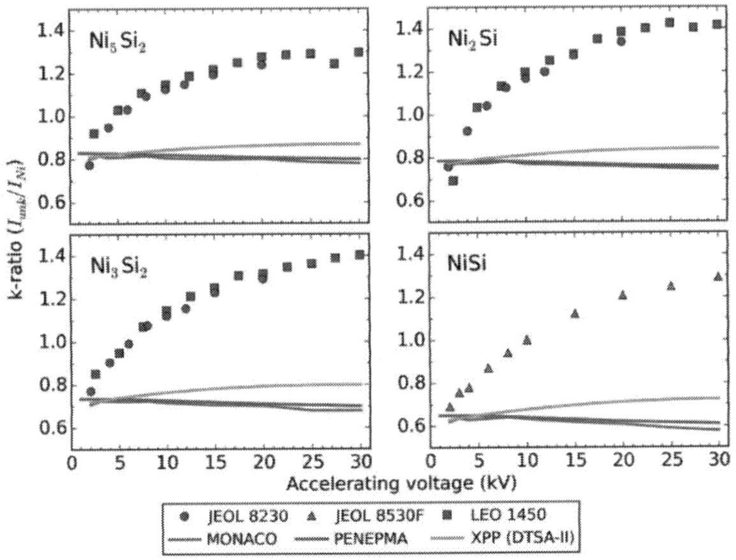

Figure 2. Experimental and calculated Ni-Lα *k*-ratios of four Ni-silicides as a function of accelerating voltage. Measurements were performed on 3 different instruments, as indicated in the legends. Calculations were performed with the MC codes MONACO, PENEPMA and the analytical model XPP using DTSA-II.

The large differences observed between experimental and calculated *k*-ratios at 6 keV are consistent with the erroneous Ni concentrations measured at this energy. The error in the concentration would be undoubtedly worse at higher beam energies.

IOP Conf. Series: Materials Science and Engineering **109** (2016) 012005 doi:10.1088/1757-899X/109/1/012005

4.2. Absorption X-ray spectra

In order to shed light on the origin of the observed *k*-ratio discrepancies, X-ray absorption spectra around the L_2- and L_3-absorption edges were acquired on pure Ni and on the Ni-silicide samples. Absorption spectra were obtained from the ratio of two normalized X-ray emission spectra measured at low and high incident electron energy [22-24]. As shown by e.g., Chopra [24], absorption spectra obtained in this way are equivalent to X-ray absorption measurements using X-ray beams and thin foil absorbers.

The X-ray emission spectra from pure Ni obtained at 2 and 30 keV are presented in figure 3a. The Lα- and Lβ-lines represent radiatiave transitions of M_5-electrons to vacancies in the L_3-shells, and of M_4-electrons to vacancies in the L_2-shells, respectively. The spectra have been normalized to yield the same peak intensity for the Lα-line. For both the Lα- and Lβ-lines, the peak positions at 30 keV are shifted towards a lower energy with respect to those at 2 keV. The shape of the lines also shows a change, the high-energy side of the Lα- and Lβ-lines as they are lower at 30 keV than at 2 keV.

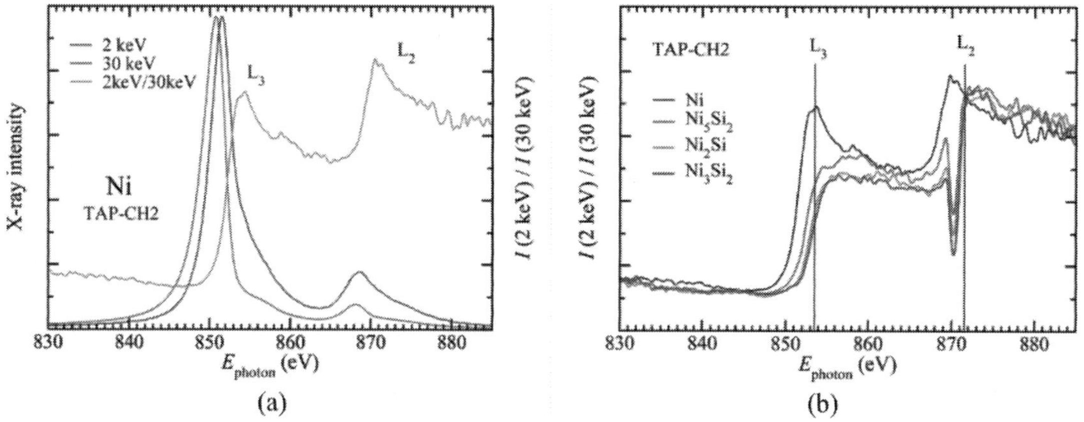

Figure 3. L X-ray emission spectra at 2 and 30 keV incident electron energy for pure Ni and $L_{2,3}$-absorption spectrum obtained from the ratio of the X-ray emission spectra at 2 and 30 keV (a) and X-ray absorption spectra of the Ni $L_{2,3}$-region for pure Ni and three Ni-silicides (b).

Figure 3a also shows the Ni $L_{2,3}$-absorption spectrum obtained from the ratio of the X-ray emission spectra at 2 and 30 keV. Two lines are observed just above the L_3- and L_2-absorption edges. These lines are generally referred to as white lines [24] or also as anomalous absorption [22]. White lines are believed to be associated with transitions from an inner-shell (in this case $L_{2,3}$-electrons) to an unoccupied level with a high density of states near the absorbing atom (in this case the 3d states). White lines are also known to depend on chemical bonding. As pointed out by Bonnelle [22], the changes in the shape and position of the Ni-Lα and -Lβ lines with increasing beam energy (see figure 3a) can be satisfactorily explained by the attenuating effect of the $L_{2,3}$-absorption spectrum on the natural Lα- and Lβ-line widths.

Figure 3b shows a comparison of the Ni $L_{2,3}$-absorption spectra obtained for Ni, Ni_5Si_2, Ni_2Si and Ni_3Si_2. The white lines are less visible for the Ni-silicides when compared to pure Ni. This suggests that the absorption of Ni-Lα X-rays (E = 851.47 eV) by Ni atoms bonded to Si atoms is lower than for Ni atoms bonded to Ni atoms. The L_3-absorption edge of Ni shows a shift towards higher energy for the Ni-silicides, with respect to that of pure Ni. The drop in intensity observed at 871 eV near the L_2-edge for the Ni-silicides is due to a 2nd-order Bragg reflection of the Si-Kα which is stronger in the 30 keV spectrum.

It therefore appears that the increase of Si concentration in the Ni-silicides decreases the absorption of Ni-Lα X-rays by Ni atoms. As the Ni concentration decreases, the total absorption of Ni-Lα X-rays by Ni atoms decreases following the linear combination of the weighted mass absorption coefficients. However, in the case of Ni-silicides, the chemical bonding between Ni and Si atoms seems to eliminate, or at least to reduce, the anomalous absorption of Ni-Lα X-rays by Ni atoms. Absorption in the Ni-silicides is consequently lower than expected by the conventional matrix correction procedures. This effect can be one explanation for the large discrepancies between the experimental and calculated *k*-ratios (figure 2) and for the experimental *k*-ratio values above 1.

As discussed by Jarrige *et al.* [25], the electronic structure of Ni-silicides is governed by the interaction of the almost "discrete" Ni d-valence states with the "continuum" of Si p-states, which depends on the environment of the Ni and Si atoms. As a consequence, the absorption coefficient, which is proportional to the density of unoccupied states, also depends on the considered compound.

4.3. Empirical MACs

In order to quantify the different absorption behaviour of Ni-Lα X-rays by Ni atoms in the Ni-silicides, empirical MACs were experimentally determined from EPMA measurements. Indeed, for a given X-ray line and material, it is possible to obtain a MAC value by considering it as a free parameter in the EPMA correction procedure and then fit the predictions of the model to relative X-ray intensities measured as a function of beam energy (or take-off angle) [21]. Obviously, this empirical MAC is expected to give reliable results only in combination with the considered model and other experimental conditions such as the spectral resolution of the spectrometer used (in the case of peak height measurements).

We have developed our own programme which allows determination of the MAC for the absorption of one or several X-ray lines by atoms of different elements within materials of known composition by using the XPP model [14].

Figure 4 shows both the experimental and fitted Ni-Lα intensities as a function of beam energy for Ni and Ni_3Si_2 (figure 4a) and the experimentally determined MACs for the Ni-Lα absorption by Ni atoms in Ni and the Ni-silicides as a function of Ni concentration (figure 4b). The agreement between the measured and fitted intensities is excellent (figure 4a). The MACs for the different Ni-silicides are significantly lower than that for pure Ni, as illustrated in figure 4b. Tabulated MACs for a given X-ray line and absorbing atom are normally supposed to be independent on the concentration of absorbing atoms and thus should have a constant value (horizontal line in figure 4b). The qualitative trends observed in the X-ray absorption spectra for Ni and the Ni-silicides (figure 3b) are confirmed by the experimentally determined MACs values.

Table 3 tabulates the MAC for Ni-Lα and -Ll X-rays in Ni from different, widely used MAC compilations. The MACs based on EPMA measurements (Pouchou and Pichoir [11] and this study), are larger than those from the other compilations, most probably because the former account for the enhanced absorption of Ni-Lα in Ni.

4.4. Improved k-ratios using the measured MACs

In order to take into account the anomalous absorption behaviour of the Ni-Lα line, the *k*-ratios were re-evaluated using the experimental MACs. To this end, the DTSA-II programme was slightly modified to allow the use of a different MAC (of the same element) for the unknown and standard.

Figure 5 shows a comparison of the new evaluated *k*-ratios (labelled "Measured MAC") with the previously obtained *k*-ratios ("Original MAC") and the experimental data as shown in figure 2 for Ni_3Si_2. Although the new *k*-ratios are closer to the measured values, large differences still remain. This suggests that other anomalies may be taking place, most probably involving X-ray emission. Pouchou [2] suggested that for Ni aluminides the probability of the Ni 2p-3d radiative transition also depends on the structure of the valence band (e.g., on chemical bonding, which implies addition or removal of valence electrons). Hence, the assumption that the radiative transition probability is the same for the unknown and the standard may no longer be valid. Unfortunately, an explanation of the dependence of the radiative transition probability on Si concentration in the case of Ni-silicides is not at hand. Furthermore, as discussed by Bonnelle [22], the intensity of an electron transition also depends on the density of states and the local partial density of Ni 3d-states largely depends on the composition of the Ni-silicide [25].

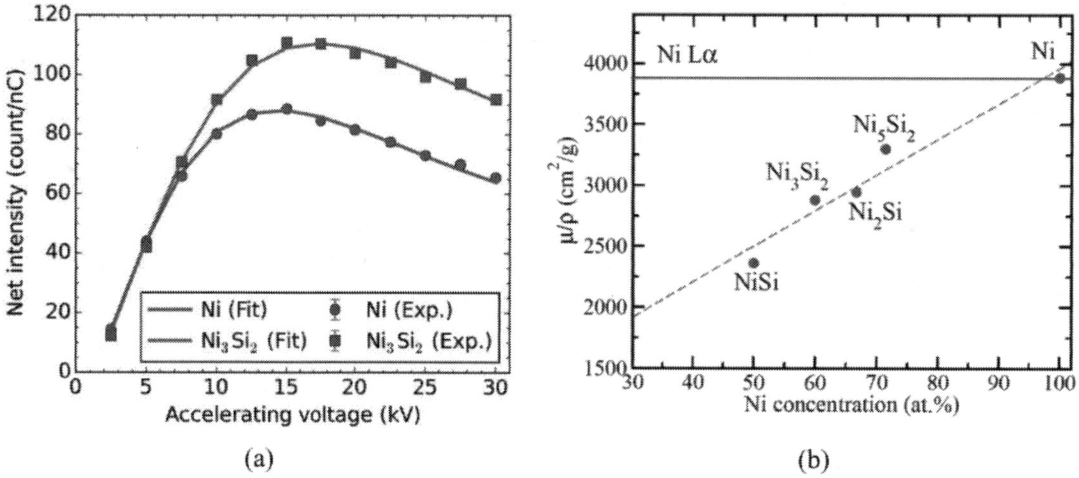

Figure 4. Experimental and fitted Ni-Lα intensities for pure Ni and Ni_3Si_2 (a) and Ni-Lα MAC by Ni atoms in pure Ni and four Ni-silicides as a function of Ni concentration (b).

Table 3. MACs for Ni-Lα and -Ll X-rays in Ni taken from different compilations (in cm^2/g). For the compilations for which the X-ray lines are not detailed, the MACs have been extracted at the photon energies of 851.47 eV (Lα) and 742.72 eV (Ll).

Source	Lα	Ll
Henke *et al.* (1985) [18]	1,819	-
Henke *et al.* (1993) [25]	1,802	2,402
Heinrich [19]	1,806	2,426
Cullen *et al.* [17] – EPDL	1,654	2,225
Chantler *et al.* [20] – FFAST	1,697	2,255
Pouchou and Pichoir [11]	3,560	-
This study	3,998	-

4.5. k-ratios using the Ni-Ll line and conventional MACs

One solution to the quantification problems of Ni-silicides at low voltage is to use the Ni-Ll line, as has already been suggested by [8-10]. The Ll-line, with an energy of 742 eV, is far from the $L_{2,3}$-absorption edges, and corresponds to a radiative transition (L_3-M_1) that does not involve the unfilled 3d-shell.

Figure 6 shows a comparison of the measured and calculated *k*-ratios for the Ni_5Si_2- and Ni_3Si_2-phases, as a function of beam energy. The calculations were performed with PENEPMA and XPP (DTSA-II), using the MAC value from EPDL and FFAST database, respectively (table 3).

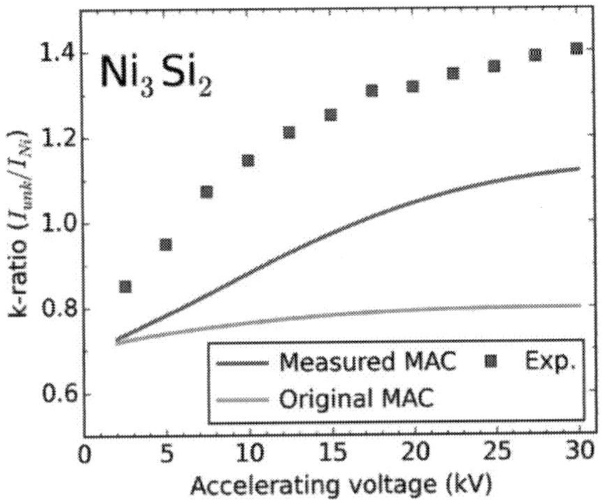

Figure 5. Experimental Ni-Lα *k*-ratios and calculated *k*-ratios using different MACs as indicated in the legends (see text for explanation).

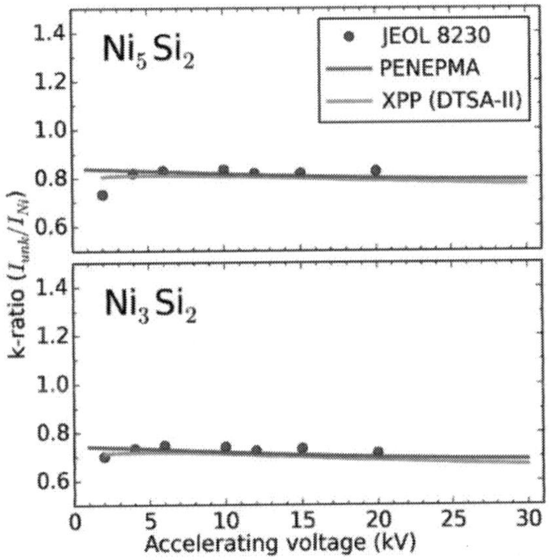

Figure 6. Experimental and calculated Ni-Ll *k*-ratios as a function of accelerating voltage.

It should be noted that Ll-line measurements are more challenging because of the much lower intensity of the Ni-Ll line when compared to the Ni-Lα line. Longer acquisition times and/or high beam currents are required which lead to problems due to contamination and other experimental issues [4]. The use of larger diffracting crystals or the simultaneous use of similar crystals in different spectrometers may help to mitigate such problems as demonstrated by the satisfactory agreement between the experimental and calculated *k*-ratios (figure 6). It should be noted that most matrix correction programmes do not allow the use of Ll-lines for quantification, since MACs are often tabulated only for the most widely used X-ray lines, namely the Kα-, Kβ-, Lα-, Lβ-, Mα- and Mβ-lines.

5. Conclusions

Systematic EPMA measurements on the Ni-Si system have been performed in order to illustrate the difficulties of using the Ni-Lα line for the quantitative analysis of Ni-silicides at low beam energy. In particular, we have shown that the behaviour of the Ni $L_{2,3}$-absorption spectra is very sensitive to chemical bonding. This means that the MACs of Ni-Lα due to Ni atoms in Ni-silicides depend on

Si concentration and they significantly differ from tabulated values. By using MACs empirically calculated from EPMA measurements, the deviation between the measured and calculated Ni-Lα *k*-ratios decreases, but the results are still far from being satisfactory. This suggests that X-ray absorption is not the only factor causing the anomalous behaviour of the Ni-Lα line in the Ni-silicides. We have finally shown that the difficulties in getting reliable accuracy for the above-studied compounds can be overcome by using the Ni-Ll line.

Acknowledgments
Financial support from the Spanish Ministerio de Economía y Competitividad (project no. FPA2013-44549-P), from the Generalitat of Catalunya (grant 2014 SGR 846) and from the Deutscher Akademischer Austauschdienst (DAAD) are gratefully acknowledged.

References

[1] Willich P and Bethke R 1996 *Mikrochim. Acta* **Suppl. 13** 63
[2] Pouchou J L 1996 *Mikrochim. Acta* **Suppl. 13** 39
[3] Pinard P T and Richter R 2014 *IOP Conf. Series: Mater. Sci. Engng.* **55** 012016
[4] Merlet C and Llovet X 2012 *IOP Conf. Series: Mater. Sci. Engng.* **32** 012016
[5] Fialin M 1990 *X-ray Spectrom.* **19** 169
[6] Llovet X, Heikinheimo E, Núñez A, Merlet C, Almagro J F, Richter S, Fournelle J and van Hoek C G 2012 *IOP Conf. Series: Mater. Sci. Engng.* **32** 012014
[7] Jonnard P, Brisset F, Robaut F, Wille G and Ruste J 2014 *X-ray Spectrom.* **44** 24
[8] Gopon P, Fournelle J, Sobol P and Llovet X 2013 *Microsc. Microanal.* **19** 1698
[9] Statham P and Holland J 2014 *IOP Conf. Series: Mater. Sci. Engng.* **55** 012017
[10] Pinard P T, Heikinheimo E, Llovet X and Richter S 2014 *Microsc. Microanal.* **20** (Suppl. 3) 700
[11] Pouchou J L and Pichoir F 1991 *Electron Probe Quantitation.* (Heinrinch K F J and Newbury D E; Eds.) (New York, NY: Plenum Press)
[12] Gopon P, Sobol P and Fournelle J 2013 *Microsc. Microanal.* **19** (Suppl. 2) 814
[13] Llovet X, Fernández-Varea J M, Sempau J and Salvat F 2005 *Surf. Interface Anal.* **37** 1054
[14] Ammann N and Karduck P 1990 *Microbeam Analysis.* (Michael J R and Ingram P; Eds.) (San Francisco, CA: San Francisco Press)
[15] Ritchie N W M 2009 *Microsc. Microanal.* **15** 454
[16] Salvat F 2015 *PENELOPE-2014. A Code System for Monte Carlo Simulation of Electron and Photon Transport.* (Issy-les-Moulineaux, France: OECD/Nuclear Energy Agency)
[17] Cullen D E, Hubbell J H and Kissel L 1997 *EPDL97 The Evaluated Data Library, Technical Report UCRL-50400.* (Livermore, CA: Lawrence Livermore National Laboratory)
[18] Henke B L, Lee T J, Tanaka R J, Shimabukuro R L and Fijikawa B K 1985 *At. Data Nucl. Data Tables* **27** 1
[19] Heinrich K F J 1986 in: *Proc. 11th International Congress on X-ray Optics and Microanalysis* (Brown J D and Packwood R H; Eds.) (Ontario: University of Western Ontario)
[20] Chantler C T, Olsen K, Dragoset R A, Chang J, Kishore A R, Kotochigova S A and Zucker D S 2005 *X-ray Form Factors, Attenuation and Scattering Tables (version 2.1)*
[21] Pouchou J L and Pichoir F 1988 *Microbeam Analysis.* (Newbury D E; Ed.) (San Francisco, CA: San Francisco Press)
[22] Bonnelle C 1987 *Ann. Rep. Progr. Chem., Sect. C, Phys. Chem.* **84** 201
[23] Rémond G, Myklebust R, Fialin M, Nockolds C, Phillips M and Roques-Carmes C 2002 *J. Res. Natl. Inst. Stand. Technol.* **107** 509
[24] Chopra D 1970 *Phys. Rev. A* **1** 230
[25] Henke B L, Gullikson E M and Davis J C 1993 *At. Data Nucl. Data Tables* **54** 181
[26] Jarrige I, Capron N and Jonnard P 2009 *Phys. Rev. B* **79** 035117

IOP Conf. Series: Materials Science and Engineering **109** (2016) 012006 doi:10.1088/1757-899X/109/1/012006

Characterisation of nanoparticles by means of high-resolution SEM/EDS in transmission mode

V-D Hodoroaba[1], S Rades[1], T Salge[2], J Mielke[1], E Ortel[1] and R Schmidt[3]

[1] BAM – Federal Institute for Materials Research and Testing, Unter den Eichen 87, 12205 Berlin, Germany

[2] Natural History Museum, Core Research Laboratories, Cromwell Road, London SW7 5BD, Great Britain

[3] Hitachi High-Technologies Europe, Europark Fichtenhain 12A, 47807 Krefeld, Germany

E-mail: dan.hodoroaba@bam.de

Abstract. Advances in scanning electron microscopy (SEM) enable the high-resolution imaging of single nanoparticles (NPs) with sizes well below 10 nm. The SEM analysis in transmission mode (T-SEM) of NPs on thin film supports has many benefits when compared to the analysis of NPs on bulk substrates. The enhanced material (mass – thickness) contrast of the T-SEM imaging mode is well suited for in-depth and, particularly valuable, to very accurate, traceable, lateral dimensional measurements of NPs. Compared to samples prepared on bulk substrates, T-SEM with energy dispersive X-ray spectroscopy (EDS) achieves a drastically improved spatial resolution of the emitted X-rays. The poor signal-to-noise ratio of the X-ray spectra emitted by a single nanoparticle (NP) can be improved by the use of high-sensitivity (high collection solid angle) silicon drift (SDD), energy-dispersive X-ray spectrometers (EDS). The EDS spectral imaging of a single NP with a spatial resolution below 10 nm has become possible. This is demonstrated by means of various examples of nanostructures. Advanced data processing of T-SEM/EDS results sets the stage for the automated classification of NPs by feature analysis. This method combines the detection of morphological structures of interest by image processing of T-SEM micrographs with the chemical classification by EDS.

1. Introduction

Due to the application of nanoparticles (NPs) within commercial products, there is an increasing demand for SEM/EDS laboratories to characterise nanoparticulate materials. Technological improvements in SEM and within energy-dispersive X-ray spectrometry (EDS) allow the analysis of sample surface morphology, the inner structure and the elemental composition with a comparable spatial resolution in the nanometre range.

The conventional Everhart–Thornley (E-T) and the *In-Lens* detector of modern SEMs are able to image sample surface morphology at high spatial resolution to a sub-nm scale. In addition, STEM (scanning transmission electron microscopy) detectors collect transmitted electrons resulting in imaged NPs with a superior material (mass and thickness) contrast. For the latter option, the specimen must be prepared on an electron transparent thin supporting film which is analogue to the preparation for transmission electron microscopy (TEM).

Imaging by means of transmitted electrons in an SEM exploits the high signal-to-noise ratio resulting from the dominant "forward" electron scattering within a nanoscale sample. In addition, the transmitted electrons suffer reduced scattering events within the thin sample, resulting in an intense electron signal in transmission mode. Surface effects, such as charging and contamination are not as critical in comparison with the secondary electron (SE) imaging mode [1]. Hence, the SEM in transmission mode (T-SEM) provides a metrological tool for dimensional (lateral) measurements of NPs, i.e., particle size distribution [1-4], but is also well suited for "in-depth" observation of structured NPs [4, 5].

The direct measurement of nanoparticle (NP) size and shape/morphology by high-resolution, electron-beam based imaging techniques can be complemented by spatially resolved chemical information as offered by EDS. Two main technological developments have enabled the improvement of the spatial resolution of EDS analysis from micro- to nano-range: (i) The development of high-sensitivity, high solid angle SDD detectors, and (ii) the application of T-SEM by preparing the NPs on standard TEM grids. Hence, the weak X-ray signal emitted by a single NP on an electron-transparent supporting film can be detected with sufficiently high signal-to-noise ratio. The excited sample volume is drastically reduced in comparison to a bulk support.

Furthermore, it has recently been demonstrated that the "bulk" EDS analysis of NPs can be supplemented by chemical imaging of ultra-thin (up to a few nm) outer shells by scanning Auger electron microscopy (SAM) [6, 7]. The lateral resolution attained by this technique is in the order of 10 nm.

2. Experimental methods

In order to achieve high-resolution imaging from the upper-most surface of a specimen using an SEM, advanced detectors are required. These are used to selectively detect so-called SE1 electrons that are emitted locally at the point of interaction between the primary electron beam and the specimen surface. In this study the detector employed was an *In-Lens* SE detector (Zeiss Supra 40, FE-SEM, Oberkochen, Germany), see figure 1a. Additional selected images for studying the inner structure of silica particles have been included, that were obtained using a FE-SEM (Hitachi SU 8230, Tokyo, Japan) equipped with different electron detectors, see later figure 6.

The proof-of-principle of T-SEM was demonstrated over thirty years ago [8-10]. The application to the characterisation of NPs however, has only recently been reported [1-5, 11]. In particular, the capability of T-SEM to the metrological measurement of lateral dimensions in the nanometre range, e.g., for the purpose of accurate NP size and distribution determination, has already been successfully exploited. Basically, the STEM detector which is placed directly under the thin specimen (figure 1b) can be used for acquiring images in transmission mode in an SEM [12].

In this study, a commercially available transmission unit has been employed [13]. This consists of a specimen holder that guides the transmitted electrons (TE) onto an electron multiplier in the form of a gold plate located below the bright-field (or alternatively dark-field) aperture (figures 1c and 1d). The multiplied transmitted electrons are then collected by the conventional E-T detector. The screening ring (see figure 1d) prevents surface sample generated SE and BSE electrons from directly entering the E-T detector. On the other side, the screening ring in operation mode prevents also the X-rays emitted by the sample to reach the EDS detector, so that it is necessary to remove manually the ring before an EDS analysis. Furthermore, the single-unit transmission setup is designed for only one TEM grid. The TSEM detector can analyse up to four grids mounted in commercial sample holders and EDS can be carried out without being necessary to (re)move parts. Contrary to the STEM detector the single-unit setup works reliably at low energies (below 7 keV). Not unimportant, the single-unit transmission sample holder can be easily taken and inserted in any other SEM having a sample stage with dove-tail mount. Summing up, T-SEM imaging with the single-unit setup can be carried out practically with similar performance when compared to imaging using a STEM detector [1-4].

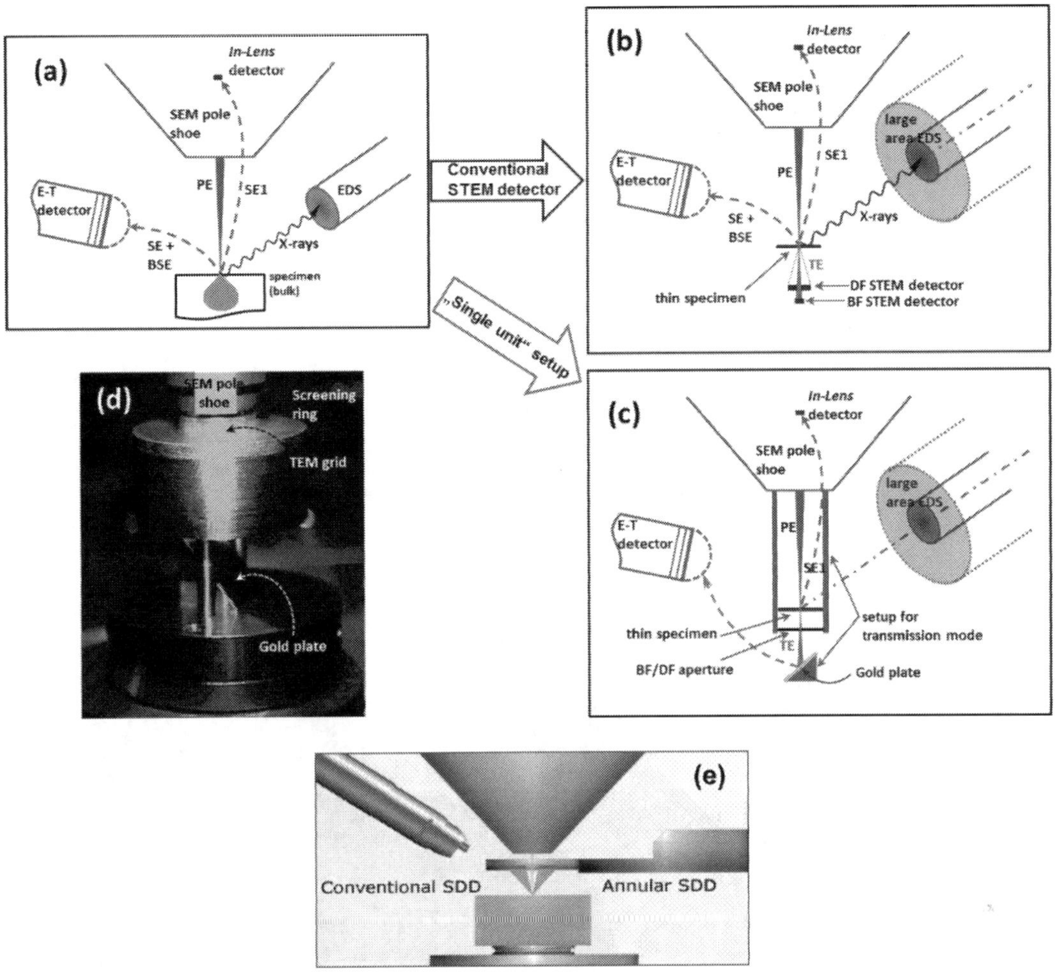

Figure 1. Schemes of various SEM/EDS configurations suited to various types of analysis: (a) Bulk sample analysis in a SEM showing a conventional E-T detector for the collection of SE/BSE electrons. SE electrons can also be collected with a high-resolution *In-Lens* SE detector. Note the relative large emission volume of X-rays which can be collected by an EDS detector. (b) Thin sample analysis: in addition to (a), the STEM detector collects only the transmitted electrons (operating either in bright or dark field mode). A larger active EDS detector area increases the collection efficiency of emitted X-rays. (c) Thin sample analysis in transmission mode (T-SEM) with the single-unit setup as used in this study. (d) Photograph of the T-SEM setup in working position (close to pole shoe). (e) Geometry of the annular SDD EDS encompassing a significantly larger solid angle (up to about 1 sr) compared to conventional SDD EDS.

In this study the following EDS detectors were used: a 10 mm^2 Si(Li) (Thermo Scientific, USA), a 10 mm^2 SDD (Bruker, Germany), a 100 mm^2 SDD (Thermo Scientific, USA), and an annular 60 mm^2 FlatQUAD SDD (Bruker, Germany). The annular SDD is inserted between the pole shoe and the sample, so that a very large solid angle of about 1 sr of the X-rays emitted by the sample can be encompassed (figure 1e). The corresponding collection solid angles of the 10 mm^2 and 100 mm^2 detectors amounted to about 6-7 msr and about 50 msr, respectively. Four detection segments of

15 mm^2 each around the central orifice for the primary electrons contribute to the total signal after parallel processing. For the classification of NPs, the automated feature analysis software of the Bruker Quantax EDS system was used with a 10 mm^2 SDD and an annular SDD, respectively.

3. Results and discussion

In order to test the analytical capabilities of high-resolution T-SEM/EDS, various materials containing NPs of known size (distribution) and chemical composition have been analysed. The first example focusses on particle size distribution (PSD) measurements of bimodal SiO_2 NPs which were prepared on carbon/*Formvar* TEM grids by aerosolisation (figure 2a). The superior material (mass- thickness) contrast of T-SEM imaging well suited to dimensional measurements can be observed in figure 2b. In addition, it reveals the presence of SiO_2 NPs also on the rear side of the thin foil (see the arrows in figure 2b and compare to figure 2a). For PSD measurements, binarisation of T-SEM micrographs, automatic counting and data reduction were performed using an open source image processing software, e.g., *ImageJ* [14]. The particles size evaluated as the PSD of the minimum Feret diameter extracted from the T-SEM micrographs as in figure 2b is presented in figure 2c. The validity of the T-SEM results has been demonstrated elsewhere by the good agreement with results obtained by TEM, SMPS and AFM analysis of the same material [3].

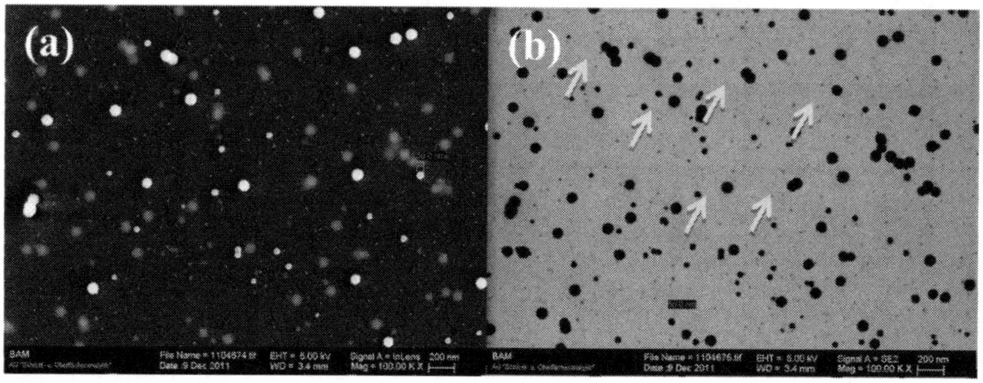

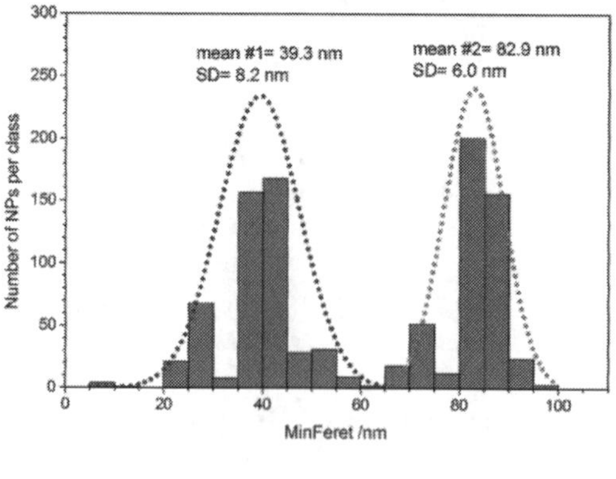

Figure 2. (a) *In-Lens* and (b) T-SEM micrographs of bimodal SiO_2 NPs prepared on carbon/*Formvar* supporting films. Arrows show examples of NPs which are located on the rear side of the film. The results of particle size distribution measurements of ~ 1,000 NPs as minimum Feret diameter from the T-SEM micrographs is shown in (c).

The second analysed sample represents a mixture of SiO_2 and TiO_2 NPs (figure 3). Imaging with a conventional E-T detector or even with an *In-Lens* detector may indicate the presence of two different NP types (see figure 3a). The high surface sensitivity of the *In-Lens* imaging only suggests the presence of some larger NPs between a rather monodisperse, smaller NPs. The evidence that there are two distinct NP types of different mid-atomic numbers, sizes (about 20 nm and 40 nm, respectively) and shapes is shown by the superior transmission imaging (mass-thickness) contrast of T-SEM (figure 3b). It should be noted that the very good transmission contrast between the two types of NPs is enhanced additionally by the somewhat larger NPs of the heavier type.

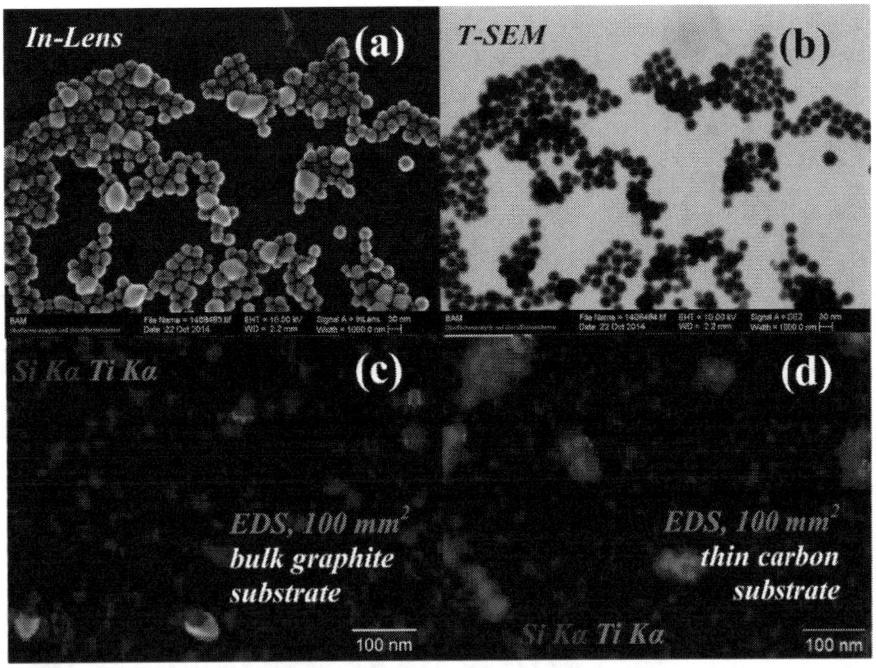

Figure 3. 10 kV imaging of a mixture of SiO_2 and TiO_2 NPs prepared on a thin, electron transparent, carbon foil TEM grid: (a) SE *In-Lens* mode (surface sensitive). (b) T-SEM mode (Z and thickness sensitive). (c) 10 kV EDS mapping overlain with SEM micrograph of the sample on a graphite bulk substrate (100 mm^2 SDD). (d) 10 kV EDS mapping overlain with SEM micrograph of the sample on a carbon foil TEM grid (100 mm^2 SDD). EDS data were enhanced with a map filter to reduce the local image noise situation.

To compare the benefit of T-SEM/EDS with conventional bulk EDS in respect to NPs elemental analysis, the material was prepared on a bulk graphite substrate and also on a thin carbon film. Figure 3c shows that on a bulk substrate, it is not possible to analyse the element distribution (figure 3c) due to the large contribution of the carbon substrate, even when a large-area, 100 mm^2 SDD is used. The discrimination between different NP types can be attained when the T-SEM/EDS approach is exploited (figure 3d).

The third example of a nanoparticulate material consists of gold NPs with an average particle size (diameter) of 17 nm. A 3 μL droplet of the initial Au suspension has been prepared on thin (lacey) carbon supporting foil (TEM grid). The T-SEM micrograph (figure 4a) shows that NPs are well dispersed on the carbon foil. The main interest of the analysis of this material was to compare the detection sensitivity of EDS detectors with various collection solid angles: (i) a conventional SDD with 10 mm^2 detector area, (ii) a new generation SDD with 100 mm^2 detector area, and (iii) and a state-of-the-art annular 60 mm^2 SDD detector. The 10 mm^2 SDD detector was not able to detect single Au NPs even after several minutes of acquisition. Longer acquisition times could not be considered due to significant image drifts. The elemental identification of the nanoparticulate material became possible by spectrum acquisition only from agglomerates (figure 4b). The 100 mm^2 SDD (figure 4c) could provide the element distribution using the Au M X-ray line after about two minutes. With the annular SDD, the acquisition time for a similar map area could be significantly reduced to only 13 seconds (figure 4d). These results prove the capability of the annular EDS detector for rapid identification of the Au NPs with sizes below 20 nm. For NP materials with a low mean atomic number, such as oxides and polymers, the detection sensitivity decreases, so that EDS identification of NPs with sizes below 20 nm is barely possible with the state-of-the-art instrumentation.

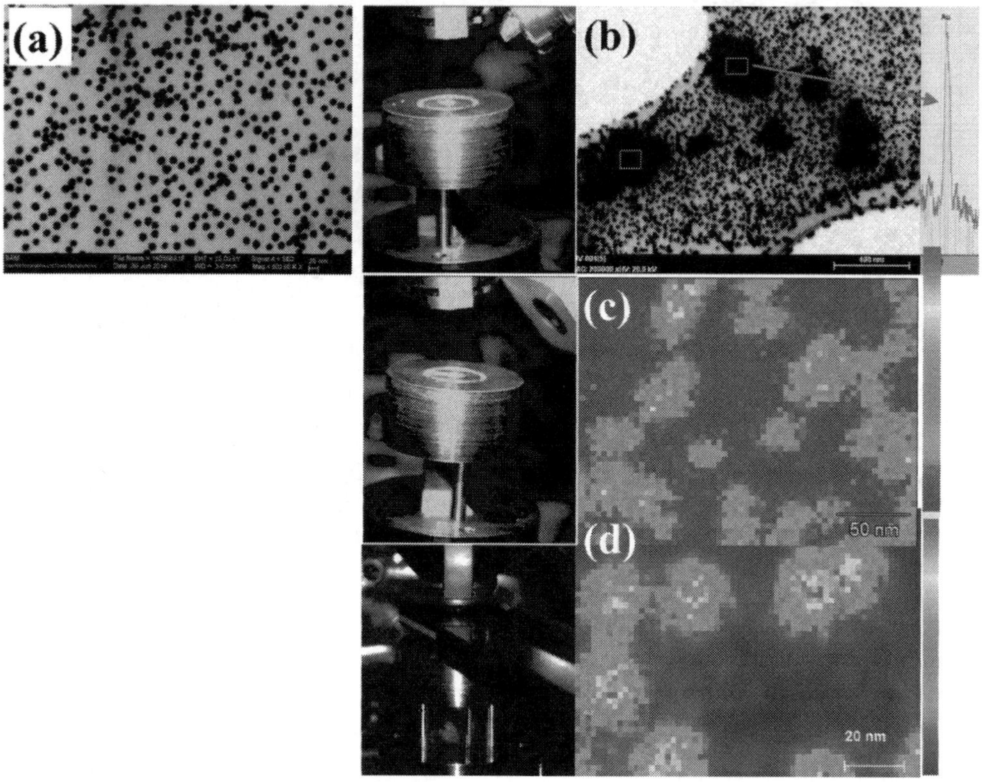

Figure 4. T-SEM/EDS characterisation of 17 nm Au NPs which were prepared on a carbon foil TEM grid. (a) T-SEM micrograph showing the contrast and separation of the individual NPs. (b-d) Chamberscope photos (left) of the various T-SEM/EDS setups and EDS maps (right) of the Au Mα X-ray line. For T-SEM/EDS the screening ring has to be removed (compare to figure 1d). (b) EDS spectrum of Au NP agglomerates acquired with a 10 mm^2 SDD (20 kV, 218 pA). (c) EDS map in false colour display acquired with a 100 mm^2 SDD (20 kV, 441 pA, 64 x 48 pixels, 237 s). (d) EDS map in false colour display acquired with an annular SDD (20 kV, 485 pA, 64 x 48 pixels, 13 s).

The fourth example focusses on the certified reference material (CRM) BAM-L200 [15, 16], a sample which is ideally suited for the determination of the spatial resolution of EDS. The material (GaAs) contains sub-micrometre to -nanometre sized Al doped GaAs stripe structures ($Al_{(0.70)}Ga_{(0.30)}As$) with certified periods and/or widths. The EDS linescan of Ga Lα and Al Kα-lines as acquired with a 10 mm^2 Si(Li) on the cross-sectioned bulk sample of BAM-L200 can be seen in figure 5a. The smallest resolved structure is the "P4" stripes period of 193 nm. Using a 100 mm^2 SDD for a FIB prepared TEM lamella of nominal thickness of 100 nm (figures 5b,c), the "P5" grating period of 136 nm is clearly resolved, which demonstrates the significant improvement in the spatial EDS resolution. The presence of "W5" stripe of 19.5 nm width is just detected close to the noise level. These studies confirm recent results on the detection of structures in the range of only a few nm in the case of FIB lamellas analysed by large-area SDDs [16]. It should be noted that the finally attained spatial resolution is a function of the geometrical collection efficiency of the EDS detector, of the specimen thickness, but also of the counting statistics. The imaging, nevertheless, at high magnification, even if drift corrected, proceeds mostly within a limited acquisition time. A higher signal-to-noise ratio of the emitted EDS spectra from single NPs can be obtained from high-sensitivity EDS detectors.

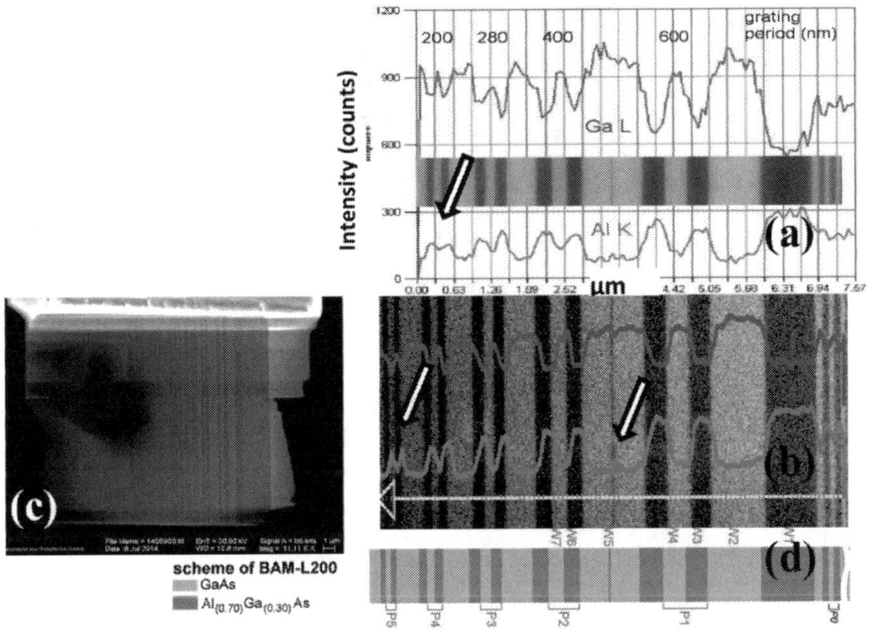

Figure 5. EDS linescans of Ga Lα (green) and Al Kα (red) over corresponding stripe structures from certified reference material BAM-L200. Different spatial resolutions correspond to (a) bulk sample and (b) TEM lamella. The arrows indicate the smallest structures resolved: P4 grating period (193 nm) in case (a), P5 grating period (136 nm), and, close to the noise limit, the stripe W5 (19.5 nm), respectively, in (b). (c) Overview micrograph of the TEM lamella as prepared by FIB of the BAM-L200 bulk sample. Variable thickness is suggested by a hole visible on the upper left side; (d) sketch with corresponding certified structures as denoted in the BAM-L200 certificate [15].

The last application example demonstrates the possibility of automated particle classification using feature analysis with stage control. The motorized sample stage of modern SEMs enables steering with a precision of several micrometres. The automatic acquisition of a large number of micrographs allows analysing a statistically relevant number of NPs. Commercially available software packages which are based on combined SEM and EDS analysis can be applied to the classification of NPs. The case study will present adapted T-SEM/EDS instrumental conditions to classify particles after appropriate deposition (clean and well dispersed) on electron-transparent supporting foils. An analytical workflow consisting of three steps was devised to classify two main classes of SiO_2 particles of different "consistencies" visible at best in the T-SEM mode (figure 6a): "bulk" (the darker ones) and "hollow" (the brighter ones).

Figure 7 illustrates the sequence of steps to be applied for measurement comprising of (i) identification of the particles according to their corresponding contrast in the transmission (T-SEM) micrographs, (ii) subsequent EDS measurement on the identified particles only, and (iii) chemical classification based on the EDS results.

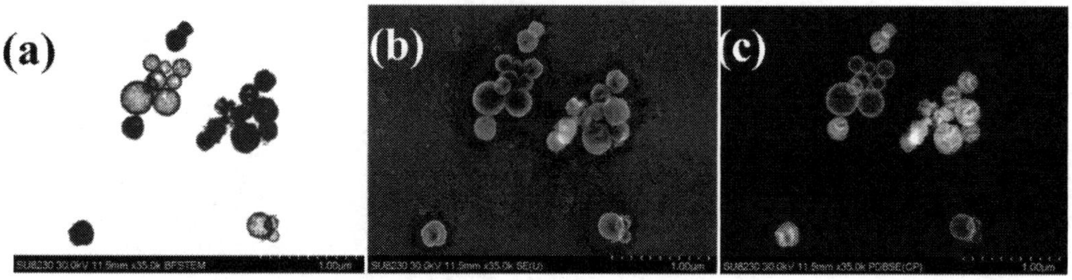

Figure 6. SEM micrographs showing two classes ("bulk" and "hollow") of SiO_2 particles as obtained (a) in the bright-field T-SEM mode, (b) with the conventional SE (upper) detector, and (c) with a BSE detector.

1. Features of interest are first detected in T-SEM micrographs using image processing (figure 7c). Instrumental parameters such as magnification, number of pixels, scan speed were optimized (see Table 1).

Table 1. Parameters of image analysis for the detection of particles in T-SEM micrographs.

Parameter	Setting
Magnification	10,000 x
Image resolution	3,000 x 2,250 pixel
Pixel size	4 nm
Grey level threshold	0 - 145
Circularity	≤ 1.37

It is important to set a grey level threshold within the T-SEM micrograph to accurately distinguish particles from the substrate according to their corresponding contrast. The "hollow" particles (consisting only of an ultra-thin shell) have almost the same T-SEM contrast in the centre as the supporting foil; the particle boundaries are somewhat darker. Therefore, the post-binary filter "Fill Hole" was used in order to improve the detection of particles. The morphological parameter:

$$\text{Circularity} = \frac{\pi \,\text{maxFeret}^2}{4\,\text{Area}}, \tag{1}$$

a dimensionless measure near 1 for round particles and above 1 for irregular particles, was applied in order to accept single NPs and reject agglomerates. By doing so, we assume that all the separated particles have same size distribution and same chemistry as those agglomerated. Extraction of information from agglomerated (or aggregated) particles is physically very little possible (at the outer boundaries).

2. Subsequently, EDS spectra on the identified single particles are automatically acquired. The acquisition time is the decisive parameter, so that sufficient signal-to-noise ratios can be attained in a short time. Due to the high solid angle of the annular SDD, a current of 216 pA and accelerating voltage of 5 kV resulted in an input count rate of 4 - 27 kcps, so that an acquisition time of just 2 seconds per spectrum was sufficient for chemical classification. This is significantly shorter than the spectrum acquisition time of 120 s with a 10 mm^2 SDD where an accelerating voltage of 10 kV and current of 115 pA led to an input count rate of just 50 - 250 cps.
3. The last step comprises reviewing the acquired EDS spectra to setup classes by evaluating the Si K / C K line net intensity ratio and the Si-K net intensities. "Bulk" particles were classified with a Si/C ratio $\geq$ 1.4 and Si net intensities > 6,000 counts, "hollow" particles correspond to a Si/C ratio $\leq$ 1.14 and Si net intensities < 4500 counts (figure 7d).

The method was developed at one mesh of the TEM grid (figure 7a) which consists of 35 T-SEM micrographs (figure 7b). This was then applied to four meshes using stage control (figure 7a and Table 2). The whole feature analysis run took 3h24 min, the most of the time was spent on T-SEM micrograph recording. Spectrum acquisition took no more than 14 min altogether.

To verify whether the net intensities are dependent on the "bulk"/"hollow" structure or the particle size, the evaluated data are shown in figure 8. Figure 8a shows that the net intensity ratio of Si K-lines to O K-lines of all classified particles are not dependent upon grey level. This indicates that the chemical composition of the SiO_2 particles does not vary between "bulk" and "hollow" particles. Similarly, the diagram of the Si K to O K net intensity ratio versus particle average diameter (figure 8b) indicates that the particle size has no relationship to the main elemental composition of the SiO_2 particles. In addition, figure 8b points out that most "bulk" particles are larger than the "hollow" ones.

Table 2. Classified particles using automated feature analysis.

Class	Number
Bulk	128
Hollow	104
Unclassified	185
Total	417

It can be summarized that without need of accurate elemental quantification, the qualitative evaluation of the chemical composition of SiO_2 particles of different inner morphologies ("consistencies") by means of the T-SEM/EDS automated analysis is possible. The range of applications of T-SEM/EDS is extended to further examples of complex NP shape and compositions such as core-shell structures. One particular application is wear particles (at both nano- and microscale) in synthetic (additivated as well as non-additivated) oil after lubricated tribological testing [17].

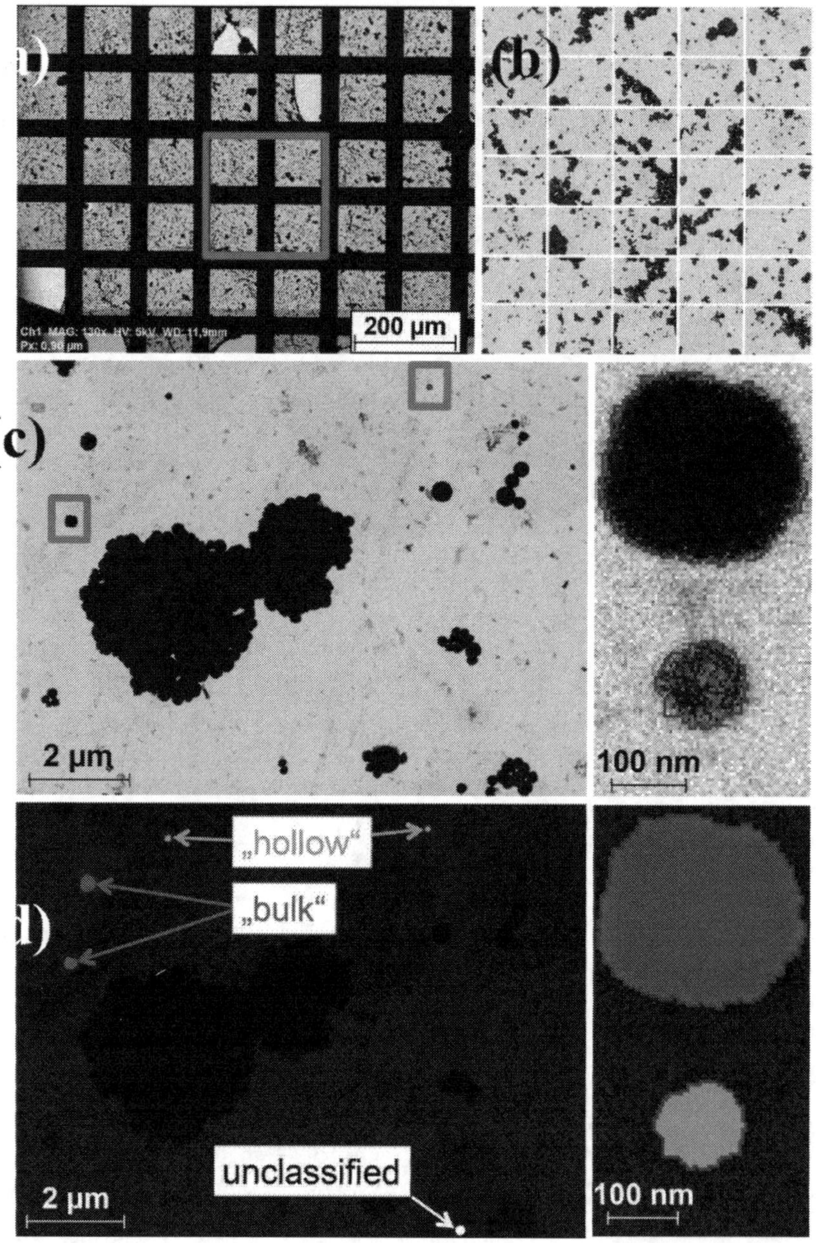

Figure 7. Classification of particles using automated feature analysis. (a) T-SEM micrograph providing an overview of the analysed area (red rectangle) with SiO_2 NPs which were deposited on a TEM grid (150 µm mesh copper) with carbon thin foil. (b) Mosaic of one analysed mesh of the TEM grid consisting of 35 fields. (c) T-SEM micrograph (left) of one analysed field. Red rectangles mark two particles (right) which were detected by image analysis using a corresponding greyscale threshold and the post-binary filter "Fill hole". (d) Feature analysis result overlain with T-SEM micrograph showing classified bulk (violet) and hollow (turquoise) particles which were classified by the Si K / C K line net intensity ratio and the Si-K net intensities.

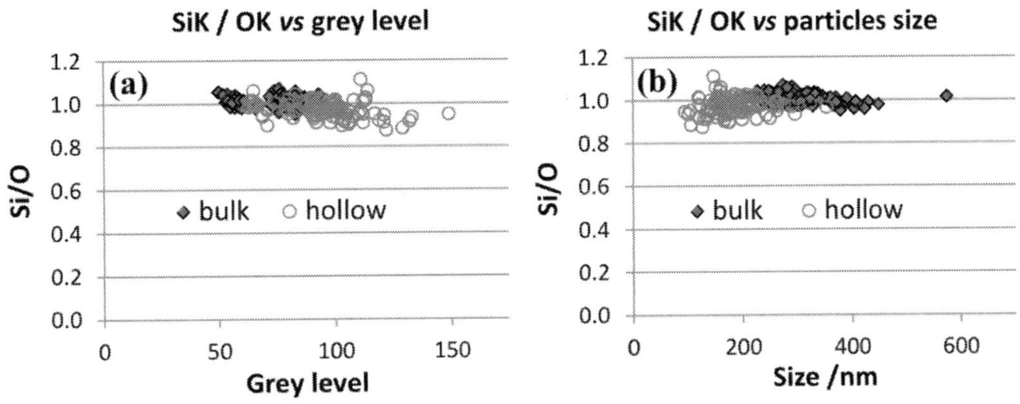

Figure 8. Representation of the chemical composition of 232 classified SiO_2 particles: (a) Si K / O K intensity ratio in dependence on the grey level, and (b) Si K / O K intensity ratio in dependence on the particles size as measured automatically in the T-SEM images.

4. Conclusions

This study provides a detailed overview of the instrumental and methodical possibilities offered by modern SEM/EDS systems for the advanced characterisation of NPs. It demonstrates that high-resolution SEM imaging provides information on the morphology, size, shape and the inner structure of NPs. In addition, advanced EDS spectrum imaging has become possible which adds essential chemical characterisation of nanoparticulate materials.

Two methodical/instrumental approaches enable this analytical undertaking: (i) T-SEM: Preparation of the nanoparticulate material onto electron-transparent foils and operation of the SEM in transmission mode, and (ii) EDS: Use of a high-sensitivity, high solid angle SDD.

A specific application was developed by combining both methods which allows the automated classification of NPs. The advanced software options included within modern analytical systems combine both methods. The image analysis of high contrast T-SEM micrographs provides numerous morphological parameters. The chemical information obtained from EDS spectra are taken only from the identified NPs. This significantly reduces acquisition time as the number of analysed particles is limited to only those that are of potential interest. The spatial resolution is furthermore addressed by T-SEM/EDS for the particular experimental conditions that were evaluated by means of a suitable certified reference material.

The potential of the methodical approach presented in this case study on relatively simple materials can serve as a basis for its application to nanoparticulate commercial products of more complex morphologies and chemical composition. Challenging studies taking into account material characteristics have to be carried out systematically.

Acknowledgements

The research leading to these results has received funding from the European Union's Seventh Framework Programme (FP7/2007-2013) under grant agreement n° 263147 (NanoValid), n° 604347 (NanoDefine) and n° 604577 (SETNanoMetro). Thanks are due to Charles Motzkus (LNE, Paris, France) for the preparation of the aerosol SiO_2 bimodale NPs from figure 2, to W. Österle and R. Hesse for the preparation of the FIB lamella of the BAM-L200 sample (figure 5) and Bruker Nano GmbH (Berlin, Germany) for the measurements with the annular SDD. John Spratt (NHM) is gratefully acknowledged for suggestions improving the manuscript.

References

[1] Buhr E, Senftleben N, Klein T, Bergmann D, Gnieser D, Frase C G and Bosse H 2009 Characterization of nanoparticles by scanning electron microscopy in transmission mode. *Meas. Sci. Technol.* **20** 084025

[2] Klein T, Buhr E, Johnsen K-P and Frase C G 2011 Traceable measurement of nanoparticles size using a scanning electron microscope in transmission mode (TSEM). *Meas. Sci. Technol.* **22** 094002

[3] Motzkus C, Macé T, Gaie-Levrel F, Ducourtieux S, Delvallee A, Dirscherl K, Hodoroaba V-D, Popov I, Popov O, Kuselman I, Takahata K, Ehara K, Ausset P, Maillé M, Michielsen N, Bondiguel S, Gensdarmes F, Morawska L, Johnson G R, Faghihi E M, Kim C S, Kim Y H, Chu M C, Guardado J A, Salas A, Capannelli G, Costa C, Bostrom T, Jäamting Å K, Lawn M A, Adlem L and Vaslin-Reimann S 2013 Size characterization of airborne SiO_2 nanoparticles with on-line and off-line measurement techniques: an interlaboratory comparison study. *J. Nanoparticle Res.* **15**, 1919

[4] Hodoroaba V-D, Motzkus C, Macé T and Vaslin-Reimann S 2014 Performance of high-resolution SEM/EDX systems equipped with transmission mode (TSEM) for imaging and measurement of size and size distribution of spherical nanoparticles. *Microsc. Microanal.* **20** 602-612

[5] Hodoroaba V-D, Akcakayiran D, Grigoriev D O and Shchukin D G 2014 Characterization of micro- and nanocapsules for self-healing anti-corrosion coatings by high resolution SEM with coupled transmission mode and EDX. *Analyst* **139** 2004-2010

[6] Rades S, Wirth T and Unger W E S 2014 Investigation of silica nanoparticles by Auger electron spectroscopy (AES). *Surf. Interface Anal.* **46** 1096-9918

[7] Rades S, Hodoroaba V-D, Salge T, Wirth T, Pilar Lobera M, Hanoi Labrador R, Natte K, Behnke T, Gross T and Unger W E S 2014 High-resolution imaging with SEM/T-SEM, EDX and SAM as a combined methodical approach for morphological and elemental analyses of single engineered nanoparticles. *RSC Adv.* **4** 49577-49587

[8] Crawford B J and Liley C R W 1970 A simple transmission stage using the standard collection system in the scanning electron microscope. *J. Phys. E: Sci. Instrum.* **3** 461-462

[9] Woolf R J, Joy D C and Tansley D W 1972 A transmission stage for the scanning electron microscope. *J. Phys. E: Sci. Instrum.* **5** 230-233

[10] Reimer L, Gentsch P and Hagemann P 1975 Anwendung eines Rasterzusatzes zu einem Rasterelektronenmikroskop. I Grundlagen und Abbildung amorpher Objekte. *Optik* **43** 431-452

[11] Hodoroaba V-D, Rades S and Unger W E S 2014 Inspection of morphology and elemental imaging of single nanoparticles by high resolution SEM/EDX in transmission mode. *Surf. Interface Anal.* **46** 945-948

[12] Vermeulen J P and Jaksch H 2005 A novel STEM detector system. *Imaging and Microscopy* **1** 22-23

[13] Golla U and Schindler B 2004 US Patent 6 815 678 B2

[14] Rasband W S 1997–2012. *Image J*, Bethesda, Maryland, USA: US National Institute of Health. Available at http://rsb.info.nih.gov/ij/ (accessed 22-05-2015)

[15] www.webshop.bam.de, Reference Materials, Layered materials, BAM-L200 (accessed 05-06-2015)

[16] Senoner M, Maassdorf A, Rooch H, Österle W, Malcher M, Schmidt M, Kollmer F, Paul D, Hodoroaba V-D, Rades S and Unger W E S 2015 Lateral resolution of nanoscaled images delivered by surface-analytical instruments: application of the BAM-L200 certified reference material and related ISO standards. *Anal. Bioanal. Chem.* **407** 3211-3217

[17] Wäsche R, Hartelt M and Hodoroaba V-D 2015 Analysis of nanoscale wear particles from lubricated steel-steel contacts. *Tribol. Lett.* **58** 49

IOP Conf. Series: Materials Science and Engineering **109** (2016) 012007 doi:10.1088/1757-899X/109/1/012007

Fundamentals of electron energy-loss spectroscopy

F Hofer, F P Schmidt, W Grogger and G Kothleitner

Graz University of Technology and Graz Centre for Electron Microscopy, Institute of Electron Microscopy and Nanoanalysis, Steyrergasse 17, 8010 Graz, Austria

E-mail: ferdinand.hofer@tugraz.at

Abstract. Electron energy-loss spectroscopy (EELS) is an analytical technique that is based on inelastic scattering of fast electrons in a thin specimen. In a transmission electron microscope (TEM) it can provide structural and chemical information about a specimen, even down to atomic resolution. This review provides an overview of the physical basis and new developments and applications of EELS in scanning transmission electron microscopy. Recent advances in elemental mapping, spectrum imaging of plasmonic structures and quantitative analysis of atomically resolved elemental maps are highlighted.

1. Introduction

The need to understand atomic processes in solids has led to an increasing demand for new imaging, diffraction and spectroscopy methods with high spatial resolution. This need has been reinforced by the growing interest in nanoscience and nanotechnology. Particularly, transmission electron microscopy (TEM) is of unrivalled value, because it can provide structural information with excellent spatial resolution (down to atomic dimensions) via high resolution TEM imaging and electron diffraction. These structural data can be supplemented by chemical information from the same specimen region, obtained using analytical techniques, such as energy-dispersive X-ray spectroscopy (EDS) and electron energy-loss spectroscopy (EELS) [1].

Due to the broad range of inelastic interactions of the high energy electrons with the specimen atoms, ranging from phonon interactions to ionisation processes, electron energy-loss spectroscopy offers unique possibilities for advanced materials analysis. It can be used to map the elemental composition of a specimen, but also for studying the physical and chemical properties of a wide range of materials and biological matter.

Beginning with the increasing use of EEL spectrometers during the eighties of the last century [1], the energy-filtering technique was a major step forward for two-dimensional mapping of the spectral features visible in an EEL spectrum [2, 3]. The wide distribution of energy-filtering TEMs (EFTEM) led to numerous practical applications both in the materials and biological sciences [4]. In the meanwhile, EFTEM is a widely used method for both overview and nanoscale characterisation of thin samples, applicable to most chemical elements and especially sensitive to light elements.

With the introduction of high brightness electron sources and spherical aberration correctors, the image resolution of the TEM reaches now the 50 picometre level [5]. Especially scanning transmission electron microscopy (STEM) equipped with a Cs-probe corrector is extremely useful for high resolution imaging [6] and the parallel chemical analysis by EELS and EDS because of the much increased probe current in the corrected beam.

Before discussing how these various methods and their recent developments provide us with information about the specimen, we will briefly examine the principles and instrumentation involved in recording energy-loss spectra in the TEM.

2. The electron energy-loss spectrum

A beam electron in a (S)TEM may be inelastically scattered when it interacts with the atomic electrons in the specimen. The electron beam loses energy and is bent through a small angle (5 - 100 milliradians). The energy distribution of all the inelastically scattered electrons provides information about the local environment of the atomic electrons which in turn relates to the physical and chemical properties of the specimen. This is the basis of electron energy-loss spectroscopy (EELS) [1]. Much of the information obtainable from EELS is similar to that of X-ray absorption spectrometry (XAS) in the synchrotron.

Figure 1 shows a typical electron energy-loss spectrum recorded up to 600 electron volts. The first peak, the most intense for a very thin specimen, occurs at 0 eV and is therefore called the zero-loss peak. It represents electrons which have not been scattered in the specimen (transmitted electrons) and which have been elastically scattered via interaction with the atomic nuclei.

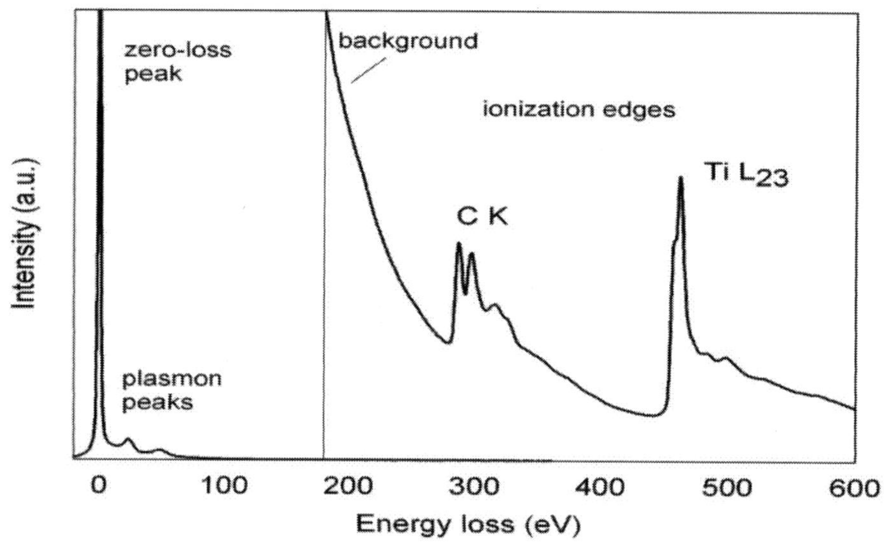

Figure 1. EELS spectrum of a 20 nm thin titanium carbide specimen recorded in a conventional 200 kV TEM equipped with an energy-filtering spectrometer.

The low-loss or valence region of an EEL spectrum (< 50 eV) provides similar information to that provided by optical spectroscopy, containing valuable information about the band structure and in particular about the dielectric properties of a material (e.g., band gap, surface plasmons). The most prominent peak, centred at 24 eV, comes from a plasma resonance of the valence atoms. Signal intensities in the low-loss region are larger than in the high-loss region of the spectrum.

At higher energy losses (> 50 eV), where the number of inelastically scattered electrons is much lower, the spectrum shows characteristic features called "ionisation edges" (due to their typical shape, a rapid rise followed by a more gradual fall). These edges are the exact equivalent of an absorption edge in XAS and arise from the same process. The edges are formed when an inner-shell electron absorbs enough energy from a beam electron to be excited to a state above the Fermi level. Not all ionisation edges are saw-toothed like the carbon K-edge in figure 1, but exhibit more complex edge shapes such as the $L_{2,3}$-edge of titanium in figure 1. This $L_{2,3}$-edge includes sharp excitations at the onset of the ionisation edge, called white-lines, which are typical for elements in the first row of transition elements and for the rare earth elements. The ionisation edges can be used for the analysis of almost all chemical elements in particular for the lighter elements, the edge onset gives the ionisation energy and allows the qualitative analysis and in case of very thin samples the edge intensities are proportional to the concentration of the corresponding elements.

3. Recent advances in EELS methodology

TEM-EELS instrumentation is based on a magnetic prism, in which a uniform magnetic field is generated by an electromagnet with specially designed pole pieces. The prism bends the inelastically scattered electrons by about 90°, disperses them according to their different kinetic energies and also has a focussing action. The spectra are recorded with a charge coupled device (CCD camera). Due to the enormous dynamic range of an EEL spectrum, spanning many orders of magnitude, and due to restrictions in dynamic range and sensitivity of CCD based spectrometers, a complete EEL spectrum normally has to be recorded in several segments by changing illumination conditions and/or acquisition times.

Energy filters are used to form images from electrons that have suffered a specific energy loss [energy filtered TEM (EFTEM) or electron specific imaging (ESI)]. In particular, two types of energy-filtering instruments are most often used: firstly post-column energy filters, with a single prism geometry and multipole lenses, can be attached and retrofitted to practically any TEM or STEM instrument [7]. The post-column filter is not only an efficient imaging system, but also a versatile instrument for acquiring EEL spectra in high quality. Secondly the in-column filter, whose spectrometer consists of four magnetic prisms that are arranged symmetrically according to the shape of a Greek Omega, is located below the projector lens system of the TEM. This type of filter can be also used for spectroscopy, but the main application field is high quality imaging such as large area elemental mapping and energy-filtered electron diffraction studies [8].

Another important development was the introduction of monochromators for the electron source, which paved the way for acquiring EEL spectra at high energy resolution, typically in the 100 - 200 meV range (e.g., the Wien filter approach) [9]. The improved energy resolution opens new possibilities for studying detailed electronic structure and bonding effects evaluated from near edge fine structures of the ionisation edges, but also accurate band gap and dielectric function measurements via the low-loss part of the spectrum [1].

Several years ago, the problem of the high dynamic range of the EEL spectrum could be addressed by acquiring the elastic and inelastic parts at nearly coincident times at identical experimental conditions, offering the advantage of using the low-loss regime as a reference for quantitative data analysis. This system works with an additional electrostatic vertical deflector and is now successfully used by several groups worldwide [10, 11].

Advances in X-ray detection with the advent of large area or four-quadrant solid state silicon drift detectors [12] and the rapid progress in EDS data processing, pushed the ideas for highly efficient elemental mapping in combination with EELS. The collaboration between Gatan, Bruker and the TU Graz enabled the fast recording of multimodal data so that STEM-images, low-loss and high-loss EELS and EDS data can now be acquired with a speed of 1,000 to 1,500 spectra per second [13].

While the conventional method of EELS mapping combines TEM/STEM images with the local concentration of the elements, the development of more powerful computers and data reduction procedures nowadays allows a "holistic" approach: the whole spectral information is gathered for each point of an image, generating a three-dimensional data set which is often called a spectrum image [14]. The spectrum-imaging technique offers various advantages: the wealth of data allows a certain amount of "post experiment microscopy" and because of the completeness of data, interpretation mistakes can be avoided. Additionally, a data-evaluation software can automatically identify and highlight the most prominent features, e.g., chemical phases [15]. Therefore, spectrum-imaging techniques are the essential basis for successful EELS mapping at the atomic scale.

IOP Conf. Series: Materials Science and Engineering **109** (2016) 012007 doi:10.1088/1757-899X/109/1/012007

4. Elemental mapping at high spatial resolution

One of the most commonly used applications of STEM-EELS or EFTEM is to derive compositional information by recording energy-filtered images using the element characteristic ionisation edges. Although advanced energy-filters or spectrometers now enable the fast acquisition of elemental maps almost on a routine basis, care must be taken due to several experimental limitations of EELS: generally, the signal-to-noise ratio of the elemental signal is very low which is mainly caused by the high uncharacteristic background below the ionisation edges and the low ionisation cross-sections for heavier elements and for elements occurring at low concentrations.

Another important limitation comes from multiple scattering of the inelastic signal in thicker specimens, restricting elemental mapping to specimen thicknesses well below the mean free path lengths of the inelastically scattered electrons (specimen thickness < 70 nm for most materials). Alternatively, EELS spectra from thicker specimens may be deconvolved by using the low-loss part of the EELS spectrum, recovering a single scattering distribution EELS spectrum. Typical procedures involve Richardson-Lucy or Fourier log methods [16, 17], which however decrease the signal-to-noise ratio of the spectrum. When crystalline specimens are studied, a frequent problem is the preservation of diffraction contrast in inelastic imaging, which can be reduced by employing advanced illumination conditions (hollow-cone or rocking-beam illumination) [2].

Firstly, detailed strategies had to be developed to optimize the quality of the experimental spectrum. Secondly, advanced spectrum imaging techniques, which include a variety of data correction or processing procedures, are a prerequisite for successful elemental mapping with EELS or EFTEM [15, 18] (see example in figure 2). Additionally, it is known for a long time that EELS spectra or elemental maps can be quantified using the standard EELS quantification procedures thus providing absolute and relative concentration values at the nanometre scale [1, 19].

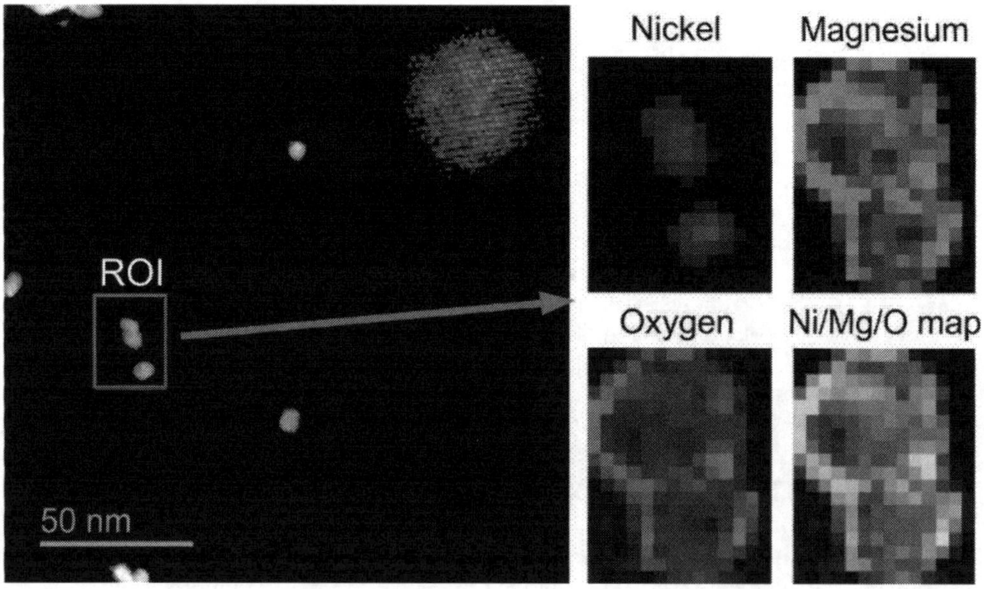

Figure 2. MgO/Ni core shell nanoparticles investigated with a C_S-corrected STEM and EELS spectrum imaging revealing the Ni core and the MgO shell, the inset shows the Z-contrast image of the Ni core (specimen courtesy: Gopi Krishnan, University of Groningen, The Netherlands).

Although EFTEM elemental mapping is now widely used for nanoscale analysis, it cannot deliver atomic resolution information even in aberration-corrected TEMs. Since the C_S-image corrector affects only spherical aberrations, but not chromatic aberrations (C_C), EFTEM mapping at atomic resolution is only possible by incorporating a chromatic aberration corrector. This so far is only realized at the NCEM in Berkeley (USA) and the Ernst-Ruska Center in Jülich (Germany) [20].

Elemental mapping at atomic resolution can be easier achieved via the scanning transmission electron microscope (STEM) which is now in wide general use, harnessing Z-contrast imaging and EELS [21]. The first elemental maps at atomic resolution have been recorded by using the STEM-EELS technique by Kimoto *et al.* [22] and they have been able to visualize the atomic columns of La, Mn and O in the layered manganite $La_{1.2}Sr_{1.8}Mn_2O_7$. In the meanwhile Allen and co-workers [23] have shown that it is also possible to record atomically resolved elemental maps by using X-ray spectroscopy in the STEM.

Two important developments enhanced the potential for successful elemental mapping at atomic resolution: Firstly, the introduction of collection-efficient solid state X-ray detectors (SDD) [12] and secondly, the simultaneous acquisition of EELS and EDS spectrum images with fast scanning rates [13]. Since then many examples of atomically resolved elemental maps have been published, but it has to be realized that coloured elemental maps only yield qualitative information. However, it is extremely important to provide quantification data in terms of volumetric densities (i.e., number of atoms per unit volume), as it allows correlating structural and chemical information with each other ("unit cell atom counting"). Using a $SrTiO_3$ crystal of known thickness and density Kothleitner *et al.* [24] could show how EELS and EDS elemental maps (figure 3) can be quantified by combining both spectroscopies and including the zeta factor method for X-ray quantification. The results show that it is possible to obtain quantitative numbers of Sr, Ti and O atoms in the individual atom columns, however elastic and thermal scattering effects can severely influence accuracy at atomic resolution [25].

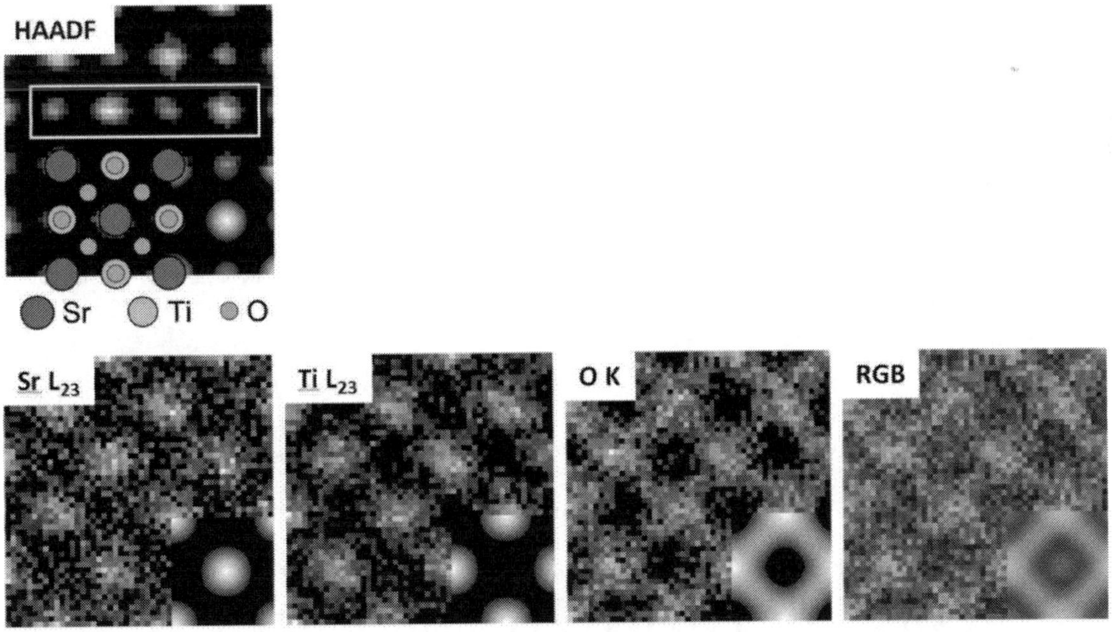

Figure 3. Experimental atomic resolution EELS elemental maps for the {001} direction in $SrTiO_3$ for the Sr $L_{2,3}$-edge, the Ti $L_{2,3}$- and the O K-edge with the simultaneously acquired HAADF image and the RGB composite of the elemental maps; quantum mechanical simulations, assuming a Gaussian incoherent source size of 0.08 nm, are inset in each figure; from [25].

5. Chemical bonding information

Edge fine structures arouse considerable interest in the application of EELS, especially in the field of materials science, because they can be used to extract information regarding local charge distributions, coordination numbers and bonding characteristics. The fine structures are divided into the near edge-fine structures (ELNES) within 50 eV of the edge onset and the extended fine structure (EXELFS) about some 100 eV above the edge onset. Most of the ionisation edges contain a more complex structure than can be explained in simple atomic terms. The ionisation edges are often modified by the solid state environment of the atom undergoing the inner-shell excitation and this information can be used as a "fingerprint" from elements in similar chemical environments [26, 27]. In the meanwhile the calculation of ELNES structures has reached a high degree of sophistication [28].

One important development for ELNES studies was the introduction of the monochromated (S)TEMs which helped to improve the instrumental energy resolution to values as low as 100 meV. In combination with advanced energy-filters and/or spectrometers it was now possible to fully exploit informational the spectral details contained in the EEL spectrum. It was not at all obvious, however, at the time of the development during the late nineties that these tools would gain such importance. The reduced energy spread of the TEM immediately opened possibilities for detailed studies of the near-edge fine structures at K- and $L_{2,3}$-ionisation edges of oxides, perovskites and similar materials [29]. The knowledge about these ELNES structures at high energy resolution is an important basis for STEM-EELS imaging of chemical bonding in materials. Figure 4 shows the ELNES structure of the V $L_{2,3}$ white lines of V_2O_5 recorded at different energy resolutions. The L_3 white line reveals the symmetry of the vanadium site being surrounded by six oxygen atoms forming a strongly distorted octahedron unit [30].

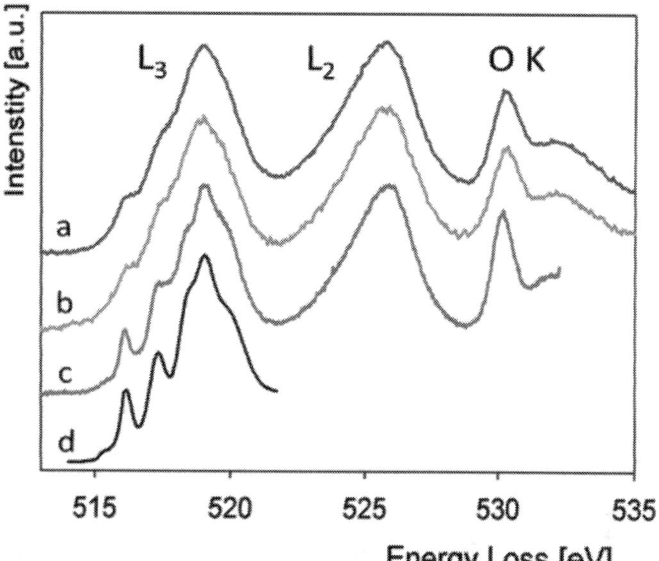

Figure 4. Comparison of the V $L_{2,3}$- and O-ionisation edges of V_2O_5 which have been recorded with different energy resolution; a) 200 kV TEM with LaB_6 cathode (0.7 eV); b) 200 kV TEM with Schottky emitter (0.6 eV); c) 200 kV TEM with a monochromator and a HR-energy-filter (0.3 eV); d) X-ray absorption spectrum recorded with an energy resolution of 0.08 eV; from [30].

6. Physical property mapping

The low-loss spectrum contains energy losses to valence or conduction electrons which provide rich information about the physical properties of a specimen e.g., inter- or intraband transitions, band gaps and the dielectric properties [31, 32]. One important application of low-loss EELS lies in the study of optical properties of metallic nanostructures which may drastically change at the nanometre scale as a function of size, morphology and environment. In particular, collective oscillations of quasi free electrons on a metallic/dielectric interface, i.e., surface plasmons, are increasingly studied in the (S)TEM, taking advantage of its unbeaten high spatial resolution (compared to the scanning optical near-field microscope). Here the introduction of monochromators, providing an energy resolution of 100 meV or even less [33], was mainly responsible for the big success of the technique [34].

Since the early seventies of the last century EELS has been used for studying surface plasmons [31, 35], but it was only in 2007 when two independent groups showed how fast electrons can be used to map localized surface plasmons of single noble metal nanoparticles [36, 37]. In these pioneering works STEM-EELS was used, but later it was also shown that a monochromated EFTEM can yield comparable results [38, 39]. It is now widely accepted that EELS or EFTEM are the most advanced methods to probe plasmonic modes and they are increasingly used to study nanostructures of increasing complexity. For example STEM-EELS was used for studying dark plasmonic breathing modes in silver nanodisks [40], for morphing a plasmonic nanodisk into a nanotriangle [41], for revealing universal dispersions of surface plasmons in flat nanostructures [42], for studying the 3D distribution of surface plasmons around a metal nanoparticle [43] and for exotic nanostructures as shown in figure 5.

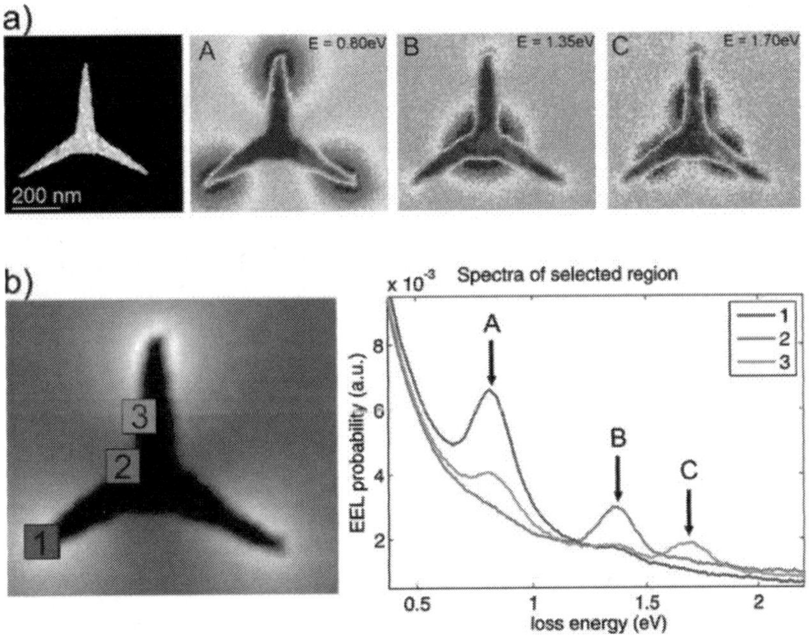

Figure 5. a) High-angle annular dark field (HAADF) image of a 30 nm thick Au-nanostar on a 15 nm thin silicon nitride membrane (left) and corresponding electron energy loss (EEL) maps at 0.8 eV (A), 1.35 eV (B) and 1.70 eV (C) integrated over an energy width of 150 meV. The EELS maps were generated using the STEM-EELS approach with a monochromated 200 keV electron beam with 150 meV energy resolution (FWHM). The sample was prepared by electron beam lithography, raw data are presented. B) The regions 1-3 in the spectrum image (left) mark the areas from which EEL spectra were extracted. The peaks labelled by A, B and C corresponds to the energies of the EELS maps shown in (a).

IOP Conf. Series: Materials Science and Engineering **109** (2016) 012007 doi:10.1088/1757-899X/109/1/012007

Recent progress in the development of monochromators with improved energy resolution in the range of 10 meV has shown that it is now possible to reveal vibrational modes (phonons) in EEL spectra recorded in a STEM [44]. This new technique allows to study the spatial variation of vibrational modes in nanostructures and nanoparticles.

7. Conclusions

Overall, inelastic excitations, probed by electron microscopy, have been a continuous source of information on composition, chemical and physical properties of nanostructured materials over the last decades. In this review we briefly describe the physical basis of electron energy-loss spectroscopy and recent developments in EELS instrumentation. We show how STEM-EELS can be used for elemental mapping of heterogeneous nanoparticles and discuss the prospects for quantitative elemental analysis at atomic resolution. We describe the advantages of monochromated STEMs for obtaining chemical bonding information from the EELS near-edge fine structures. Furthermore, we explain how the low-loss region of the EEL spectrum can be used for physical property mapping. Here we concentrate on surface plasmon imaging of noble metal nanoparticles. In summary, EELS imaging and spectroscopy with aberration corrected STEMs offer many new opportunities to study fundamental questions in physics, chemistry and materials science at atomic resolution.

Acknowledgements

This work has received funding from the European Union within the 7th Framework Programme under Grant Agreement no. 312483 (ESTEEM2). This work was financially supported by the FFG in Vienna, Austria (COIN-project ASTEM).

References

[1] Egerton R F 2011 *Electron energy-loss spectroscopy in the electron microscope, 3rd edition.* (New York-Dordrecht-Heidelberg-London: Springer)
[2] Shuman H, Chang C F and Somlyo A P 1986 *Ultramicroscopy* **19** 121
[3] Mayer J 1992 in: *Proc. 50th EMSA Meeting.* (San Francisco, CA: San Francisco Press) 1198-1199
[4] Hofer F and Warbichler P 2004 in: *Transmission electron energy loss spectrometry in materials science and the EELS atlas. 2nd edition.* (Ahn C C; Ed.) (Weinheim: Wiley-VCH) 159-222
[5] Haider M, Uhlemann S., Schwan E, Kabius B and Urban K 1998 *Nature* **392** 768
[6] Batson P E, Dellby N and Krivanek O L 2002 *Nature* **418** 617
[7] Krivanek O L, Gubbens A J, Dellby N and Meyer C E 1991 *Microsc. Microanal. Microstruct.* **2** 315
[8] Mayer J, Deininger C and Reimer L 1995 in: *Energy-filtering transmission electron microscopy.* (Reimer L; Ed.) (Berlin-Heidelberg-New York: Springer Verlag) 291-346
[9] Tiemeijer P C 1999 *J. Phys.: Conf. Ser.* **161** 191-194
[10] Scott J, Thomas P J, MacKenzie M, McFadzean S, Wilbrink J, Craven A J and Nicholson W A P 2008 *Ultramicroscopy* **108** 1586
[11] Angseryd J, Albu M, Andren H O and Kothleitner G 2011 *Micron* **42** 608
[12] Von Harrach H S, Dona P, Freitag B, Soltau H, Niculae A and Rohde M 2010 *J. Phys.: Conf. Ser.* **241** 012015
[13] Kothleitner G, Fladischer S, Grogger W, Haber T, Meingast A and Hofer F 2012 *Microsc. Microanal.* **18** (Suppl. 2) 972
[14] Jeanguillaume C and Colliex C 1989 *Ultramicroscopy* **28** 252
[15] Schaffer B, Kothleitner G and Grogger W 2004 *Ultramicroscopy* **106** 1129
[16] Gloter A, Douiri, Tence M. Colliex C 2003 *Ultramicroscopy* **96** 385
[17] Thomas P J and Midgley P A 2001 *Ultramicroscopy* **88** 187
[18] Kothleitner G and Hofer F 1998 *Micron* **29** 349

[19] Hofer F, Grogger W, Kothleitner G and Warbichler P 1997 *Ultramicroscopy* **67** 83
[20] Urban K W, Mayer J, Jinschek J R, Neish M J, Lugg N R and Allen L J 2013 *Phys. Rev. Lett.* **110** 185507
[21] Pennycook S J and Nellist P D 2011 *Scanning transmission electron microscopy.* (New York-Dordrecht-Heidelberg-London: Springer)
[22] Kimoto K, Asaka T, Nagai T, Saito M, Matsui Y and Ishizuka 2007 *Nature* **450** 702
[23] Allen L J, D'Alfonso A J, Freitag B. and Klenov D O 2012 *MRS Bull.* **37** 47
[24] Kothleitner G *et al.* 2014 *Microsc. Microanal.* **20** 678
[25] Kothleitner G, Neish M J, Lugg N R, Findlay S D, Grogger W, Hofer F and Allen L J 2014 *Phys. Rev. Lett.* **112** 085501
[26] Hofer F and Golob P 1987 *Ultramicroscopy* **21** 379
[27] Keast V J, Scott A J, Brydson R, Williams D B and Bruley J 2001 *J. Microscopy* **203** 135
[28] Rez P, Alvarez J R and Pickard C 1999 *Ultramicroscopy* **78** 175
[29] Mitterbauer C, Kothleitner G, Grogger W, Zandbergen H, Freitag B, Tiemeijer P and Hofer F 2003 *Ultramicroscopy* **96** 469
[30] Su D S, Zandbergen H W, Tiemeijer P C, Kothleitner G., Hävecker M, Hebert C, Knop-Gericke A, Freitag B H, Hofer F and Schlögl R 2003 *Micron* **34** 235
[31] Geiger J 1967 *Phys. Stat. Sol.* **24** 457
[32] Schattschneider P 1986 *Fundamentals of inelastic electron scattering.* (Vienna, Austria: Springer)
[33] Schaffer B, Riegler K, Kothleitner G, Grogger W and Hofer F 2009 *Micron* **40** 269
[34] Grogger W, Hofer F, Kothleitner G and Schaffer B 2008 *Top. Catal.* **50** 200
[35] Batson P E 1980 *Solid State Comm.* **34** 477
[36] Nelayah J, Kociak M, Stephan O, de Abajo G, Javier F, Tence, Henrard L, Taverna D, Pastoriza-Santos, Liz-Marzan L M and Colliex C 2007 *Nature Phys.* **3** 348
[37] Bosman M, Keast V J, Watanabe M, Maarof A I and Cortie M B 2007 *Nanotechnology* **18** 165505
[38] Schaffer B, Hohenester U, Trügler A and Hofer F 2009 *Phys. Rev. B* **79** 041401
[39] Schaffer B, Grogger W, Kothleitner G and Hofer F 2010 *Ultramicroscopy* **110** 1087
[40] Schmidt F P, Ditlbacher H, Hohenester U, Hohenau A, Hofer, F and Krenn J R 2012 *Nano Lett.* **14** 5780
[41] Schmidt F P, Ditlbacher H, Hofer F, Krenn J R and Hohenester U 2014 *Nano Lett.* **14** 4810
[42] Schmidt F P, Ditlbacher H, Hohenester U, Hohenau A, Hofer F and Krenn J R 2014 *Nature Comm.* **5** 3604
[43] Nicoletti O, de la Peña F, Leary R K, Holland D J, Ducati C, Midgley P A 2013 *Nature* **502** 80
[44] Krivanek O L, Lovejoy T C, Dellby N, Aoki T, Carpenter R W, Rez P, Soignard E, Zhu J, Batson P E, Lagos M J, Egerton R F and Crozier P A 2014 *Nature* **514** 209

Quantitative ADF STEM: acquisition, analysis and interpretation

L Jones

University of Oxford, Department of Materials, STEM Group, Parks Road, Oxford OX1 3PH, Great Britain

E-mail: lewysbjones@gmail.com

Abstract. Quantitative annular dark-field in the scanning transmission electron microscope (ADF STEM), where image intensities are used to provide composition and thickness measurements, has enjoyed a renaissance during the last decade. Now in a post aberration-correction era many aspects of the technique are being revisited. Here the recent progress and emerging best-practice for such aberration corrected quantitative ADF STEM is discussed including issues relating to proper acquisition of experimental data and its calibration, approaches for data analysis, the utility of such data, its interpretation and limitations.

1. Introduction

Aberration correction in the STEM allows for the atomic resolution imaging of samples, while the incoherent nature of ADF allows for direct interpretation with a Z^n contrast relationship [1]. More than only taking 'pretty pictures', since the 1970's [2, 3] there have been efforts to quantify this image intensity. Generally, quantitative ADF is taken to mean intensity information and often it is atomic-resolution data that is studied. Spatial information such as peak-position shifts and strain mapping are closely related areas but are only discussed here in the context of error analysis. The mathematics of image normalisation is given in the highlighted literature and instead this review will concentrate on areas of best-practice, experiment design and error analysis in quantitative ADF.

2. The annular dark-field detector

The design of most ADF detectors is fairly similar. To facilitate retraction when not in use, these are usually mounted horizontally with a 45° inclined scintillator attached to a light guide, figure 1 [4]. A hole through the scintillator and light guide (often lined with a metal tube) allows the bright-field electrons to pass to another detector or spectrometer.

Scattered electrons hit the scintillator and are converted to optical photons; this light is guided to a photomultiplier tube (PMT) which amplifies the output signal. The output from the photomultiplier has some dark-level, set by the amplifier brightness (or offset) which is added to the experimental signal; this combined signal is then amplified further by an amount depending on the amplifier's gain (or contrast) setting. Finally this is integrated over the dwell-time specified by the user and passed through a so-called analogue-to-digital converter (ADC) [5]. The output units are often displayed in arbitrary counts; importantly this should not be confused with electron-counts. Modern detector and amplifier systems are sufficiently sensitive to register single electron scattering impacts [6].

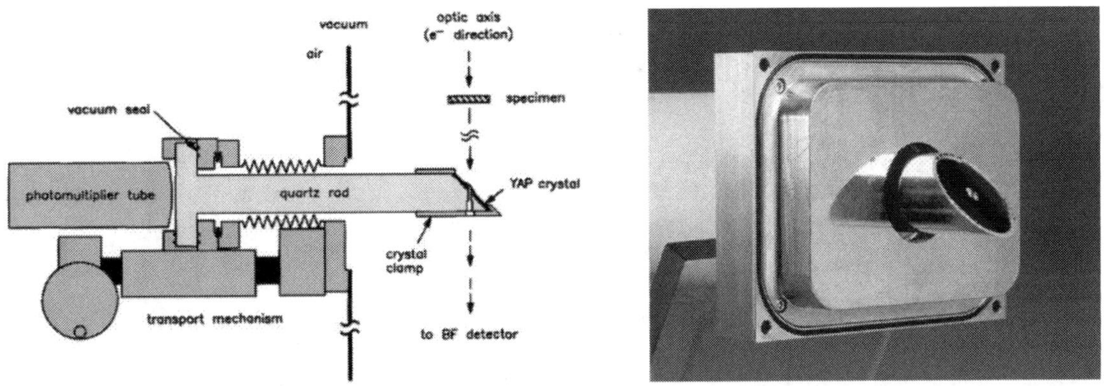

Figure 1. Left: schematic of a typical ADF detector (diagram modified from [4]). Right: photograph of a Fischione 3000 ADF detector on a retractable mounting (image credit Fischione).

3. Nomenclature and choice of analysis strategy

Quantitative ADF is a rapidly evolving field being developed in several groups; so first a brief mention of some of the most common terms in use. The most basic unit is the 'arbitrary-counts' output direct from the ADC. If the detector linearity is acceptable (see later) and a suitable vacuum level is subtracted, then these units will be proportional to the scattered signal. With this arbitrary-counts data either 'peak-intensities' can be measured or, because single-pixel measurements are often corrupted by noise, 'averaged intensities' from within some small area around the peak can be taken. For isolated atoms mask choice is flexible [7, 8], but in periodic structures this becomes more difficult and may be circles [9], squares [7], or Voronoi cells [10-12]. This last approach of Voronoi cells has the advantage that all image scattering is associated to some atomic column.

Averaged intensities are more robust to noise, but the result depends on the size of the mask used for averaging. For peaked signals (resolved columns) a wider mask results in a lower average, alternatively if units of 'integrated-intensity' are used up to a certain point a wider mask gives an *increased* value beyond which it plateaus [7]; in some cases then choice of mask size may be an issue. To eliminate choosing any mask, fitted 'Gaussian-volumes' can be used where peaks in the intensity data are fitted with Gaussian peaks [13]. For thin samples which do not experience excessive beam-broadening, cells and fitted-Gaussians are equivalent [14]. While Voronoi cells are quick to calculate, in thicker samples or where large probe-tails exist fitted-Gaussians may become more accurate.

If contrast ratios [15, 16] or statistical treatments are used [13], no further pre-processing is required, however for comparison with simulation the data must be normalized by the detector sensitivity.

3.1. Comparison with reference image simulation

Perhaps the most direct way to interpret experimental data is through comparison with image simulation. First, experimental data will be normalized by the mean detector efficiency and then some metric such as 'fractional-intensities' [9] or 'locally-averaged intensities' [10] is measured. Reference data are simulated (carefully matching accelerating-voltage, convergence angle, and detector angles) for a range of sample conditions including thickness, composition, or both, before the same metric is extracted from the simulated images. Comparing experiment to the reference data then allows an assignment on a column by column basis, figure 2.

This direct method is intuitive but has one main drawback; systematic errors are difficult to detect. For example, some errors yield global scaling factors (see Section 9). This means that thickness or composition assignments made purely by comparison with simulated libraries can always find a match

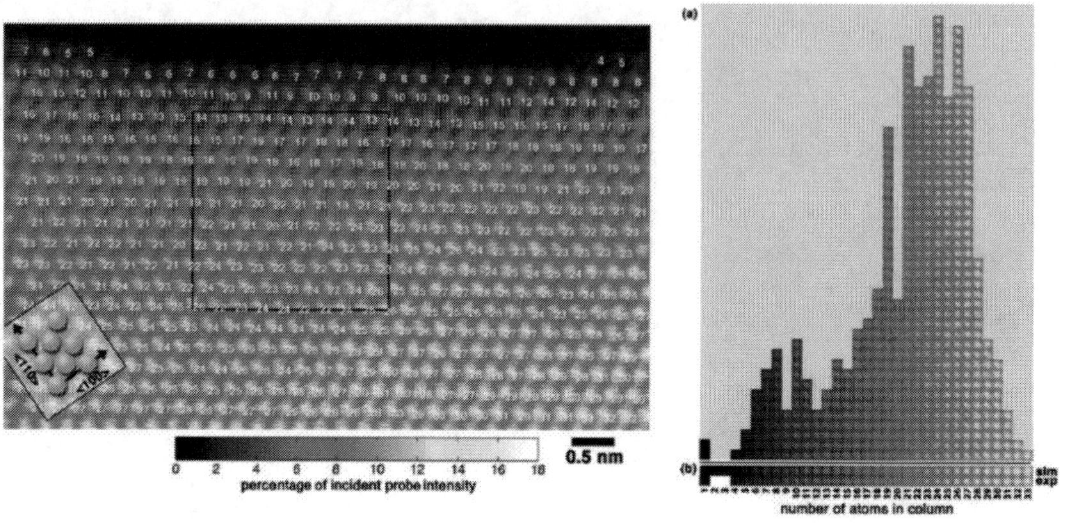

Figure 2. ADF image of [110] oriented Au wedge. Intensity at each atomic-column was averaged within a circular mask and compared with simulated values (right). Adapted with permission from [9]. Copyright American Chemical Society (2010).

and the analysis reliability depends purely on the accuracy of the data normalisation. An additional complication is the wide range of simulation software packages to choose from [17-21] (and others) each of which give very similar but sometimes slightly differing results for identical conditions.

Finally, as image simulation may never capture all the subtleties of the experimental data, care must be taken in what metric to perform the matching on. Perhaps the 'best-practice' in quantitative ADF is the unit of the 'scattering cross-section' [2, 3, 7, 22]. This parameter is robust to magnification, defocus and convergence angle [7], source-size [7], astigmatism and other aberrations [23], scan-noise [24], and small sample mis-tilt [25]. These scattering cross-sections are then principally only dependent on accelerating voltage and detector inner- and outer-angle.

3.2. Statistical interpretation of intensities

Where a large number of nominally identical atomic-columns are within the field of view, it may be possible to statistically decompose the intensities into their underlying grouping (sorted by atomic thickness) using, for example the Integrated Classification Likelihood (ICL) [13]. Figure 3 shows the main stages in this process.

The applicability of this approach is limited by two considerations; the need for a large number of observations per unique thickness, and for the spread of intensities from equal thicknesses to be significantly less than the difference between those of differing thicknesses [27, 28]. However, when these conditions are met, this approach is robust against systematic errors.

4. Variations in detector response

In Section 2 the hardware of the ADF detector was introduced, here the response of the detector and amplifier combination to electron illumination is explored.

4.1. Asymmetry in detector efficiency

Ideally electron scattering to all angles (all points on the detector) should be recorded with equal sensitivity. Unfortunately this is not always the case; the central hole in the scintillator means that light from beyond this less efficiently reaches the PMT. This generally results in a reduced efficiency at low inner angles and can be accommodated via the comparison simulations [20], flux-weighting

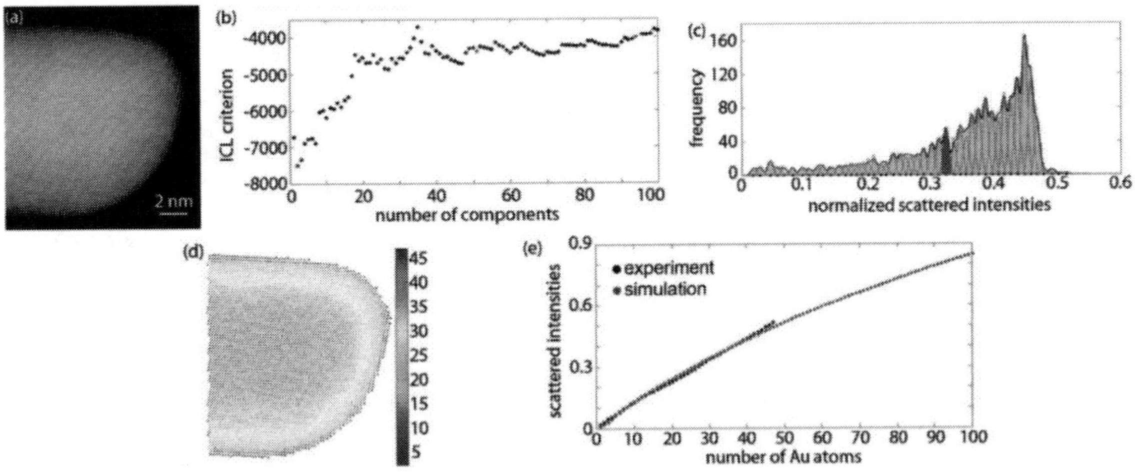

Figure 3. Stages in a statistical analysis of ADF data; a) the raw image, b) the ICL analysis of the intensities, c) the intensity data re-plotted as a histogram with the appropriate number of Gaussian components fitted, d) the atom-count assignments based on b) and c), and e) the validation test comparing the experimental assigned counts with simulated intensity data. Figure modified from [26].

[29], or through use of a reduced effective outer-angle [30]. Figure 4 shows a comparison of six of the most common detectors currently used.

Manufacturer	Company 1			Company 2	Company 3	Company 4
	Detector A	Detector B	Detector C	Detector D	Detector E	Detector F
Detector Map	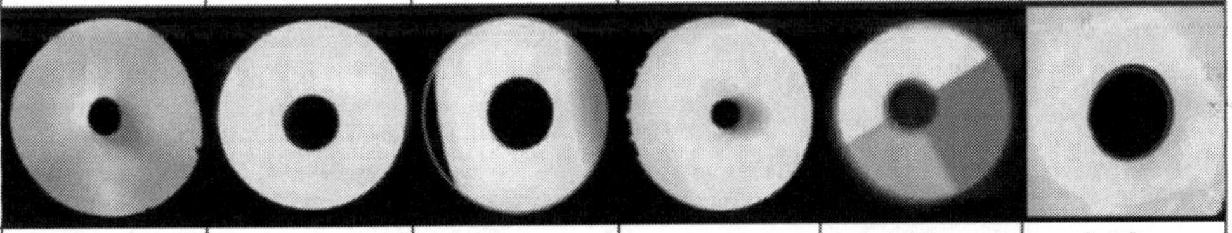					
Angle Ratio	5.47 x	3.09 x	2.89 x	5.91 x	2.90 x	3.50 x
'Flatness'	8.9 %	6.8 %	24.9 %	10.4 %	9.7 %	14.5 %
'Roundness'	8.2 %	5.4 %	10.6 %	5.4 %	28.1 %	2.4 %
'Smoothness'	30.0 % *	15.1 %	16.3 %	18.0 %	87.2 %	23.2 % *
'Ellipticity'	19.6 %	4.8 %	8.9 %	0.5 %	4.3 %	13.3 %
Average	16.7 %	8.0 %	15.2 %	8.6 %	32.3 %	13.4 %

Figure 4. Comparison of the sensitivity response of six commercially available ADF detectors, evaluated for various performance metrics. Smoothness values marked with an asterisk indicate probable overestimation from sample occlusion or detector burn-in damage. See reference [31] for metric interpretation.

Of course the ideal solution is to improve the detector design and some progress has already been made towards this by physically separating the light-guide from the scintillator and through the introduction of two-material detectors [32].

4.2. Detector response linearity
The term 'detector linearity' is a somewhat ambiguous term used to describe several slightly different things. Fundamentally we need to evaluate the 'detector-plus-amplifier' combination as a whole, and this depends greatly on the amplifier settings. Ideally the output should respond linearly with respect to brightness [33], but also with respect to gain [6]; lastly, and most importantly, it must be linear with respect to beam-current [34-35]; these three individual linearities are illustrated by figure 5. Also remember, that as scattered electrons falling on the detector are Poisson distributed, not only does the mean-intensity of thick regions need to be below the amplifier's saturation limit but also the high-count tail of this distribution [5].

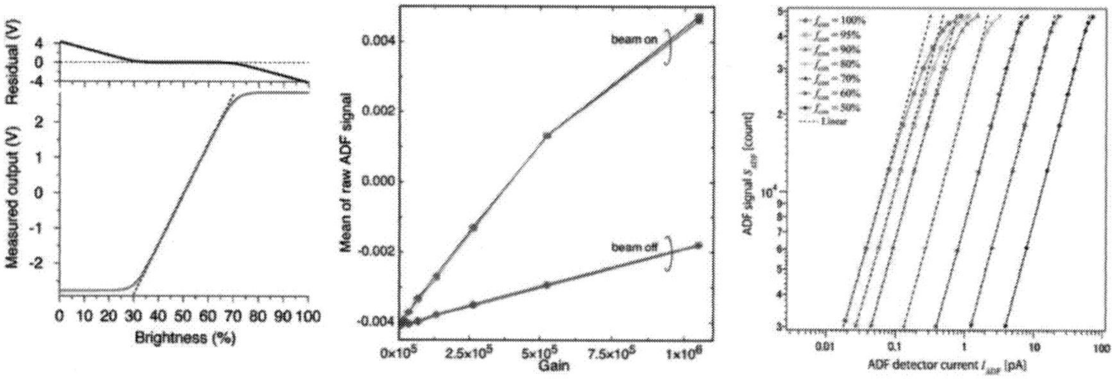

Figure 5. Evaluation of amplifier output linearity with brightness setting (left [33]), with gain setting (middle [6]), and with the current at the detector itself at various gain settings (right [34]).

Once the bounds of linearity are known, and working within these, the operator can set brightness towards the low end and contrast slightly higher (say 35% and 60% respectively for the above plots). This maximises the available dynamic range without risking saturation in thick sample areas.

5. Asymmetry in post-specimen flux-distributions

Having discussed the asymmetry in ADF detectors above, next we must consider the electron distribution which falls on them. Most STEMs are equipped with a CCD in the detection plane and with the ADF retracted these can be used to observe the flux which falls in this plane.

Figure 6 shows examples of these patterns for the most commonly used instrument types. Some show convergent-beam electron-diffraction or Kikuchi like patterns or no texture and this is only a result of the specimen thickness used to generate the high-angle scattering. In all cases the logarithm of the intensity is shown to improve the visibility of the dynamic range which spans around three orders of magnitude. The red circles indicate typical convergence angles of around 10 mrad for uncorrected, and 20 mrad for corrected systems. In all cases an 'outer cut-off' can be seen and depending on the instrument is between 130 - 200 mrad [29, 34]. This means that below a certain camera-length ADF outer-illumination-angle will not increase any further [36]. In some cases the edge is bounded with a slightly enhanced intensity; this is thought to arise from the 'folding-back in' of even higher angle scattering by the uncorrected residual aberrations in the post-specimen optics.

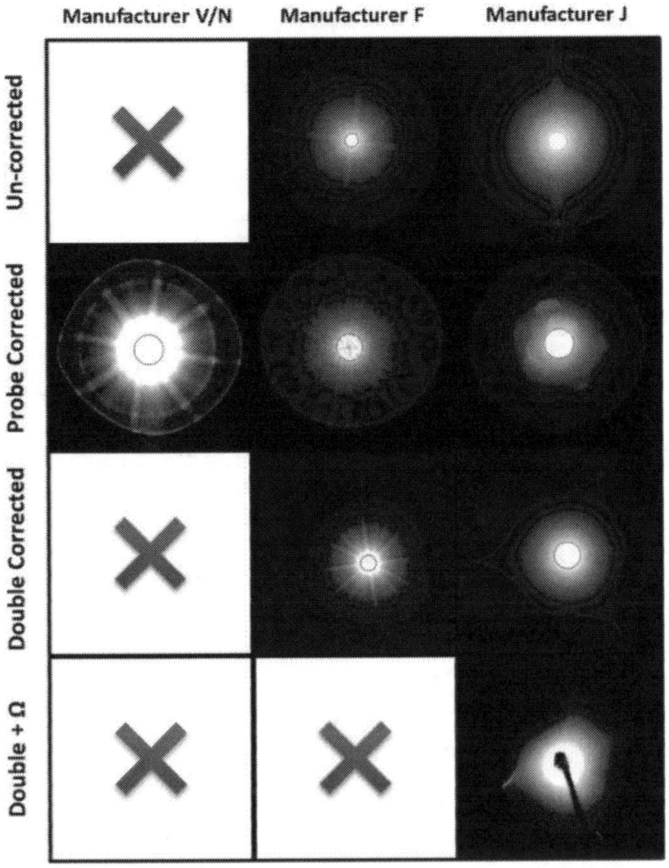

Figure 6. Various experimental STEM flux maps. Red circles indicate the BF disk (≈ 10 mrad for uncorrected, 20 mrad otherwise).

For uncorrected and probe-corrected instruments this outer cut-off is round; TEM-corrected instruments yield more complicated patterns owing to the additional non-round post specimen lenses in the image corrector [34]. Here the 6-fold hexapoles yield a 3-fold parasitic aberration resulting in a triangular flux pattern in the detection plane; with the addition of an in column energy-filter this is further distorted in one direction (figure 6). For this reason, and because no image simulation packages include the effects of non-round post-specimen lenses, probe-corrected-only instruments are expected to give the best quantitative ADF accuracy.

6. Strategies for recording detector scans

When recording detector efficiency scans two choices need to be made; what current/gain regime to use, and whether to map in real-space or angle-space. In all cases, to avoid any possible effects of variable integration time [6], the same dwell-time should be used for the detector scan and experiment.

6.1. Dropped gain versus dropped current

When imaging total high-angle scattering may be around 10% of the primary beam and is spread across the whole ADF detector. However, ADF normalisation requires a detector sensitivity scan equivalent to 100 % of beam-current. This leaves the operator with a dilemma and a choice, should the detector scan be recorded in so-called 'dropped gain' or 'dropped current' methods.

In the dropped gain method, the full primary beam is focussed onto the ADF detector [33], where to avoid massive amplifier saturation the gain settings are set very low. To prevent errors arising from gain-non-linearity, this same low gain must then be used for the imaging experiment which may reduce sensitivity. There is also uncertainty about the behaviour of the scintillator crystal (afterglow, depletion, saturation) with such differing illumination conditions between the imaging and mapping.

In the dropped current method, the amplifier is first optimized for the imaging experiment; then, by either choosing a smaller objective aperture or reducing the strength of the extraction anode, the beam-current is reduced by a factor of around 10. The detector is then mapped with this weaker-beam, and the ratio of beam currents (as measured by either the CCD, or better, a Faraday cup / EELS drift-tube) incorporated in the normalisation analysis [35].

Both are thought to yield roughly similar accuracies; with the dominant error in the dropped gain method being the reliance on the linearity of output across such a wide range of electron dose, and in the dropped current method the precision with which the current ratio can be measured.

6.2. Confocal or swung-beam detector scanning

There are also two choices of geometry for recording detector efficiency scans, the 'confocal' or 'swung-beam' methods; both are recorded without a sample present.

In the confocal method the post specimen optics are set to refocus the probe in the detector plane (FEI 'image mode' or JEOL 'STEM-align mode'), the scan-deflectors then scan the probe both in the sample plane and the detector plane [33]. Because of post-specimen magnification, confocal detector scans have dimensions of real-space millimetres. Whilst easier to record, scanning in this way passes only a narrow beam through the objective lens (OL) and does not faithfully replicate genuine scattering to the ADF.

Alternatively, the condenser-lens system can be adjusted to produce a cross-over in the back focal-plane of the objective lens and a pencil-beam at the specimen plane; the scan-coils now swing this pencil beam in angle-space through the OL. The scanning angle is calibrated by comparing the scanned-square as seen on the CCD (with ADF detector retracted) against a known diffraction pattern.

Conventional confocal method examples are shown in figure 4; while examples of swung maps for various camera-lengths are shown in figure 7.

In the swung-beam method maps are returned with units of mrad and the effect of the OL bore constriction is directly observed (compare with figure 6). This effect becomes apparent in figure 7 for camera-lengths $\leq$ 10 cm where the outer-angle is now imposed by the OL-bore rather than the detector hardware [29, 34, 36]. To verify the origin of this 'cut-off' effect we can also observe the behaviour of these swung-beam type detector maps while changing various deflectors in the microscope. The bottom row of figure 7 shows composite figure showing the effect of varying image-shift (CLA) and projection-shift (PLA), these were made my recording separate scans at different deflector settings and adding the scans together. Varying CLA (at 8 cm camera-length) the detector and cut-off ring move together, while varying PLA (at 40 cm camera-length) only the detector moves but this is always bounded by the same cut-off. This, taken with the cut-off's invariance with camera-length, confirms that the origin of the cut-off lies after the CLA deflectors but before intermediate-lenses and PLA, that is, it is caused by the objective lens.

7. Inner and outer angle optimisation

To compare ADF data against simulations, inner and outer detector angles must be precisely measured [29, 33] for each of the required camera-lengths [36]. However, image simulations can also guide experiment design to maximise the precision of the recorded data, figure 8.

For compositional selectivity at fixed thickness the Z^n relationship of ADF scattering [1] improves with higher inner angles [37, 38]. However, for light elements scattering to high angles is so weak the finite SNR becomes the dominant source of error so for thickness measurements at fixed composition lower inner angles are preferable [14, 39, 40].

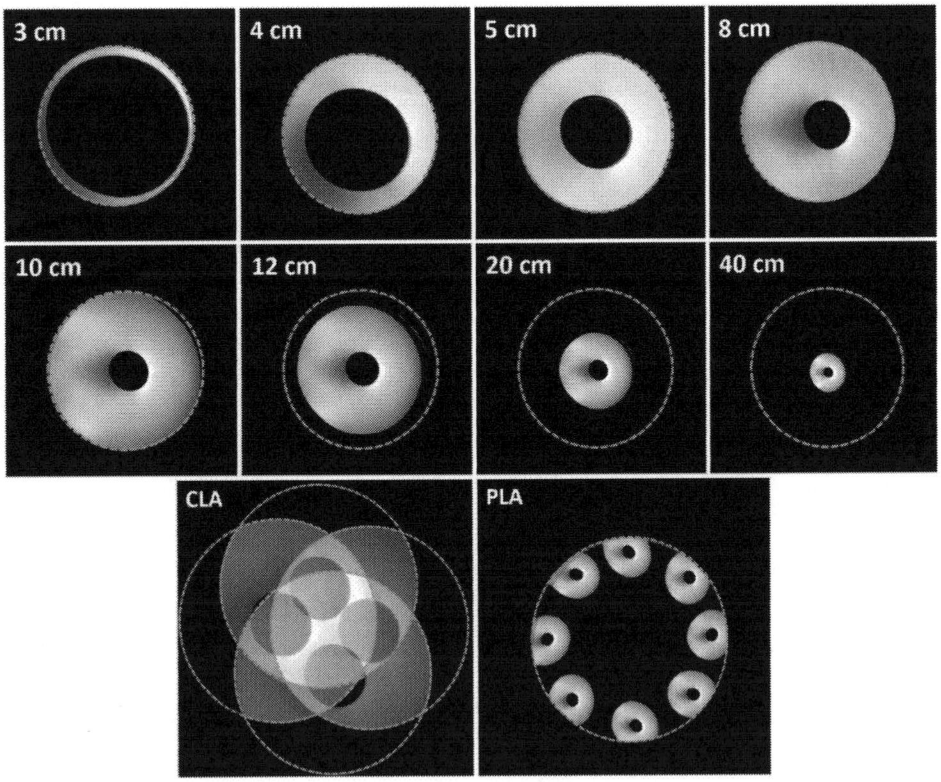

Figure 7. Effect of camera-length (top) and beam-deflectors (bottom) on the swung-beam detector scans. For camera-lengths ≤ 10 cm the outer-angle is truncated and this must be reflected in any reference simulations. Changing image-shift (CLA) translates the apparent position of the whole OL and detector set, while changing projection-shift (PLA) changes only the apparent position of the detector. Green circle is constant in all figures and represents the observed shadow of the inside of the objective-lens bore.

8. Choice of normalisation approach

The significant asymmetry in the performance of ADF detectors cannot be ignored when comparing experiment with simulations; most importantly the effect of the reduced efficiency at the centre where the majority of scattering falls. This introduces a choice for the experimentalist as to whether this hardware defect should be accounted for in the reference simulations or in the experimental data normalisation.

Some simulation packages can accommodate the detector sensitivity profile (DSP) in the creation of the library data [20]; this allows for a simple experimental normalisation to be used [33]. However, this precludes using other simulation codes that cannot introduce the DSP [17-19, 21]. In this case the asymmetry must be accounted for in the normalisation stage by either imposing some artificial effective-outer-angle [30], or more phenomenologically, by weighting the observed detector efficiency (figure 4) by the electron flux distribution (figure 6) [11, 29]. A more in depth discussion of this issue is given in Martinez *et al.* [29].

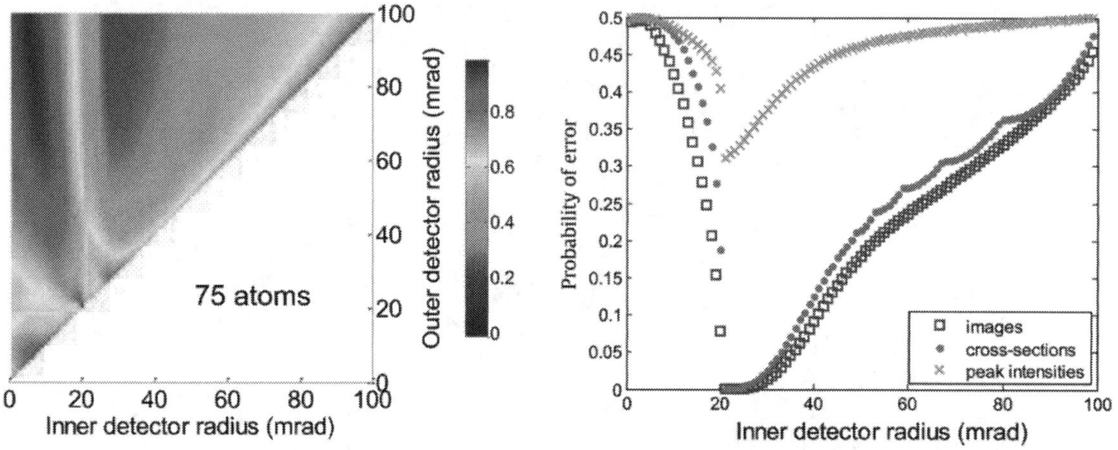

Figure 8. Experiment design optimisation study. Left: the probability of error in thickness measurement for a 75 cell slab of $SrTiO_3$ for a fixed electron dose. Right: plot for a fixed 100 mrad outer angle detector were only inner angle can be varied (compare with figure 7). Note the probability of error for fitted image-Gaussians, cross-sections and peak-intensities. Figures modified from reference [14].

9. Considerations on the accuracy of experimental intensities and peak-positions

9.1. Sample tilt

With the exception of small nano-clusters, to obtain atomic-resolution ADF STEM must be recorded close to low-order crystallographic axes. In the theory of electron channelling, where atoms in aligned crystals act as small lenses for the beam (focussing it for those that follow), we can understand this 'wave-guiding' effect as keeping intensity on an atomic-column as it propagates through samples far thicker than the probe's depth of field [41, 42]. The counterpoint then, is that small mis-tilts away from this condition have marked detrimental effects on both image resolution and scattered ADF intensity, figure 9.

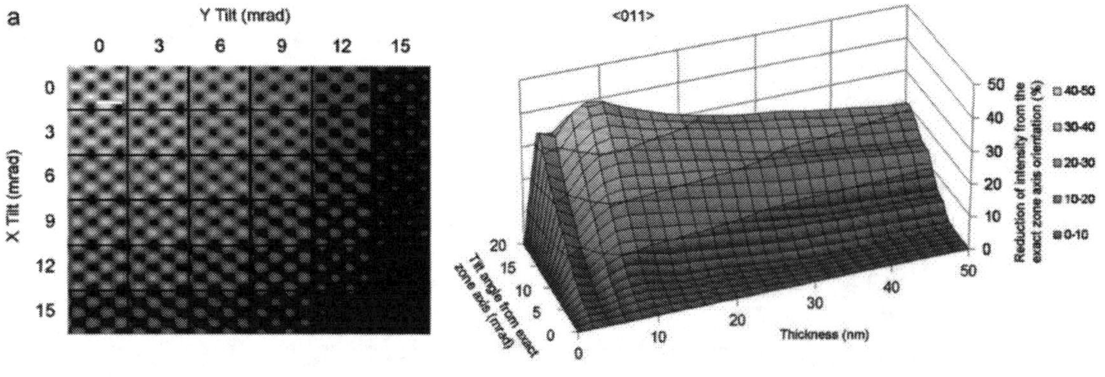

Figure 9. Left: simulated ADF images of silicon [110] with increasing tilts. All images expressed in fractional beam current (simulation details and image credit, reference [43]). Right: quantitative assessment of the loss of scattering for mis-tilted crystals of [110] oriented gold. Here a tilt of only 20 mrad yields a 40% drop in scattering (simulation details and image credit, reference [44]).

Experimentally, sample tilts of up to 15 mrad can still yield an atomic resolution image [25, 43] but can lead to a loss of ADF intensity of up to 25% or more [25, 44]. Fortunately it is possible to design the experimental conditions to also be robust to these intensity losses. If an integrated cross-section approach is followed [7], and if the ADF detector inner angles are chosen suitably [25] it is possible to create a 'tilt plateau' where the scattering cross-section remains almost invariant to sample tilts for miss-tilts of up to one probe convergence angle [25].

Sample mis-tilt not only affects image-intensity, but can also affect the apparent position of atomic-columns within an image. Often, this effect is neglected, however, where precise peak-position shifts are being measured (oxygen octahedral tilts, cation shifts, etc.) this becomes essential. Consider the data shown in figure 10 [45]. Here the apparent positions of Sr and Ti-O columns shift, but moreover they shift differently from each other. This is shown in the lower graphs for various sample thicknesses; if we consider the 11.3 mrad tilted sample (blue line) an apparent shift of 15 pm may be observed between the two different sub-lattices that is purely a tilt-artefact.

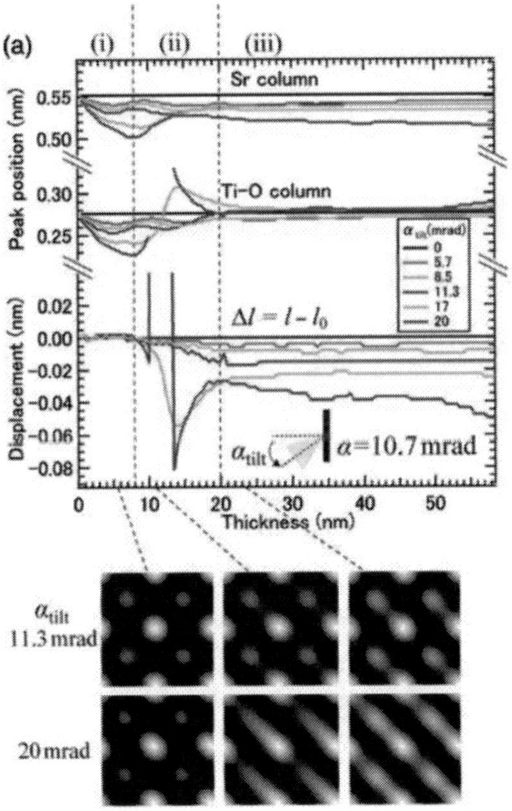

Figure 10. Absolute peak-positions of Sr and Ti-O columns in tilted simulated images. The relative shift between the two column types is also shown. Figure credit and simulation details: reference [45].

Further if we consider the constant composition with increasing thickness case, such as the Sr column data alone; here a tilt of only 8.5 mrad can yield an apparent shift of up to 20 pm that is approximately proportional to thickness up to around 8 nm. This is relevant for strain mapping of nanostructures, if a reference lattice is defined in a thick region from which lateral strains are measured at a thinner surface region [46].

From figure 10 we might consider a 'rule of thumb' that for column positions measurements better than 20 pm, sample tilt must be less than the convergence angle, and for accuracies better than 5 pm this should be less than one-half convergence angle [45].

9.2. Atomic-column 'cross-talk'

For thicker specimens, beam-broadening within the sample can lead to scattering reaching the detector originating from several Ångstroms to the side of the probe position, figure 11 [7, 15]. It should be noted that while STEM images are recorded as a function of incident probe-position, signal(s) reaching their respective detectors are integrated over the volume of any electron interaction. Figure 11 also highlights the improved robustness to the integrated column (or cell) versus the peak intensity.

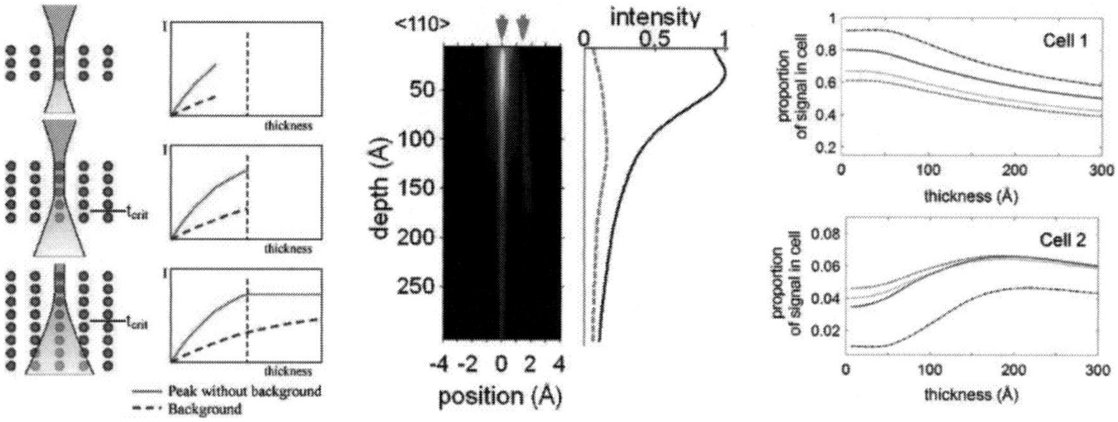

Figure 11. Left: Schematic of beam broadening and its effect on column and background intensity (figure credit: [15]). Middle: x-z intensity plot through [110] oriented GaAs sample with the probe positioned over a Ga column (black intensity plot) showing the intensity cross-talk to the nearest As column (dashed red line). Right: equivalent signal integrated laterally over an atomic-column (cell) for the nearest and neighbouring columns for different forms of probe coherence (figure modified from, and probe coherence details in, reference [12]).

While larger objective apertures can improve the diffraction limited probe size, de-channelling begins sooner meaning that optimal experiment design for quantitative ADF requires a trade-off of these considerations [16, 42, 47]. In some cases balancing of the elastic and inelastic contributions to image-intensity can lead to a sample mis-tilt invariant 'plateau' in the scattered cross-sections for tilts less than or equal to the convergence-angle [25]. This gives the experimentalist a very useful operating range of acceptable mis-tilt and again should be considered in the experiment design.

9.3. Strain contrast

Many materials are interesting because of their imperfections such as interfaces, grain-boundaries [48] and dislocations [49], many of which are accompanied by appreciable strain. These strains distort crystal planes and affect the electron-channelling behaviour and hence image contrast; this often increases image intensity in low-angle ADF images at the expense intensity in the HAADF, figure 12.

The precise role of strain in ADF contrast is not yet fully understood, but it is important to be aware of because it is often a manifestation of some other underlying effect such as varying mis-tilt or composition at interfaces [50, 51], each of which would also need to be reflected in any reference simulations.

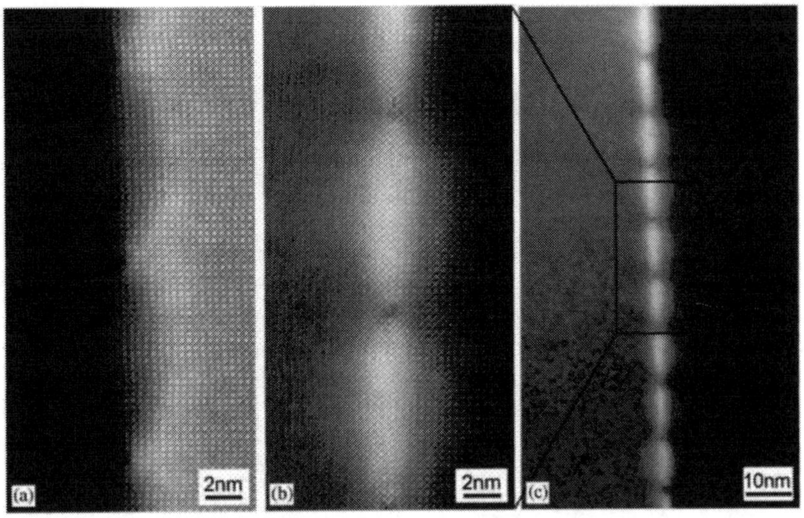

Figure 12. Effect of strain on HAADF (a) and LAADF (b & c) images of $SrTiO_3$. The HAADF image (inner-angle = 35 mrad) is very sensitive to mis-tilt across the interface but also shows darkness in the strained regions between dislocations, while the LAADF (inner angle = 17 mrad) shows brighter contrast in these places (figure credit: reference [48]). Probe convergence angles was 9 mrad in both cases.

9.4. Amorphous layers and carbon background

For free-standing samples amorphous layers can often persist with even the most careful sample preparation, whether they are from focussed ion beam damage or from carbon contamination from within the microscope itself. Equally for samples supported on thin carbon (or other) membranes, the effect of this 'second sample' cannot be ignored [28]. In either case, and depending on defocus, they may appear invisible (figure 13) but still contribute to the quantitative image intensity [47, 52].

In figure 13, though it appears that little resolution is lost by under-focussing through the amorphous layers (last row of panels), this amorphous material remains and will increase beam-broadening.

If nanostructures are supported on a thin membrane, it may be possible to improve the quantitative analysis by subtracting some image intensity contribution extrapolated from the region outside the particle [28]; however, in other situations amorphous carbon may lead to an unavoidable overestimate of the sample thickness [47] or a loss of statistical counting precision.

9.5. Electron dose and pixel size

Excessive electron dose will usually cause sample damage [53], however too little dose and the low signal-to-noise ratio caused by excessive shot-noise may make quantitative measurements less precise. Statistical studies are useful for predicting the expected error as a function of dose for different applications such as nanometrology [14, 26, 28], or light element imaging [39, 40]. Recall from figure 8 that for a given fixed dose, the expected probability of error is always bigger for peak-intensity measurements compared with integrated cross-sections. It is often perceived that finer pixel-sampling gives 'better' pictures, however these may contain no additional quantitative information content. So long as the sampling is sufficiently high to meet the instrument resolution (say equal or better than twice Nyquist) then additional pixels do not improve the reliability of measurements [7, 28]. Up to 4x Nyquist may be useful when onward reregistration or processing is planned but again, beyond that no additional benefit is conveyed.

IOP Conf. Series: Materials Science and Engineering **109** (2016) 012008 doi:10.1088/1757-899X/109/1/012008

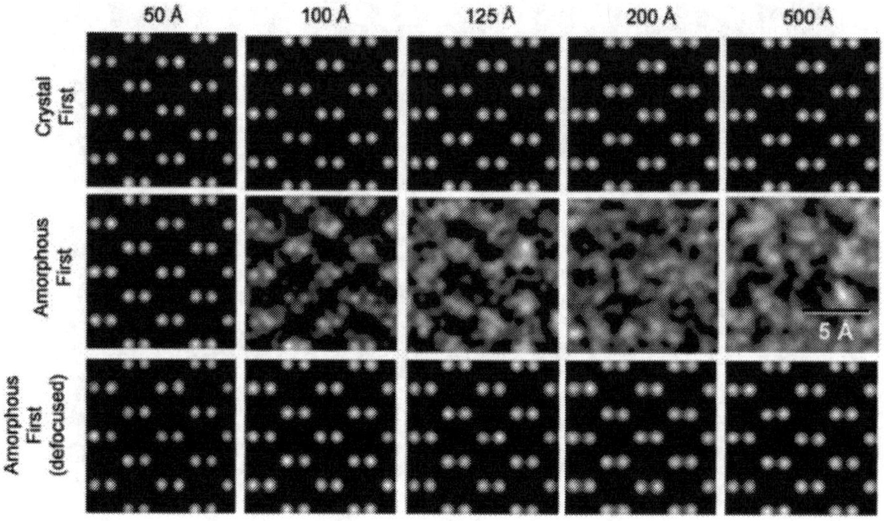

Figure 13. Effect of 50% thickness amorphous surface-layers on the contrast recorded in ADF STEM for various sample thicknesses. Note that amorphous layers on the rear-face of the crystal are invisible at all thicknesses. For the amorphous-first case small under-focus again renders this layer near invisible. Full simulation conditions, and figure credit, reference [52].

9.6. Scan noise and scanning distortion

Environmental instability can affect the quality of scanned image data dramatically [24, 54]. The important consideration here is to understand the frequency domain over which any environmental disturbance operates (see [55] for a more detailed discussion of frequency ranges). For data recorded sufficiently above the Nyquist sampling limit, high frequency scan-noise in the kHz range may corrupt peak-positions and peak-intensities, however it does not affect the *integrated-intensity* over an atomic-column. The same cannot be said of low frequency distortions ($\approx$ 0.5 - 3Hz range) and the effect of these on quantitative ADF image intensities remains an important area of research [55].

One approach is to record multiple frames and realign these by cross-correlation [56-58], or to precess the slow-scan direction orientation and apply affine transformations to correct image drift [59]; in either case after averaging these data both improves the SNR and yields a column-position precision of several picometres or better [56, 60, 61]. However, rigid registration and averaging may not mitigate residual low-frequency time-varying environmental distortion, and artefacts often remain visible in strain maps [57, 60].

To correct scanning-distortions non-rigid registration must be used, where different regions may be moved by different registration vectors [46, 55, 62]. Approaches like these are able to locally compensate for distortions and to converge a frame-series of data to a 'reduced-scan-distortion' single frame. Using non-rigid techniques peak-position precision has improved to $\approx$ 1 pm levels [46].

9.7. Cold field-emission tip current fluctuation and emission decay

In general, emission fluctuations from cold field-emission type electron sources are monitored and compensated in real time. Where this fails fluctuations in image contrast may result and users should be aware of this when using such systems. Perhaps more important with these systems is to be aware of any emission current decay between the detector mapping and the experimental acquisition [11].

9.8. Summary of sources of error

Table 1 summarises the dominant sources of relative error in quantitative ADF. Not all will apply in all instances and magnitudes will vary with individual systems and experiments.

Table 1. Dominant sources of error in quantitative ADF.

Error Source	Severity	Cause	Solution	Comments
Detector asymmetry	Global error, up to 15% if ignored with worst case detector [29]	Detector geometry / design [31]	Improve design / symmetry [32]	Can be as low as 2% if corrected during normalisation or simulation [29], [30]
Sample tilt	Local error, up to 10% for peak-intensities, ≈2% for cross-sections, at 5 mrad mis-tilt [7], [44]	Loss of electron channelling condition	Optimise convergence angle to be more robust to tilt [25], keep tilt $\leq 0.5\alpha$	Varies by column, depends on sample thickness
Detector angle mis-measurement	Global error, ≈1% error per 1% error in inner-angle, ≈0.2% error per 1% error in outer angle	Errors lead to incorrect comparison simulations	Careful detector angle measurement, including outer-angle / flux cut-off evaluation [29], [33], [34]	Camera-length dependant. Statistical analysis approaches not affected [13]
Detector or flux 'cut-off' mis-alignment	Global error, up to 3% with shifts of around 10 mrad [30]	Incorrect projection lens settings or de-scan calibration; incorrect image-shift setting	Improve column alignment verify anode wobble centre in Ronchigram	May also cause appearance of astigmatism / peak-shifts [30]
Magnification (pixel width)	Global error, generally <2%; cross-section error scales with mag error squared [7]	Imperfect scan coil calibration	Verify calibrations (in x and y) from bulk reference material	Statistical intensity analysis approaches not affected [13]
Scan noise / distortion	Local error, ≈0-5%, frequency and dwell-time dependant [24], [55]	Environmental instability [54]	Room design [54], fast-scan and post-processing techniques [6], [55]–[57]	May also affect peak-positions and strain mapping [57], [60]
Shot noise	Global error, generally <1% for cross-sections, worse for peak-intensities	Fundamental limit of Poisson distributed intensities [28]	Increase beam-current, dwell-time or number of frames [44]	Error scales with $1/\sqrt{\text{dose}}$ [44]
Amplifier brightness / contrast problems	Local error depending on sample thickness, <1% if amplifier adjusted properly [5], [6]	Poisson distribution of counts can lead to 'pre-clipping' in thick sample regions [5]	Observe waveform during acquisition, increase brightness or reduce amplifier gain	Up to 10% if brightness too low or gain is too high [5]
Emission current fluctuation	Local error, generally <1%, affects bands parallel to fast-scan	Power supply and tip stability	Fast-scan averaging techniques [55]–[57]	Cold-FEGs only. Error may become 5-10% if compensation fails

10. Example applications of quantitative ADF

Extracting reliable information from quantitative ADF relies on being able to fix as many experimental factors as possible, ideally to leave only one free parameter to explore.

Where thickness is constant (or presumed from extrapolation) quantitative ADF intensity can be used, with extensive simulation, to reveal local compositional variations [10]. It should be noted that in this case the limiting error will be some combination of the reliability of thickness extrapolation, the effects of de-channelling, effects from carbon contamination reducing the image contrast (up to 4% error), and from the simulated reference data (generally better than 1% error contribution). However, for the measurement shown in figure 14, the relatively good SNR of quantitative ADF still yields a lower error than that of atomic resolution spectroscopic chemical mapping approaches which may be as high as 10%.

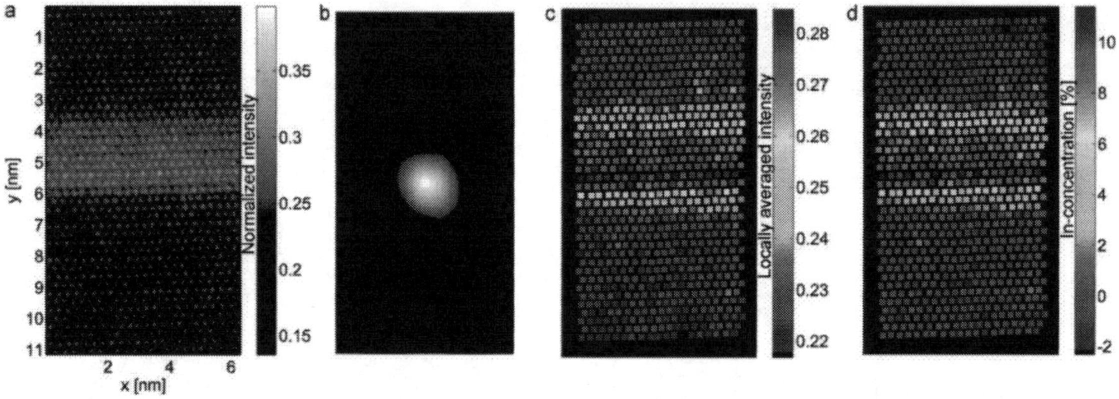

Figure 14. Experimental ADF image (a) and the integration polygons (b) used to locally average the image intensity (c). Adjusting for thickness, and through comparison with extensive simulation, the resultant In-concentration map (d) at atomic resolution. Figure modified from reference [10].

For fixed composition it is possible to observe changes in thickness or structure, figure 15 [11, 63].

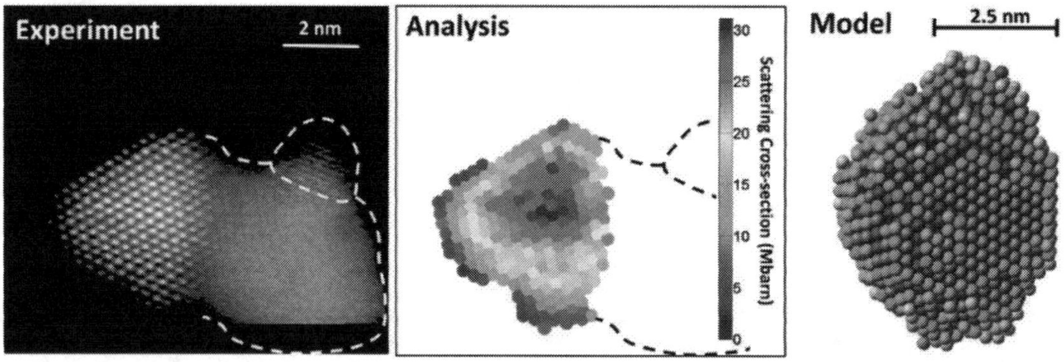

Figure 15. Pt nanoparticle (left) and the scattering cross-section analysis (middle). Atom counts were then used to rebuild a 3D atomic model of the sample (right). Adapted with permission from [11]. Copyright American Chemical Society (2014).

In rare cases, where both thickness and composition can be presumed to remain constant, the quantitative ADF signal can be used to infer the column-occupancy of heavy species [64]. Dynamic processes are special cases of quantitative ADF [11, 63, 64], but make use of the principle that individual images from a frames-series may be scrutinized separately to yield temporal resolution instead of tomography which sacrifices dynamic information to retrieve spatial resolution.

11. Future possibilities in quantitative ADF

The basic geometry of the ADF detector has changed little since its invention; however, several possibilities exist for the future of quantitative ADF with some new detector geometries recently being explored. Perhaps the most simple of these is the 'quadrant-detector' ADF, where the annulus has been divided into four quarters. These can be added together to give a classical ADF image or signals from opposite sides subtracted to give a differential phase-contrast signal [65]. Taking this concept further, multiple annular rings (themselves each divided into quadrants) can be used, figure 16 [66]. Such a detector allows for a host of sum-, or difference-, images to be synthesized after the scattering data has been recorded, moving experiment design from a pre-imaging to a post-processing regime.

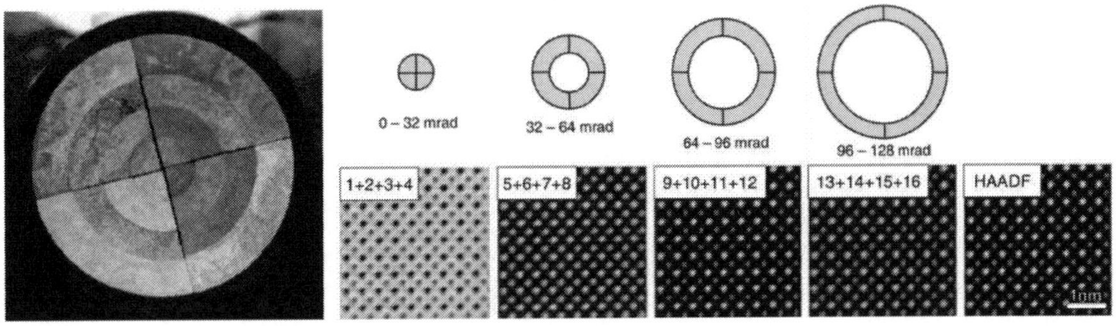

Figure 16. Left: photograph of a fibre-optically coupled 16-segment detector. Right: example detector geometries available and their associated images of $SrTiO_3$. Far right: the simultaneously acquired conventional HAADF image. Figure modified from reference [66].

This may allow light elements to be seen in the ABF images, or for strain and thickness / composition effects to be separated.

Taking this concept to its extreme is the use of pixelated detectors which can be used to record all the scattering in the detector plane (figure 17). This yields flux-patterns similar to figure 6, including Kikuchi patterns for thicker samples [67], but has the advantage of a homogenous detection efficiency through applying a conventional CCD gain-reference.

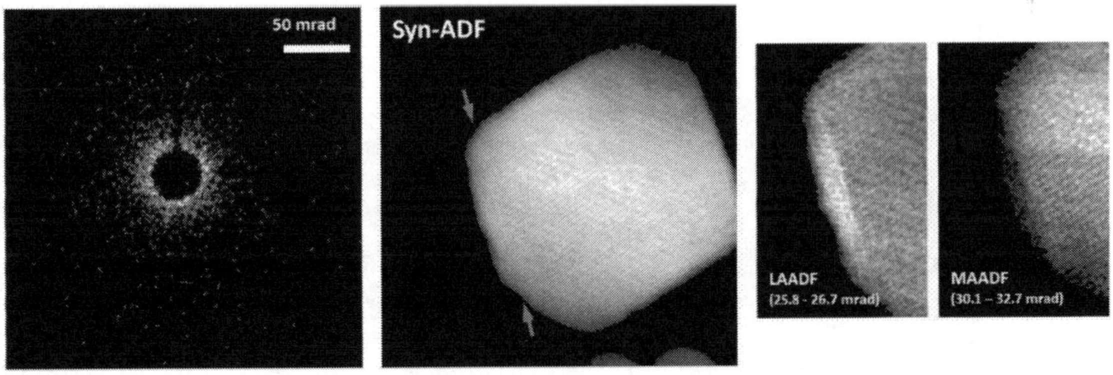

Figure 17. Left: example pixelated detector readout. A beam-stopper was used to mask the BF-disk. Bright points indicate individual electron scattering events. Right: example images synthesized from the 4D dataset, the integrated ADF signal (arrows indicate the position of a twin-boundary) and synthetic LAADF and MAADF images where the edge region appears bright and dark respectively.

With CCD technology dynamic range is an issue, making it difficult to record both the BF and DF regime; also frame-speed, as the whole detector-plane-image must be read-out at every real-space probe position. In figure 17 a beam-stopper was used to mask the BF disk, however this contains much useful sample information [68]. There also exist possibilities for detectors in other planes [69].

In the coming years the development of quantitative cross-sections for analytical signals such as EELS [22] and EDX [12, 70, 71] is expected, as well as hybrid solution engines that combine the complimentary robustnesses of simulation-comparison and statistical-interpretation based quantification techniques. These advances may allow for the combined measurement of both composition and thickness variation at atomic resolution. Further improvements may also be possible in the detector and objective lens hardware design. It is certainly expected that error-analysis driven experiment design will mature as a field to deliver the best possible results from the powerful aberration corrected STEM.

Acknowledgments
The author would like to thank G. Martinez, A. Béché, Q. Ramasse and J. Kim for assistance collecting some of the data shown in figure 6; H. Sawada for assistance developing the swung-mapping method used in figure 7; and pnDetector for the camera used in figure 17. This work was supported by the European Union Seventh Framework Programme under Grant Agreement 312483 - ESTEEM2 (Integrated Infrastructure Initiative-I3), and EPSRC grant code, EP/K040375/1, for the 'South of England Analytical Electron Microscope'.

References

[1] Krivanek O L, Chisholm M F, Nicolosi V, Pennycook T J, Corbin G J, Dellby N, Murfitt M, Own C S, Szilagyi Z S, Oxley M P, Pantelides S T and Pennycook S J 2010 Atom-by-atom structural and chemical analysis by annular dark-field electron microscopy. *Nature* **464**, no. 7288, 571-574

[2] Retsky M 1974 Observed single atom elastic cross sections in a scanning electron microscope. *Optik (Stuttg)*. 127-142

[3] Isaacson M S S, Langmore J, Parker N W W, Kopf D and Utlaut M 1976 The study of the adsorption and diffusion of heavy atoms on light element substrates by means of the atomic resolution stem. *Ultramicroscopy* **1** 359-376

[4] Kirkland E J and Thomas M G 1996 A high efficiency annular dark field detector for STEM. *Ultramicroscopy* **62** 79-88

[5] Grillo V 2011 An advanced study of the response of ADF detector. *J. Phys.: Conf. Ser.* **326** 012036

[6] Ishikawa R, Lupini A R, Findlay S D and Pennycook S J 2014 Quantitative annular dark field electron microscopy using single electron signals. *Microsc. Microanal.* **20** 99-110

[7] E H, MacArthur K E, Pennycook T, Okunishi E, D'Alfonso A J J, Lugg N R R, Allen L J and Nellist P D 2013 Probe integrated scattering cross sections in the analysis of atomic resolution HAADF STEM images. *Ultramicroscopy* **133** 109-119

[8] Singhal A, Yang J C and Gibson J M 1997 STEM-based mass spectroscopy of supported Re clusters. *Ultramicroscopy* **67** 191-206

[9] Lebeau J M, Findlay S D, Allen L J and Stemmer S 2010 Standardless atom counting in scanning transmission electron microscopy. *Nano Lett.* **10** 4405-4408

[10] Rosenauer A, Mehrtens T, Müller K, Gries K, Schowalter M, Satyam P V, Bley S, Tessarek C, Hommel D, Sebald K, Seyfried M, Gutowski J, Avramescu A, Engl K and Lutgen S 2011 Composition mapping in InGaN by scanning transmission electron microscopy. *Ultramicroscopy* **111** 1316-1327

[11] Jones L, MacArthur K E, Fauske V T, van Helvoort A T J and Nellist P D 2014 Rapid estimation of catalyst nanoparticle morphology and atomic-coordination by high-resolution Z-contrast electron microscopy. *Nano Lett.* **14** 6336-6341

[12] Nguyen D T, Findlay S D and Etheridge J 2014 The spatial coherence function in scanning transmission electron microscopy and spectroscopy. *Ultramicroscopy* **146** 6-16

[13] Van Aert S, Batenburg K J, Rossell M D, Erni R and Van Tendeloo G 2011 Three-dimensional atomic imaging of crystalline nanoparticles. *Nature* **470**, no. 7334 374-377

[14] De Backer A, De wael A, Gonnissen J and Van Aert S 2015 Optimal experimental design for nano-particle atom-counting from high-resolution STEM images. *Ultramicroscopy* **151** 46-55

[15] Klenov D O and Stemmer S 2006 Contributions to the contrast in experimental high-angle annular dark-field images. *Ultramicroscopy* **106** 889-901

[16] Kotaka Y, Lehmann M and Kotaka Y 2010 Essential experimental parameters for quantitative structure analysis using spherical aberration-corrected HAADF-STEM. *Ultramicroscopy* **110** 555-562

[17] Kirkland E J 1998 *Advanced computing in electron microscopy*. (New York, NY: Springer Science & Business Media)

[18] Ishizuka K 2002 A practical approach for STEM image simulation based on the FFT multislice method. *Ultramicroscopy* **90** 71-83

[19] Koch C T 2002 *Determination of core structure periodicity and point defect density along dislocations*. (Tempe, AZ: Arizona State University)

[20] Rosenauer A and Schowalter M 2008 STEMSIM - a New Software Tool for Simulation of STEM HAADF Z-Contrast Imaging. *Microsc. Semicond. Mater.* **120** 170-172

[21] Forbes B D, Martin A V, Findlay S D, D'Alfonso A J and Allen L J 2010 Quantum mechanical model for phonon excitation in electron diffraction and imaging using a Born-Oppenheimer approximation. *Phys. Rev. B* **82** 104103

[22] Rez P 2002 Scattering Cross Sections in Electron Microscopy and Analysis. *Microsc. Microanal.* **7** 356-362

[23] Martinez G T, De Backer A, Rosenauer A, Verbeeck J and Van Aert S 2013 The effect of probe inaccuracies on the quantitative model-based analysis of high angle annular dark field scanning transmission electron microscopy images. *Micron* **63** 57-63

[24] Jones L and Nellist P D 2013 Identifying and Correcting Scan Noise and Drift in the Scanning Transmission Electron Microscope. *Microsc. Microanal.* **19** 1050-1060

[25] Macarthur K E, D'Alfonso A J, Ozkaya D, Allen L J and Nellist P D 2015 Optimal ADF STEM imaging parameters for tilt-robust image quantification. *Ultramicroscopy* **156** 1-8

[26] Van Aert S, De Backer A, Martinez G T, Goris B, Bals S, Van Tendeloo G and Rosenauer A 2013 Procedure to count atoms with trustworthy single-atom sensitivity. *Phys. Rev. B - Condens. Matter Mater. Phys.* **87** 1-6

[27] De Backer A, Martinez G T, Rosenauer A and Van Aert S 2013 Atom counting in HAADF STEM using a statistical model-based approach: methodology, possibilities, and inherent limitations. *Ultramicroscopy* **134** 23-33

[28] De Backer A, Martinez G T, MacArthur K E, Jones L, Béché A, Nellist P D and Van Aert S 2015 Dose limited reliability of quantitative annular dark field scanning transmission electron microscopy for nano-particle atom-counting. *Ultramicroscopy* **151** 56-61

[29] Martinez G T, Jones L, De Backer A, Béché A, Verbeeck J, Van Aert S and Nellist P D 2015 Quantitative STEM normalization : the importance of the electron flux. *Ultramicroscopy* **159** 46-58

[30] Findlay S D and Lebeau J M 2013 Detector non-uniformity in scanning transmission electron microscopy. *Ultramicroscopy* **124** 52-60

[31] MacArthur K E, Jones L and Nellist P D 2014 How flat is your detector? Non-uniform annular detector sensitivity in STEM quantification. *J. Phys.: Conf. Ser.* **522** 012018

[32] Kaneko T, Saitow A, Fujino T, Okunishi E and Sawada H 2014 Development of a high-efficiency DF-STEM detector. *J. Phys.: Conf. Ser.* **522** 012050

[33] Lebeau J M and Stemmer S 2008 Experimental quantification of annular dark-field images in scanning transmission electron microscopy. *Ultramicroscopy* **108** 1653-1658

[34] Yamashita S, Koshiya S, Ishizuka K and Kimoto K 2015 Quantitative annular dark-field imaging of single-layer graphene. *Microscopy (Oxford)* **64** 143-150

[35] E H, Nellist P D, Lozano-Perez S and Ozkaya D 2010 Towards quantitative analysis of core-shell catalyst nano-particles by aberration corrected high angle annular dark field STEM and EDX. *J. Phys.: Conf. Ser.* **241** 012067

[36] Lotnyk A, Poppitz D, Gerlach J W and Rauschenbach B 2014 Direct imaging of light elements by annular dark-field aberration-corrected scanning transmission electron microscopy. *Appl. Phys. Lett.* **104**. 4-7

[37] Walther T 2006 A new experimental procedure to quantify annular dark field images in scanning transmission electron microscopy. *J. Microsc.* **221** 137-144

[38] Wang A, Li Z, Park S, Abdela A, Tang D and Palmer R 2011 Quantitative Z-contrast imaging in the scanning transmission electron microscope with size-selected clusters. *Phys. Rev. B* **84** 073408

[39] Gonnissen J, De Backer A, den Dekker A J, Martinez G T, Rosenauer A, Sijbers J and Van Aert S 2014 Optimal experimental design for the detection of light atoms from high-resolution scanning transmission electron microscopy images. *Appl. Phys. Lett.* **105** 063116

[40] Hovden R and Muller D A 2012 Efficient elastic imaging of single atoms on ultrathin supports in a scanning transmission electron microscope. *Ultramicroscopy* **123** 59-65

[41] Van Dyck D and Opdebeeck M 1996 A simple intuitive theory for electron diffraction. *Ultramicroscopy* **64** 99-107

[42] Rossouw C J, Dwyer C, Katz-Boon H and Etheridge J 2014 Channelling contrast analysis of lattice images: conditions for probe-insensitive STEM. *Ultramicroscopy* **136** 216-223

[43] Maccagnano-Zacher S E, Mkhoyan K A, Kirkland E J and Silcox J 2008 Effects of tilt on high-resolution ADF-STEM imaging. *Ultramicroscopy* **108** 718-726

[44] Katz-Boon H, Rossouw C J, Dwyer C and Etheridge J 2013 Rapid measurement of nanoparticle thickness profiles. *Ultramicroscopy* **124** 61-70

[45] So Y-G and Kimoto K 2012 Effect of specimen misalignment on local structure analysis using annular dark-field imaging. *J. Electron Microsc. (Tokyo)* **61** 207-215

[46] Yankovich A B, Berkels B, Dahmen W, Binev P, Sanchez S I, Bradley S A, Li A, Szlufarska I and Voyles P M 2014 Picometre-precision analysis of scanning transmission electron microscopy images of platinum nanocatalysts. *Nature Comm.* **5** 4155

[47] Lefebvre W, Hernandez-Maldonado D, Moyon F, Cuvilly F, Vaudolon C, Shinde D and Vurpillot F 2015 HAADF–STEM atom counting in atom probe tomography specimens: Towards quantitative correlative microscopy. *Ultramicroscopy* in press

[48] Fitting L, Thiel S, Schmehl A, Mannhart J and Muller D A 2006 Subtleties in ADF imaging and spatially resolved EELS: A case study of low-angle twist boundaries in $SrTiO_3$. *Ultramicroscopy* **106** 11-12

[49] Grillo V and Rossi F 2011 A new insight on crystalline strain and defect features by STEM-ADF imaging. *J. Cryst. Growth* **318** 1151-1156

[50] Yu Z, Muller D A and Silcox J 2004 Study of strain fields at a-Si/c-Si interface. *J. Appl. Phys.* **95** 3362-3371

[51] Wu X and Baribeau J M 2009 Composition and strain contrast of $Si_{1-x}Ge_x$ (x=0.20) and $Si_{1-y}C_y$ (y=0.015) epitaxial strained films on (100) Si in annular dark field images. *J. Appl. Phys.* **105** 043517

[52] Mkhoyan K A, Maccagnano-Zacher S E, Kirkland E J and Silcox J 2008 Effects of amorphous layers on ADF-STEM imaging. *Ultramicroscopy* **108** 791-803

[53] Egerton R, Li P and Malac M 2004 Radiation damage in the TEM and SEM. *Micron* **35** 399-409

[54] Muller D A, Kirkland E J, Thomas M G, Grazul J L, Fitting L and Weyland M 2006 Room design for high-performance electron microscopy. *Ultramicroscopy* **106** 11-12
[55] Jones L, Yang H, Pennycook T, Marshall M, Van Aert S, Browning N, Castell M and Nellist P D 2015 Smart Align – A new tool for Robust Non-rigid Registration of STEM Data. *Adv. Struct. Chem. Imaging* **1** 8
[56] Kimoto K, Asaka T, Yu X, Nagai T, Matsui Y and Ishizuka K 2010 Local crystal structure analysis with several picometer precision using scanning transmission electron microscopy. *Ultramicroscopy* **110** 778-782
[57] Couillard M, Radtke G and Botton G A 2013 Strain fields around dislocation arrays in a Σ9 silicon bicrystal measured by scanning transmission electron microscopy. *Philos. Mag.* **93** 1250-1267
[58] Spurgeon S R, Balachandran P V, Kepaptsoglou D M, Damodaran A R, Karthik J, Nejati S, Jones L, Ambaye H, Lauter V, RamasseQ M, Lau K K S and Martin L W Polarization Screening-Induced Magnetic Phase Gradients at Complex Oxide Interfaces. *Nature Comm.* **6** 6735
[59] Sang X and Lebeau J M 2013 Revolving scanning transmission electron microscopy: Correcting sample drift distortion without prior knowledge. *Ultramicroscopy* **138C** 28-35
[60] Zuo J-M, Shah A B, Kim H, Meng Y, Gao W and Rouviére J-L 2014 Lattice and strain analysis of atomic resolution Z-contrast images based on template matching. *Ultramicroscopy* **136** 50-60
[61] Sang X, Grimley E D, Niu C, Irving D L and LeBeau J M 2015 Direct observation of charge mediated lattice distortions in complex oxide solid solutions. *Appl. Phys. Lett.* **106** 061913
[62] Berkels B, Binev P, Blom D A, Dahmen W, Sharpley R C and Vogt T 2014 Optimized imaging using non-rigid registration. *Ultramicroscopy* **138** 46-56
[63] Katz-Boon H, Walsh M, Dwyer C, Mulvaney P, Funston A M and Etheridge J 2015 Stability of Crystal Facets in Gold Nanorods. *Nano Lett.* **15** 1635-1641
[64] Pennycook T J, Jones L, Pettersson H, Coelho J, Canavan M, Mendoza-Sanchez B, Nicolosi V and Nellist P D 2014 Atomic scale dynamics of a solid state chemical reaction directly determined by annular dark-field electron microscopy. *Sci. Rep.* **4** 7555
[65] Shibata N, Findlay S D, Kohno Y, Sawada H, Kondo Y and Ikuhara Y 2012 Differential phase-contrast microscopy at atomic resolution. *Nature Phys.* **8** 611-615
[66] Shibata N, Kohno Y, Findlay S D, Sawada H, Kondo Y and Ikuhara Y 2010 New area detector for atomic-resolution scanning transmission electron microscopy. *J. Electron Microsc. (Tokyo)* **59** 473-479
[67] Kimoto K and Ishizuka K 2011 Spatially resolved diffractometry with atomic-column resolution. *Ultramicroscopy* **111** 1111-1116
[68] Pennycook T J, Lupini A R, Yang H, Murfitt M F, Jones L and Nellist P D 2014 Efficient phase contrast imaging in STEM using a pixelated detector. Part 1: Experimental demonstration at atomic resolution. *Ultramicroscopy* **151** 160-167
[69] Lazar S, Etheridge J, Dwyer C, Freitag B and Botton G A 2011 Atomic resolution imaging using the real-space distribution of electrons scattered by a crystalline material. *Acta Crystallogr. Sect. A Found. Crystallogr.* **67** 487-490
[70] Rez P 2003 Electron Ionization Cross Sections for Atomic Subshells. *Microsc. Microanal.* **9** 42-53
[71] Zhu Y and Dwyer C 2014 Quantitative Position-Averaged K-, L-, and M-Shell Core-Loss Scattering in STEM. *Microsc. Microanal.* **20** 1070-1077

IOP Conf. Series: Materials Science and Engineering **109** (2016) 012009 doi:10.1088/1757-899X/109/1/012009

PENEPMA: a Monte Carlo programme for the simulation of X-ray emission in EPMA

X Llovet[1] and F Salvat[2]

[1] Universitat de Barcelona, Scientific and Technological Centres, Lluis Solé i Sabaris 1-3, 08028 Barcelona, Spain
[2] Universitat de Barcelona, Facultat de Fisica (ECM), Diagonal 645, 08028 Barcelona, Spain

E-mail: xavier@ccit.ub.edu

Abstract. The Monte Carlo programme PENEPMA performs simulations of X-ray emission from samples bombarded with electron beams. It is both based on the general-purpose Monte Carlo simulation package PENELOPE, an elaborate system for the simulation of coupled electron-photon transport in arbitrary materials, and on the geometry subroutine package PENGEOM, which tracks particles through complex material structures defined by quadric surfaces. In this work, we give a brief overview of the capabilities of the latest version of PENEPMA along with several examples of its application to the modelling of electron probe microanalysis measurements.

1. Introduction

The Monte Carlo (MC) method is widely recognized as the most reliable tool for the simulation of X-ray emission in electron probe microanalysis (EPMA) [1]. One of the advantages of the method is its ability to handle geometrical configurations that would be difficult to treat by other means and, therefore, it is well suited for the analysis of heterogeneous samples such as small particles, inclusions, interfaces or multilayer films. MC methods are also useful for deducing compositions where current matrix corrections are unable to provide reliable results. Another advantage of MC simulation is that it can incorporate realistic interaction cross-section models, and atomic relaxation parameters, usually in the form of extensive numerical databases, and thus provide highly accurate results. One drawback of MC methods is that simulation results are affected by statistical uncertainties which, in principle, can only be reduced at the expense of increasing the simulation time. However, significant progress has been made in recent years on the application of variance-reduction techniques which, combined with the increasing availability of fast computers, has enhanced the attractiveness of MC simulation.

Various MC programmes aiming at facilitating the application of MC simulation to EPMA, with different capabilities and degrees of sophistication, have been developed over the years (e.g., [2-5]). Our group at the University of Barcelona has developed and maintains the general purpose MC package PENELOPE [6] for the simulation of coupled electron-photon transport in complex material structures. After more than two decades work, PENELOPE has become a mature simulation tool, and has been applied to a wide variety of problems in electron microscopy, radiation metrology, radiotherapy, and other fields. In 2006 we introduced the programme PENEPMA [7], a dedicated version of an earlier main programme called PENMAIN [3], to allow occasional users to perform simulations of X-ray emission without requiring any programming work. Since then, a number of upgrades have been made to extend the capabilities of the programme and improve its versatility. Nevertheless, a detailed description of PENEPMA has not yet been published.

In this article we provide a brief overview of the latest version of PENEPMA along with some examples of its application to the modelling of EPMA measurements. A more detailed description of the code and its application to EPMA will be published elsewhere. Among the latest extensions of the programme we mention the addition of N x-ray lines of elements with $Z > 50$ and of X-ray lines with energies as low as 50 eV, the incorporation of the variance-reduction techniques of bremsstrahlung and X-ray splitting, and the possibility of combining independent runs of the same problem to take advantage of multi-core processors. In Section 2 we give a brief description of the physics and structure of the PENELOPE package. The programme PENEPMA is presented in Section 3. Specific examples that illustrate applications of PENEPMA are presented in Section 4. Finally, Section 5 contains some concluding remarks.

2. Brief overview of PENELOPE

As indicated above, the current code PENEPMA is based on the version 2014 of PENELOPE, released in early 2015 [6]. PENELOPE is a general-purpose MC package for the simulation of the coupled transport of electrons and photons in arbitrary materials. This means that the code not only tracks the primary electrons, but also all secondary electrons and photons produced in interactions of the transported particles (i.e., the complete particle shower originated by each primary electron). PENELOPE uses a combination of numerical and analytical differential cross-sections (DCS) to describe the different interaction mechanisms of electrons and photons with matter. For electrons, these include elastic scattering, inelastic collisions (which includes inner-shell ionisation) and bremsstrahlung emission, while for photons the considered interactions mechanisms are coherent (Rayleigh) and incoherent (Compton) scattering, photoelectric absorption, and electron-positron pair production (which is not active at the energies of interest in EPMA). The cross-section models and the electron tracking algorithm implemented in PENELOPE are described in detail in the manual of PENELOPE [6]. Although the interaction databases cover the interval between 50 eV and 109 eV, the accuracy of the physics models worsens when the energy decreases below about 1 keV (see Ref. [6]).

Electron elastic scattering is simulated by using numerical DCSs obtained from Dirac partial-wave calculations for the electrostatic potential derived from Dirac–Fock atomic electron densities [8], with the exchange potential of Furness and McCarthy. Inelastic collisions are described by means of DCSs obtained from the Sternheimer-Liljequist generalized oscillator strength model [9, 10], including the density-effect correction. In the 2014 version of PENELOPE ionisations of inner subshells by electron impact are considered as proper inelastic collisions, with the DCSs re-normalized so as to yield a total ionisation cross-section identical to the numerical value obtained from a combination of the distorted-wave Born approximation and the planewave Born approximation [11]. A global renormalisation is also applied to the cross-sections of outer subshells to ensure that the collision stopping power remains unaltered. The energy of bremsstrahlung photons is sampled from numerical energy-loss spectra derived from the scaled cross-section tables of Seltzer and Berger [12], while the angular distribution of radiated quanta is described by an analytical expression consisting of an admixture of two "boosted" dipole distributions [13], with parameters determined by fitting the benchmark partial-wave shape functions of Kissel, Quarles and Pratt [14].

Coherent (Rayleigh) scattering of photons is simulated by using the Born DCS with an analytical atomic form factor and theoretical anomalous scattering factors; incoherent (Compton) scattering uses DCSs calculated from the relativistic impulse approximation with analytical one-electron Compton profiles [15]; and photoelectric absorption is described by using total atomic cross-sections and partial cross-sections for the K-shell and L-, M- and N-subshells from the LLNL Evaluated Photon Data Library [16]. The initial direction of photoelectrons is sampled from Sauter's K-shell hydrogenic DCS.

Electron tracks and photon histories are simulated from the initial energy down to the absorption energies selected by the user, at which the particles are considered to be effectively absorbed in the medium. Secondary particles (electrons and photons) emitted with initial energy larger than the corresponding absorption energy are simulated after completion of each primary electron track. PENELOPE accounts for the emission of characteristic X-rays and Auger electrons that result from vacancies produced in K-shells and L-, M- and N-subshells by photoelectric absorption and Compton scattering of photons and by electron impact. The relaxation of these vacancies is followed until all vacancies have migrated to O- and outer shells. PENELOPE uses radiative and non-radiative transition probabilities extracted from the LLNL Evaluated Atomic Data Library [17]. The energies of X-rays emitted in radiative transitions resulting from vacancies in the K-shell and L-subshells are taken from the compilation by Deslattes *et al.* [18]. X-ray energies for transitions of vacancies in M-subshells are taken from Bearden's [19] review. In the case of transitions not included in these compilations, and also in transitions resulting from vacancies in N-subshells, the energy of the X-ray is approximated as $E_x = U_{S0} - U_{S1}$, where U_{S0} and U_{S1} are the ionisation energies of electrons in shell $S0$ where the vacancy has been produced and in shell $S1$ where the vacancy has migrated, respectively. U_{S0} and U_{S1} are taken from the LLNL Evaluated Atomic Data Library [17].

Electrons can be simulated by following the usual detailed procedure, i.e., all the interaction events along an electron track are sampled in chronological order, or alternatively, by using a mixed (class II) simulation algorithm [6]. Mixed simulation algorithms are useful when the electrons undergo a very large number of interactions and detailed simulation becomes impractical (e.g., for primary electron energies above ~ 100 keV). Photon histories are simulated in detail, i.e., interaction after interaction.

The practical usefulness of MC simulation lies in its ability to handle samples with complex geometries. PENELOPE contains the subroutine package PENGEOM which automatically tracks particles through arbitrary material systems consisting of homogeneous bodies limited by quadric surfaces. Limiting surfaces can be specified either through their implicit equation, $\boldsymbol{f(r)} = 0$, or by means of their reduced form and a few simple geometrical transformations. A Java graphical user interface (GUI) is available [20] to ease the definition and allow the visualisation of the geometrical structure of the sample (see figure 1)

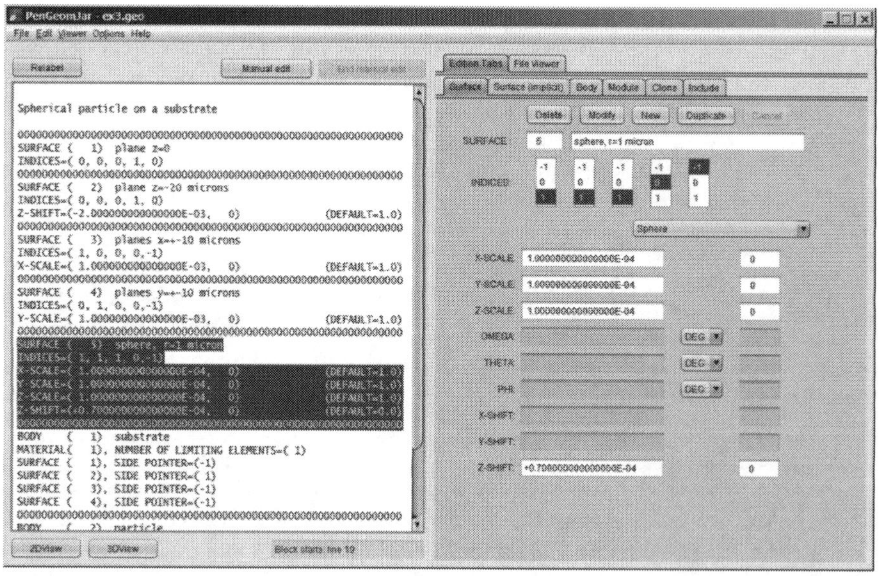

Figure 1. Main window of the GUI [20] provided to help defining and visualising the sample geometry in PENELOPE.

IOP Conf. Series: Materials Science and Engineering **109** (2016) 012009 doi:10.1088/1757-899X/109/1/012009

3. The programme PENEPMA

Because of its specificity, PENEPMA is not distributed with the PENELOPE package, but it is publicly available from the authors. It is pertinent to also mention that older versions of the code, with limited capabilities, have been incorporated by other researchers into free microanalytical platforms such as CALCZAF, PYPENELOPE or PYMONTECARLO.

As mentioned above, PENEPMA was designed to allow occasional users to simulate EPMA experiments without having to write a PENELOPE main programme. The user can define the details of his or her experiment (electron-beam characteristics, geometrical structure and composition of the target, photon detectors) and set the simulation control parameters by simply editing a single input text file. PENEPMA assumes that primary electrons are emitted from a point source, at a certain position, $\mathbf{r}_0 = (x_0, y_0, z_0)$, with fixed energy. The initial direction of primary electrons is sampled isotropically within a cone of given semi-aperture α and central axis in the direction (θ_0, ϕ_0) (see figure 2a). Notice that $\alpha = 0$ defines a parallel pencil beam, and $\alpha = 180°$ corresponds to an isotropic source.

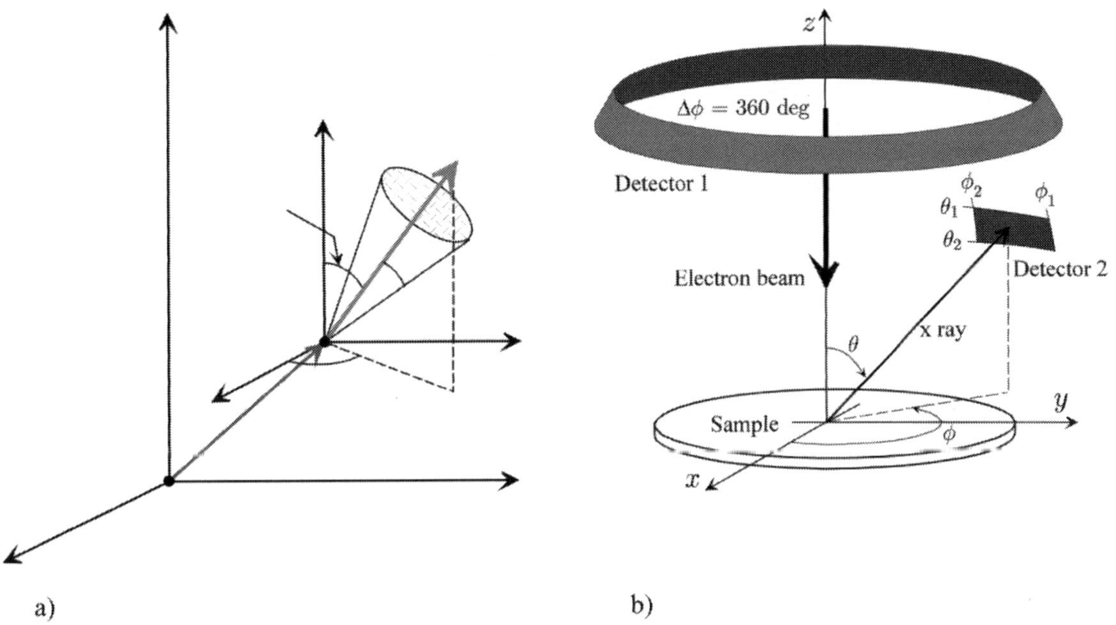

Figure 2. a) Definition of the electron-beam geometry. Electrons start off from the position $\mathbf{r}_0$. The beam is specified by giving the polar and azimuthal angles, θ_0 and ϕ_0, of its axis direction, and the semi-aperture α, all in degrees. b) Geometry of the experimental set-up assumed in PENEPMA. Each photon detector is defined by specifying the limiting values of the polar and azimuthal angles, $(\theta_1, \theta_2) \times (\phi_1, \phi_2)$. Notice that θ is the complement of the traditional take-off angle χ, i.e., $\chi = \pi - \theta$. The annular detector 1 is useful to obtain fast simulation results in cases where the X-ray flux is axially symmetric. (Reprinted from [7], with kind permission from Springer Science and Business Media).

The user can define a number of ideal photon detectors (up to 25). Each detector covers a solid angle corresponding to a "rectangle", $(\theta_1, \theta_2) \times (\phi_1, \phi_2)$, on the surface of the unit sphere centred at the origin of the reference frame (see figure 2b). This particular shape of the active detector surface is useful for normalising the simulated spectrum. The programme delivers energy spectra of photons that enter the various detectors, as well as characteristic line intensities, with fluorescence contributions given separately. It is worth mentioning that fluorescence effects are consistently described. The programme also offers the option of scoring the spatial distribution of X-ray emission within the volume of a parallelepiped (the scoring box) with a Cartesian mesh, for selected characteristic lines or for a given interval of X-ray energies.

In actual EPMA measurements, X-ray detectors cover a very small solid angle about a direction (θ_d, φ_d) with a take-off angle χ relative to the sample surface, i.e., $\chi = \pi - \theta_d$, of 40°. Simulations with this configuration would be very inefficient, because of the small collection solid angle. In cases where the flux of emerging photons has axial symmetry about the z-axis (e.g., for plane-surface samples that are homogeneous within the interaction volume, and at normal incidence), the efficiency of the simulation can be increased by using an annular detector (with $\phi_1 = 0$ and $\phi_2 = 360°$) that covers a finite interval of polar emission angles θ (see figure 2b).

PENEPMA has a dump/resume option which allows the user to stop the simulation at any time, and to resume it from the last dumping point in a completely consistent way. Furthermore, PENEPMA allows the possibility of combining independent runs of the same simulation problem to take advantage of multi-core processors.

The straight simulation of EPMA spectra may be very inefficient, even when using mixed simulation algorithms for electron transport. The reason is that the processes leading to the emission of X-rays (namely, inner-shell ionisation and/or bremsstrahlung emission) occur with exceedingly small probabilities. That is, on average, we need to generate many electron histories to obtain one emerging photon. In addition, as the solid angle subtended by the detector is usually very small, only a tiny fraction of the emerging photons hits the detector and contributes to the spectrum. To cope with this difficulty, PENEPMA takes full advantage of the variance-reduction techniques implemented in PENELOPE. The most effective of these are interaction forcing and photon splitting. The method of interaction forcing consists of artificially increasing the probability of inner-shell ionisation and bremsstrahlung emission and, at the same time, assigning appropriate statistical weights to the generated secondary particles so that simulation results remain unbiased. Photon splitting consists of considering that during each radiative event several photons are emitted in various directions, with appropriate weights to avoid biasing.

A simulated X-ray spectrum corresponds to an ideal detector, and hence, characteristic lines are very narrow, i.e., each line is fully contained in a single channel of the output histogram. In order to obtain a realistic spectrum, which can be compared, e.g., to an experimental spectrum, the effect of the finite energy resolution of the detector must be introduced. The PENEPMA package contains a Fortran programme called CONVOLG that performs this calculation by assuming that the energy-resolution function of the detector is a Gaussian with energy-dependent full-width-at-half-maximum.

4. Examples of the application of PENEPMA

The reliability of PENEPMA and PENELOPE has been assessed by our group as well as by other researchers through systematic comparisons of simulated and experimental X-ray spectra from bulk samples [21-23], X-ray yields from bulk samples [24-27], *k*-ratios from thin films on substrates [28], and *k*-ratios from material couples [29, 30]. Examples of recent applications of PENEPMA or PENELOPE in EPMA include the study of secondary fluorescence effects in bulk samples and near phase boundaries [31, 32], the analysis of inclusions in steel [33], the simulation of energy-dispersive spectra for automated materials identification [34, 35], the determination of the detector efficiency for standard-less analysis [36] or the examination of carbon contamination effects in high-resolution EPMA [37].

In this Section, we present several examples of application of PENEPMA in EPMA that illustrate the capabilities of the code. These examples refer to the simulation of X-ray spectra including N-lines, the determination of the spatial resolution at low energies, the analysis of thin films on substrates and the calculation of secondary fluorescence near phase boundaries.

4.1. X-ray spectra

X-ray spectra consist of characteristic peaks superimposed on a continuous background. The background results mostly from bremsstrahlung photons (with a lesser contribution from Compton-scattered characteristic X-rays), while characteristic peaks arise from X-rays emitted in radiative transitions of atoms that have been ionized in an inner shell, either by electron impact, photon absorption or incoherent photon scattering. As already mentioned, the simulation of X-ray spectra is generally difficult because both bremsstrahlung emission and inner-shell ionisation occur with very small probability when compared with the dominant interactions of elastic scattering and outer-shell ionisation (or excitation). Only with the use of variance-reduction techniques [22] can X-ray spectra be generated in reasonably short computer times.

Figure 3a shows an X-ray spectrum from a Pt bulk sample irradiated with a 20 keV electron beam measured with a Si(Li)-detector, compared with the simulation result. The experimental data were taken from Ref. [22]. The spectra are given in absolute units, i.e., as the probability density for detecting a photon per unit photon energy, per unit solid angle and per incident electron. Also shown in figure 3a is the result of convolving the simulated spectrum with the energy-resolution function of the detector, represented by a Gaussian distribution with an energy-dependent width. We see that simulation agrees closely with experiment. Significant differences are visible only at low energies, where the efficiency of the detector, which was considered to be independent of the photon energy, becomes less than its geometrical efficiency.

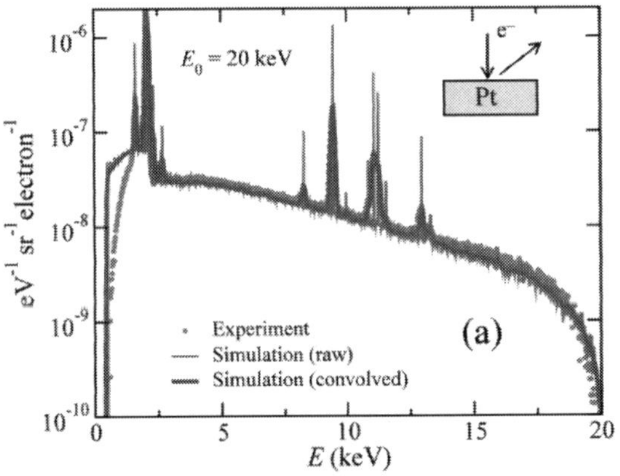

a)

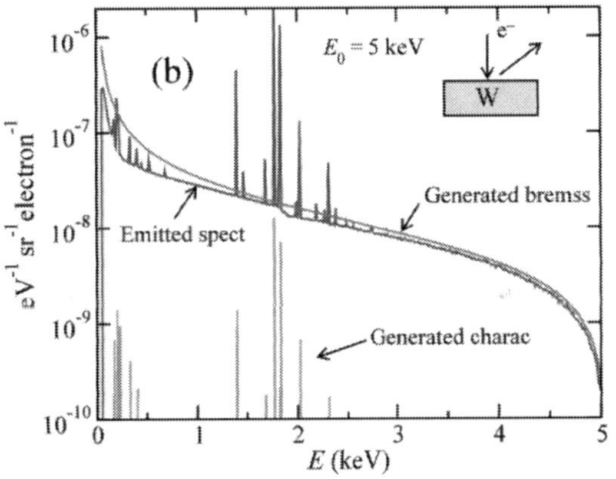

b)

Figure 3. a) Comparison of simulated and experimental spectra from Pt with a 20 keV electron beam. The dots represent the experimental data and the solid curves represent the simulated raw spectrum (green) and the result of its convolution with the energy-resolution function of a Si(Li)-detector (blue); notice the logarithmic scale on the vertical axis. b) Emitted X-ray spectrum from W with a 5 keV electron beam. Also displayed are the generated bremsstrahlung and characteristic X-ray spectra, the latter is shown at the bottom of the plot. See text for details.

Figure 3b shows the spectrum of x rays produced by a 5 keV electron beam, emerging from a W sample for an annular detector covering the angular range $45° < \theta < 55°$. The spectrum is given in absolute units. Since PENEPMA keeps track of the interaction mechanism that originates each secondary particle, it is able to provide the intensity of characteristic X-rays that reach a detector (i.e., with the bremsstrahlung background removed), along with the different contributions (primary photons resulting from interactions of the beam electrons, characteristic fluorescence and bremsstrahlung fluorescence). Also plotted in figure 3b is the energy spectrum of bremsstrahlung and

IOP Conf. Series: Materials Science and Engineering **109** (2016) 012009 doi:10.1088/1757-899X/109/1/012009

characteristic X-rays "generated" within the sample, i.e., at the start of their motion. The characteristic lines observed below 1 keV correspond to N-lines while those above 1 keV are M-lines. A list of the radiative transitions corresponding to the N-lines for W, along with the corresponding X-ray energies and detected line intensities (for the 5 keV incident electron beam) are tabulated in table 1.

Table 1. List of radiative transitions corresponding to the N-lines for W implemented in PENEPMA. The column *S*0 corresponds to the subshell where the vacancy has been produced, *S*1 is the shell to which the vacancy migrates and *E* is the energy of the N X-ray line. The columns *I* and '*unc*' are the intensity and statistical uncertainty (at 3σ level), respectively, of the N X-ray lines detected by an annular detector covering the angular range 45° < θ < 55° (i.e., a take-off angle χ of 40°) for an incident electron beam of 5 keV. This simulation took 2.5 h on an Intel Core i7 2.93 GHz processor with a RAM size of 8 Gb.

*S*0	*S*1	*E* (eV)	*I* (1/(sr×electron))	unc
N1	N2	96.09	2.14×10^{-8}	4.9×10^{-9}
N3	N4	159.57	4.96×10^{-8}	7.4×10^{-9}
N1	N3	165.08	1.65×10^{-7}	1.3×10^{-8}
N3	N5	172.28	5.91×10^{-7}	2.6×10^{-8}
N5	O3	201.69	2.66×10^{-7}	1.7×10^{-8}
N5	N6	201.73	5.88×10^{-8}	8.0×10^{-9}
N4	O2	204.03	1.76×10^{-7}	1.4×10^{-8}
N5	N7	204.09	1.13×10^{-6}	3.5×10^{-8}
N4	O3	214.40	2.62×10^{-8}	5.4×10^{-9}
N4	N6	214.44	7.15×10^{-7}	2.8×10^{-8}
N2	N4	228.56	8.14×10^{-7}	3.0×10^{-8}
N3	O1	334.48	4.22×10^{-7}	2.2×10^{-8}
N2	O1	403.47	2.34×10^{-7}	1.6×10^{-8}
N3	O4	411.35	6.43×10^{-9}	2.7×10^{-9}
N3	O5	412.21	2.56×10^{-8}	5.3×10^{-9}
N2	O4	480.34	5.66×10^{-9}	2.5×10^{-9}
N1	O2	528.68	1.33×10^{-7}	1.2×10^{-8}

Published quantitative measurements of N-line X-ray intensities emitted by electron impact are very scarce. Scheffel *et al.* [38] have reported measurements with an energy-dispersive X-ray spectrometer of the intensity ratio of the $N_{4,5}N_{6,7}$-line $I(N_{4,5}N_{6,7})$ to the $M\alpha\beta$ line $I(M\alpha\beta)$. Although an absolute comparison of the measured $I(N_{4,5}N_{6,7})/I(M\alpha\beta)$ ratio with results from MC simulation would require knowledge of the detector efficiency at the energies of both X-ray lines, it is possible to compare the relative shape of the $I(N_{4,5}N_{6,7})/I(M\alpha\beta)$ curve versus the electron beam energy. This comparison is presented in figure 4. We see that the agreement between simulation and experiment is quite good in relative terms. Multiplication of the simulation results by a factor of 1.46 to account for the different detector efficiencies at the energies of the $N_{4,5}N_{6,7}$ (E ~ 0.2 keV) and $M\alpha\beta$ (E ~ 1.7 keV) lines would give a perfect match between simulation and experiment. There is obviously a need for absolute N-line intensity measurements to assess the reliability of simulations of such X-ray lines.

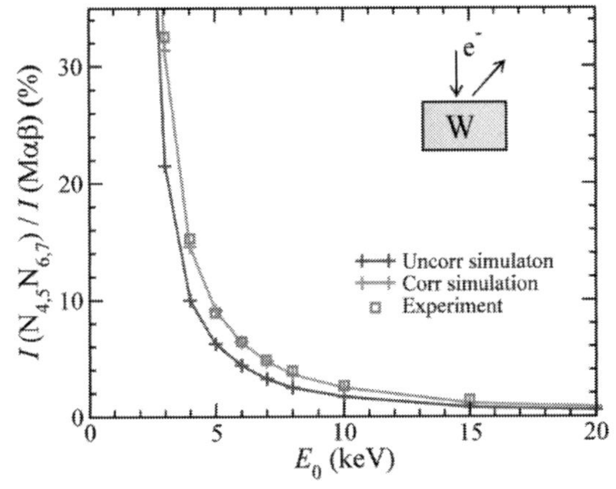

Figure 4. Comparison of the simulated (blue line) and experimental (symbols) ratio of intensities $I(N_{4,5}N_{6,7})/I(M\alpha\beta)$ from a W-target. The green line corresponds to the simulation results multiplied by a factor of 1.46. See text for details.

4.2. Spatial resolution of EPMA

The spatial (lateral) resolution of EPMA is generally estimated by means of the penetrative range of primary electrons, R_x, which is defined as the average path length (distance) travelled by a primary electron before slowing down to rest (or more precisely to an energy equal to the ionisation energy E_i of the atomic shell which originates the X-rays of interest). R_x is the difference between the ranges of electrons with energies E_0 and E_i:

$$R_x = \int_{E_0}^{E_i} \frac{dE}{dE/ds} \tag{1}$$

where dE/ds is the stopping power of the material. Approximate analytical expressions for R_x obtained from experimental measurements have been developed which are used in practice. Eq. (1) assumes that the beam has a negligible size, which is plausible for moderately high-beam energies. However the width of the electron beam may become comparable to the electron range at low-beam energies, and therefore its effect should be accounted for at such energies.

Experimentally, the lateral resolution of EPMA can be measured as edge resolution, i.e., by means of an integrated X-ray intensity profile obtained by scanning the electron probe across the boundary of two materials separated by a plane interface perpendicular to the surface of the specimen. The edge resolution is then quoted as the distance between points corresponding to 10 % and 90 % of the total intensity (or sometimes the 16 % to 84 % distance). Figure 5 shows a comparison between measured and simulated integrated X-ray profiles for the Ni-Lα line across the boundary of a Fe-Ni couple, obtained with a 5 keV electron beam having a diameter d of 54 nm. The experimental data are from [39]. Agreement between simulation and experiment is satisfactory. Measurements of this kind, although scarce, are very useful for the purpose of assessing the reliability of MC simulations.

It is possible to investigate the effect of the beam size on the spatial (lateral) resolution of EPMA by simulating the X-ray distribution as a function of the radial r distance from the beam centre (radial X-ray distribution). Figure 6 compares radial X-ray distributions for Si-Kα X rays generated in Si and 4 keV electron beams with different beam diameters. The full-width-at-half-maximum of a radial distribution is often regarded as a measure of the lateral resolution of a given imaging technique. In this case, the X-ray distributions are seen to consist essentially of two components, a central sharp peak superimposed onto a much broader contribution. Figures 6b-6d display X-ray distributions as functions of both the radial distance r and the depth z within the specimen, and allow a 2D visualisation of the beam spread within the sample. In the studied cases, the 2D X-ray distributions have been generated by scoring the number of Si-Kα X-rays that have been generated within a 0.4×0.4×0.4 μm^3 scoring box, whose edges are parallel to the axes of the laboratory frame, and integrating over the y coordinate.

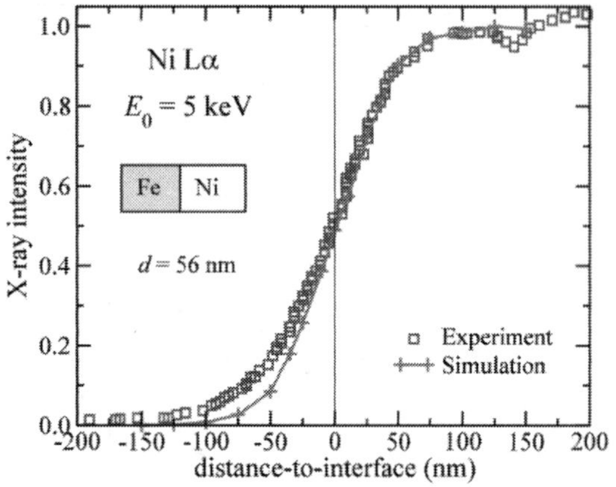

Figure 5. Comparison of experimental and simulated x-ray line scans of Ni-Lα across the boundary of a Fe-Ni couple. *d* represents the beam diameter. The experimental data are from [39].

4.3. Thin films

EPMA is conventionally used for the analysis of samples which are homogeneous at the micrometre-scale; however, it can also be used for determining the thickness and composition of thin films deposited on substrates and multilayer films, with thicknesses ranging from a few nanometres up to few micrometres. In order to analyze these kinds of samples, *k*-ratio measurements are performed at varying incident electron energy and they are processed with the help of a suitable thin-film analysis programme. The latter determines the composition and thickness of the film or multilayer by fitting the predictions of an X-ray emission model to the measured *k*-ratios.

Extensive experimental databases have become available in recent years for the purpose of assessing the reliability of thin film programmes [40]. Evidently, they can also be used to assess the performance of MC simulation programmes [28]. Figure 7 shows one such example, in which a comparison between simulated and experimental Al and Pd *k*-ratios emitted from Al and Pd films deposited on a wide variety of single-element substrates spanning the periodic table, as a function of substrate atomic number. The experimental data were taken from Bastin and Heijligers's database [41, 42]. The results from PENEPMA agree satisfactorily with the measured *k*-ratios.

4.4. Secondary fluorescence

As mentioned previously, the penetrative range R_x of an electron beam with an energy in the keV range is of few micrometres, depending on the beam energy and the material. However, the range of both characteristic and bremsstrahlung X-rays can be one-to-two orders of magnitude greater than that of primary electrons. Consequently, X-rays can also be generated by fluorescence of characteristic and bremsstrahlung X-rays absorbed in a non-excited material far away from the point of impact of the electron beam. This contribution is known as secondary fluorescence (SF) and it can be significant in certain situations, both quantitatively and qualitatively. We would like to mention that this contribution is not accounted for by conventional matrix correction methods.

Figure 8a compares simulated and measured *k*-ratio profiles for a non-diffused couple consisting of two homogeneous alloys, namely Co (4.1 wt% Cu) and Cu (2.1 wt% Co), as a function of the distance from the beam to the boundary. The electron beam impacts on the Cu (2.1 wt% Co) alloy. Material couples of this kind have been used as starting materials for diffusion experiments using the diffusion couple technique. The measured Co-Kα X-ray intensity has a primary contribution (from electron impact on the Cu (2.1 wt% Co) alloy itself) and a contribution from secondary fluorescence due to the interaction of both Co characteristic X-rays and bremsstrahlung within the adjacent Co (4.1 wt% Cu) alloy. Simulated and measured fluorescence intensities agree very well with each other.

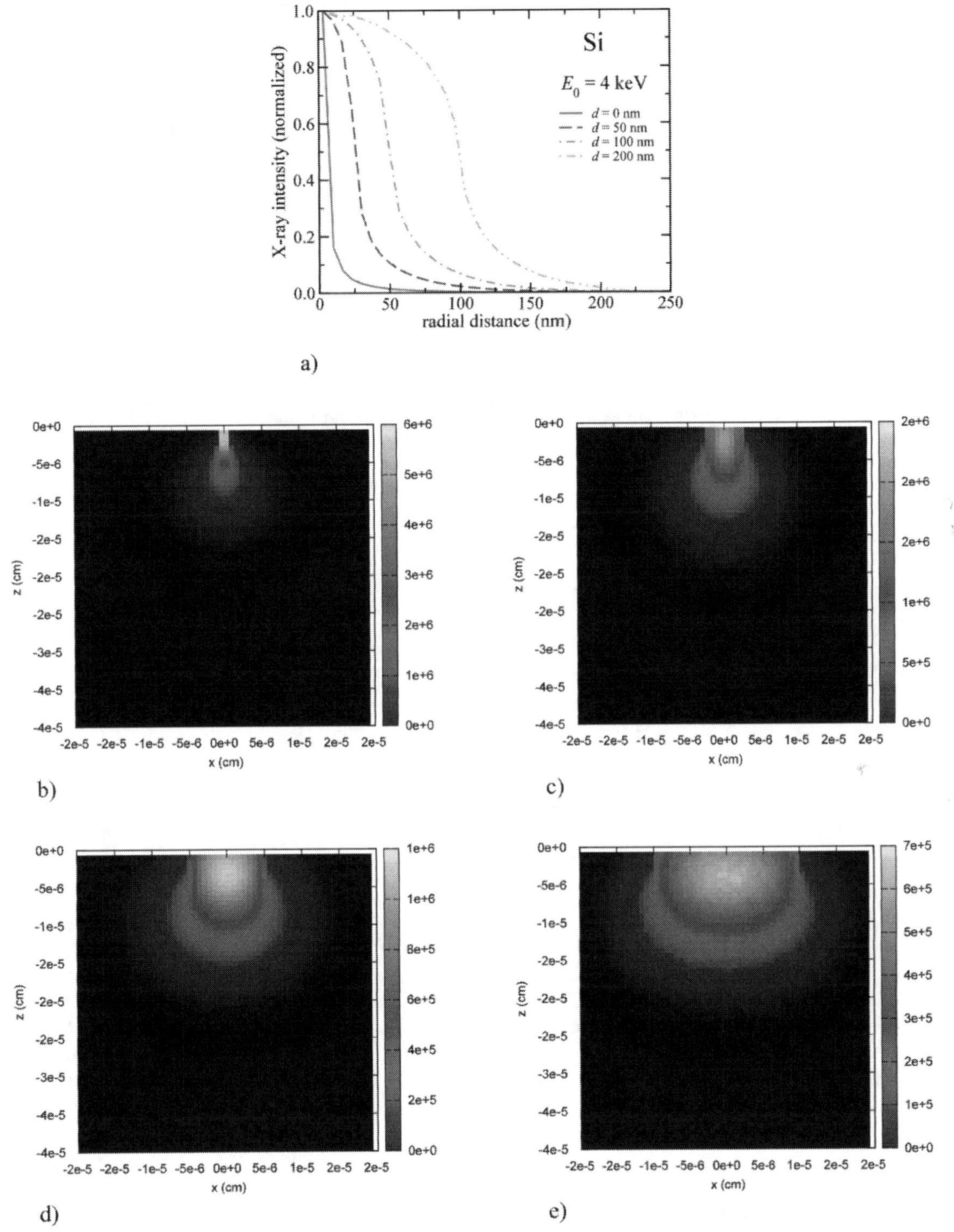

Figure 6. a) Radial distribution of Si-Kα X-ray emission in Si for a 4 keV electron beam with increasing beam radii, as indicated in the legends. b) 2D Si-Kα X-ray distribution in the $x - z$ plane generated within a Si sample with a 4 keV electron beam with diameters $d = 0$ nm (upper left), 50 nm (upper right), 100 nm (lower left), and 200 nm (lower right).

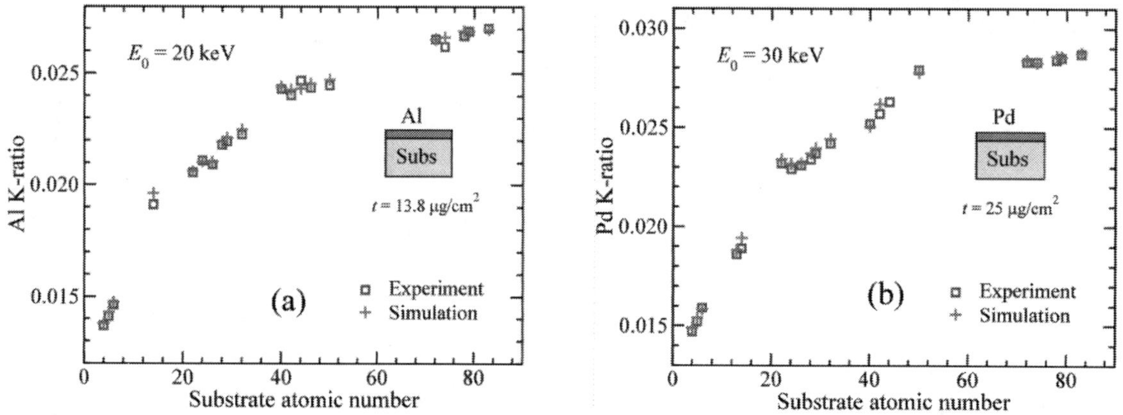

a) b)

Figure 7. Comparison of measured and simulated *k*-ratios for Al (a) and Pd (b) films deposited on different single-element substrates, versus substrate atomic number. *t* is the mass thickness of the films. Experimental data were taken from [41, 42].

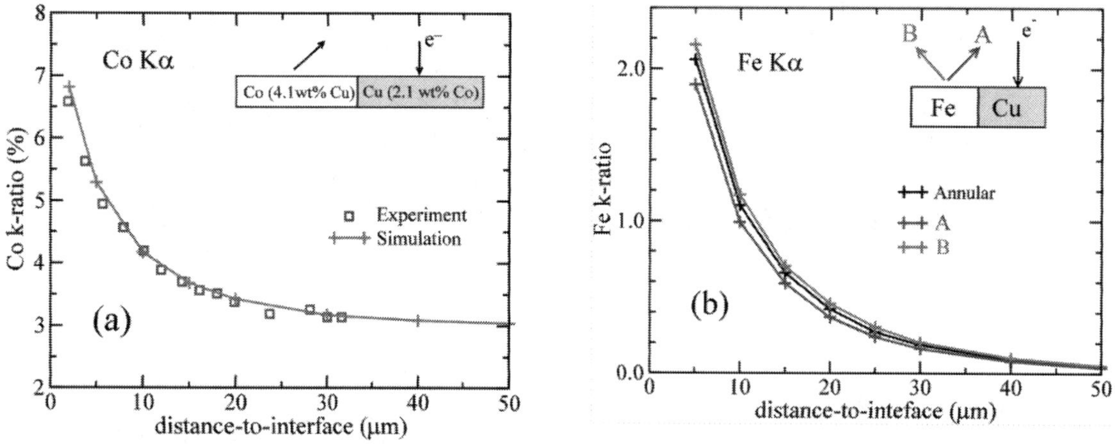

a) b)

Figure 8. a) Comparison of measured and simulated Co-Kα *k*-ratios versus electron beam distance to the interface for a Co (4.1 wt% Cu)-Cu (2.1 wt% Co) couple. The beam impacts on the Cu (2.1 wt% Co) alloy (right-hand side of the couple). The measured and simulated values are represented by open squares and crosses joined by continuous lines for visual aid. The experimental data were taken from [43]. Notice that these two slabs of alloy have been joined together under cold conditions and no atoms have diffused. b) Effect of the position of the spectrometer on the SF profile for a Fe-Cu couple. See text for details.

Because of the much higher range of primary X-rays, absorption of fluorescent X-rays within the sample may be sensitive to the relative position of the X-ray spectrometer, as discussed by Fournelle *et al.* [44]. Figure 8b illustrates this dependence for a Fe-Cu couple, with the electron beam impacting on the Cu side of the couple. Three simulated SF profiles are compared, namely one obtained by using an annular X-ray detector, i.e., with limiting polar and azimuthal angles of ($\theta_1 = 45°$, $\theta_2 = 55°$) × ($\phi_1 = 0°$, $\phi_2 = 360°$), a second one with a detector with polar and azimuthal angles (θ_1 = 45°, θ_2 = 55°) × (ϕ_1 = -90°, ϕ_2 = 90°), i.e. facing towards the Cu-side of the couple (labelled A in figure 8b), and a third one placed with polar and azimuthal angles (θ_1 = 45°, θ_2 = 55°) × (ϕ_1 = 90°, ϕ_2 = 270°), i.e. facing towards the Fe-side of the couple (labelled B in figure 8b). The differences between the different profiles are mainly caused by the different absorption of Fe-Kα X-rays in Fe and in Cu. The observed differences between the three profiles indicate that accurate correction of SF requires knowledge of the position of the spectrometer with respect to the couple boundary.

5. Conclusion

We have provided an overview of the MC simulation programme PENEPMA. The programme uses the general-purpose MC simulation package PENELOPE, which implements the most realistic interaction cross-section models available today. These models are generally described by extensive numerical databases. Examples have been provided to demonstrate the reliability of the programme in simulations of X-ray spectra including N-lines, for the determination of the spatial resolution of EPMA at low energies, for the analysis of thin films and in the calculation of secondary fluorescence across phase boundaries.

Acknowledgements

Financial support from the Spanish Ministerio de Econom´ıa y Competitividad and ERDF (project no. FPA2013-44549-P) and from the Generalitat de Catalunya (grant 2014 SGR 846) is gratefully acknowledged.

References

[1] Rinaldi R and Llovet X 2015 *Microsc. Microanal.* **21** 1053
[2] Hovington P, Drouin D and Gauvin R 1997 *Scanning* **19** 1
[3] Llovet X, Fernández-Varea J M, Sempau J and Salvat F 2005 *Surf. Interface Anal.* **37** 1054
[4] Ritchie N W M 2005 *Surf. Interface. Anal.* **37** 1006
[5] Gauvin R, Lifshin E, Demers H, Horny P and Campbell H 2006 *Microsc. Microanal.* **12** 49
[6] Salvat F 2015 *PENELOPE-2014. A code system for Monte Carlo simulation of electron and photon transport.* (Issy-les-Moulineaux, France: OECD/NEA Data Bank)
[7] Salvat F, Llovet X, Fernández-Varea J M and Sempau J 2006 *Microchim. Acta* **155** 67
[8] Salvat F, Jablonski A and Powell C J 2005 *Comput. Phys. Commun.* **165** 157
[9] Sternheimer R M 1952 *Phys. Rev.* **88** 851
[10] Liljequist D 1983 *J. Phys. D: Appl. Phys.* **16** 1567
[11] Bote D and Salvat F 2008 *Phys. Rev. A* **77** 042701
[12] Seltzer S M and Berger M J 1985 *Nucl. Instrum. Meth. B* **12** 95
[13] Acosta E, Llovet X and Salvat F 2002 *Appl. Phys. Lett.* **80** 3228
[14] Kissel L, Quarles C A and Pratt R H 1983 *At. Data Nucl. Data Tables* **28** 381
[15] Brusa D, Stutz G, Riveros J A, Fernández-Varea J M and Salvat F 1996 *Nucl. Instrum. Meth. A* **379** 167
[16] Cullen D E , Chen M H, Hubbell J H, Perkins S T, Plechaty E F, Rathkopf J A and Scofield J H 1989 *Report UCRL-50400 vol. 6, rev. 4, parts A and B.* (Livermore, CA: Lawrence Livermore National Laboratory)
[17] Perkins S T, Cullen D E, Chen M H, Hubbell J H, Rathkopf J and Scofield J 1991 *Report UCRL-50400 vol. 30.* (Livermore, CA: Lawrence Livermore National Laboratory)
[18] Deslattes R D, Kessler E G, Indelicato P, de Billy L, Lindroth E and Anton J 2003 *Rev. Mod. Phys.* **75** 35
[19] Bearden J A 1967 *Rev. Mod. Phys.* **39** 78
[20] Almansa J, Salvat-Pujol F, Díaz-Londoño G, Carnicer A, Lallena A M and Salvat F 2015 *Comput. Phys. Commun.* (in press)
[21] Acosta E, Llovet X, Coleoni E, Riveros J A and Salvat F 1998 *J. Appl. Phys.* **83** 6038

[22] Llovet X, Sorbier L, Campos C S, Acosta E and Salvat F 2003 *J. Appl. Phys.* **93** 3844
[23] Tian L, Zhu J, Liu M and An Z 2009 *Nucl. Instrum. Meth. B* **267** 3495
[24] Bote D, Llovet X and Salvat F 2008 *J. Phys. D: Appl. Phys.* **14** 105304
[25] Li X L, Zhao J, Tian L X, An Z, Zhu J J and Liu M T 2014 *Nucl. Instrum. Meth. B* **333** 106
[26] Moy A, Merlet C and Dugne O 2015 *Anal. Chem.* **87** 7779
[27] Zhao J L, Tian L X, Li X L, An Z, Zhu J J, Liu M T 2015 *Radiat. Phys. Chem.* **107** 47
[28] Statham P, Llovet X and Duncumb P 2012 *IOP Conf. Ser.: Mater. Sci. Engng.* **32** 012024
[29] Llovet X and Galán G 2003 *Am. Mineral.* **88** 121
[30] Llovet X, Pinard P T, Donovan J J and Salvat F 2012 *J. Phys. D: Appl. Phys.* **45** 225301
[31] Petaccia M, Segui S and Castellano G 2015 *Microsc. Microanal.* **21** 753
[32] Wade J and Wood B J 2012 *Phys. Earth Planet. Inter.* **192-193** 54
[33] Pistorius P C and Verma N 2011 *Microsc. Microanal.* **17** 963
[34] Weber C F, Bekar K B, Patton B W and Miller T M 2014 *Simulation of SEM-EDSB Spectra: Monte Carlo Code Development and Inverse Analysis.* (Oak Ridge, TN: Oak Ridge National Laboratory) Report ORNL/LTR-2014/557
[35] Miller T M, Patton B W and Weber C F 2014 *Trans. Am. Nucl. Soc.* **110** 497
[36] Moy A, Merlet C, Llovet X and Dugne O 2014 *IEEE Trans. Nucl. Sci.* **61** 1977
[37] Buse B and Kearns S 2015 *Microsc. Microanal.* **21** 594
[38] Scheffel A, Dellith J and Wendt M 2008 *Microchim. Acta* **161** 471
[39] Berger D and Nissen J 2012 *IOP Conf. Ser.: Mater. Sci. Engng.* **55** 012002
[40] Llovet X and Merlet C 2010 *Microsc. Microanal.* **16** 21
[41] Bastin G F and Heijligers H J M 2000 *X-ray Spectrom.* **29** 212
[42] Bastin G F and Heijligers H J M 2000 *X-ray Spectrom.* **29** 373
[43] Bastin G F, van Loo F J, Vosters P J and Vroljik J W 1983 *Scanning* **5** 172
[44] Fournelle J H, Kim S and Perepezko J H 2005 *Surf. Interface Anal.* **37** 1012

SEM and FIB-SEM investigations on potential gas shales in the Dniepr-Donets Basin (Ukraine): pore space evolution in organic matter during thermal maturation

D Misch[1], F Mendez-Martin[2], G Hawranek[2], P Onuk[3], D Gross[1] and R F Sachsenhofer[1]

[1] Montanuniversität Leoben, Chair of Petroleum Geology, Peter-Tunner-Strasse 5, 8700 Leoben, Austria
[2] Montanuniversität Leoben, Chair of Physical Metallurgy and Metallic Materials, Max-Tendler-Strasse 9, 8700 Leoben, Austria
[3] Montanuniversität Leoben, Chair of Geology and Economic Geology, Peter-Tunner-Strasse 5, 8700 Leoben, Austria

E-mail: david.misch@unileoben.ac.at

Abstract. Porosity and permeability are essential parameters for reservoir rocks. Techniques developed for conventional reservoir rocks characterized by large pores, cannot be applied to study gas shales. Therefore, high resolution techniques are increasingly used to determine reservoir quality of shale gas plays. Within the frame of the recent study, Upper Visean black shales ("Rudov Beds") from the Dniepr-Donets-Basin (DDB, Ukraine) were characterized by X-ray diffraction, conventional SEM imaging and FIB/BIB-SEM.

According to SEM and FIB/BIB-SEM data, nanopores are not abundant in primary macerals (e.g., vitrinite) even in overmature rocks, whereas they develop within secondary organic matter (bitumen) formed mainly at gas window maturity. Frequently occurring sub-micrometre porosity, probably related to gas generation from bituminous organic matter, was detected within mudstones at a vitrinite reflectance > 2.0 % Rr. However, such pores have also been detected occasionally in solid bitumen at oil window maturity (0.9 % Rr). Authigenic nanoscale clay minerals and calcite occur within pyrobitumen at gas window maturity.

Furthermore, Rudov Beds can be subdivided into mineralogical facies zones by SEM imaging and X-ray diffraction. A basin-centred, brittle siliceous facies is most likely caused by increased contribution from deeper water radiolaria and is separated from a marginal clayey and carbonate-rich facies.

1. Introduction

Despite major achievements in the investigation of organic matter-hosted pores in the nanometre scale [1], storage capacity and production behaviour of gas shales are still poorly understood [2]. Visualisation of nanopores from conventionally (mechanically) polished surfaces is impossible, whereas FIB/BIB-SEM techniques generate undisturbed surfaces without preparation-induced artefacts, allowing imaging of nanopores in the range of > 10 nm in equivalent diameter. Such pores might contribute largely to storage capacity and release behaviour, although recent findings support a significant influence of nanopores even below the resolution of modern FE-SEM devices (< 5 nm) [1].

Great progress has been made in the understanding of the evolution of nanopores in organic matter. Nevertheless, it is still unclear if, at which maturity, and to which amount, nanopores form within primary or secondary organic matter. Furthermore, most previous studies focussed on kerogen type II, whereas the behaviour of vitrinite-rich kerogen (mixed type III-II) is poorly investigated.

Taking into account the high total organic carbon (TOC) contents (average 5.5 %) and original hydrogen index values of 270 mg HC/ g TOC [3], the Upper Visean "Rudov Beds" (V-23) in the Dniepr-Donets Basin (DDB; Ukraine; figure 1) offer a unique possibility to study organic-rich rocks with a kerogen type III-II at different maturities. While their significance as source rocks for a high number of conventional hydrocarbon deposits found in Upper Visean clastic rocks [4-6] is apparent, their potential for unconventional production of shale gas/liquids is currently underinvestigated. Apart from the general generation potential, production parameters like permeability and fraccability are largely influenced by the mineralogical facies as well as type and distribution of pores within the fine-grained rocks. Furthermore, types of organic matter porosity and its evolution during thermal maturation are of great interest for the evaluation of an unconventional hydrocarbon play.

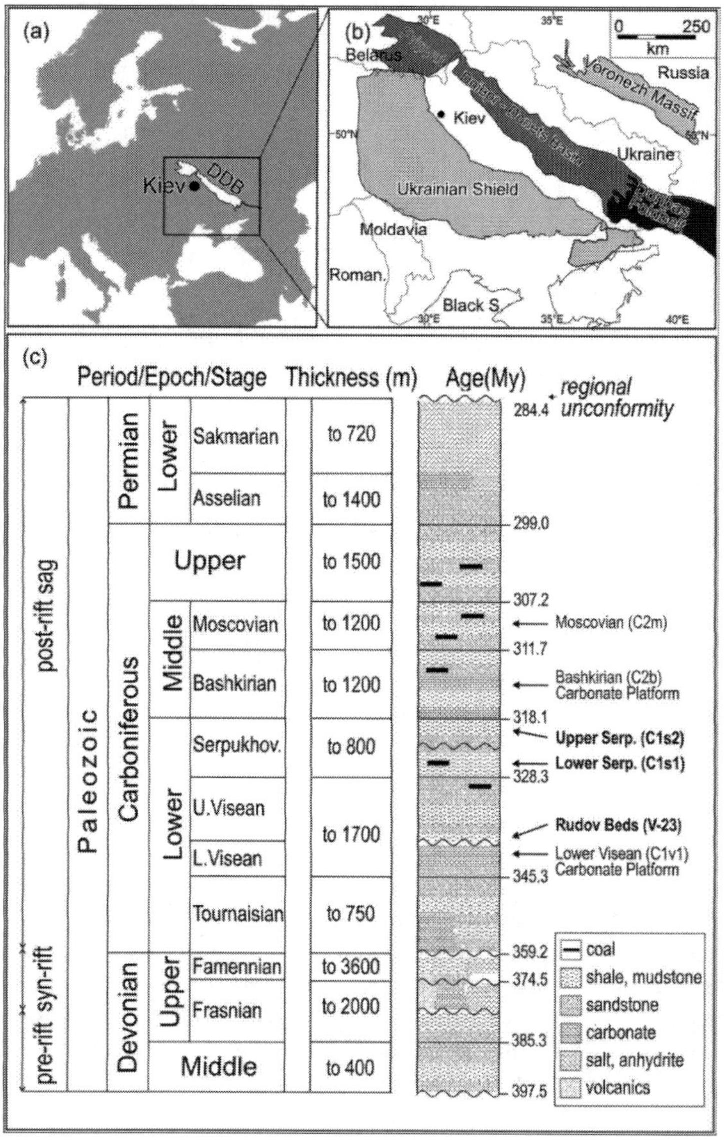

Figure 1. (a), (b) Regional setting of the Dniepr-Donets-Basin (DDB) in Eastern Europe. (c) Chrono- and lithostratigraphy of the Palaeozoic succession [7]. Age data follow [8].

The Rudov Beds cover a large area in the DDB. The best source rocks, however, are found within the Srebnen Bay, a low-energy basin surrounded by a reef belt (figures 2a,b; [9]). Siliceous, clayey, and carbonate-rich facies zones have been defined by Gavrish *et al.* [10] and Machulina and Babko [11]. However, apart from bulk mineralogical parameters, understanding the habitus and distribution of brittle and ductile mineral phases is important. Apart from that, the visualisation of organic matter and its arrangement within the mineral matrix was a main interest of the current study. For that purpose, SEM imaging of freshly broken surfaces was performed to characterize both organic and inorganic facies changes.

2. Geological setting

The DDB is an about 650 km long, Late Devonian rift-basin, located within the East-European Craton (figure 1a, b). Chrono- and lithostratigraphy of the DDB are shown in figure 1c.

Middle Devonian pre-rift deposits represent the oldest strata in the DDB (figure 1c). The Late Devonian syn-rift succession is up to 4 km thick and includes clastic deposits, carbonate rocks, extensive salt deformed to salt domes, and volcanic rocks [12].

The Carboniferous post-rift succession with a maximum thickness in excess of 10 km is represented by cyclic successions of siliciclastic and minor carbonate rocks deposited in fluvial, shallow-marine, and lagoonal environments. Water depth was typically low and exceeded 200 m only in the axial part of the basin. Typically each cycle, up to 50 m in thickness, includes a sand interval ("productive unit") and a shallow marine, organic-rich shale interval ([13]; see figure 1c). The Rudov Beds (up to 70 m thick) form the base of the Upper Visean section and mark the most prominent black shale horizon.

Basin-wide Lower Visean and Bashkirian carbonate platforms were the result of major transgressions from the south-east. The percentage of fluvial and deltaic sands increases in the Upper Serpukhovian succession. Serpukhovian to Moscovian coal seams, extensively mined in the Donbas Foldbelt, extend into the DDB. Clastic deposits, carbonate rocks, and evaporite rocks prevail in the Permian section. The Mesozoic and Cenozoic sediments unconformably overlie the Paleozoic series.

Petroleum systems in the DDB [6] are based on multiple source rocks [7, 14]. Visean and Serpukhovian black shales (incl. Rudov Beds) together with Serpukhovian to Moscovian coaly layers are potential source rock horizons. Oil window maturity (~0.65 % Rr; random vitrinite reflectance in oil) in the north-western part of the basin is reached at around 3,500 m depth, but at shallower depth in the inverted SE part of the DDB [5, 15, 16]. Hydrocarbon generation occurred predominantly during Permian deep burial, according to previous basin modelling studies [16]. Additionally, a potential generation of hydrocarbons is suggested for Mesozoic to Cenozoic burial stages. Carboniferous and Permian clastic rocks are the dominant reservoir lithology [6]. Small quantities of hydrocarbons occur in Lower Carboniferous and Lower Permian carbonates. Oil fields prevail in the north-western part of the basin as well as along the basin flanks.

3. Samples and methodology

Samples were obtained from wells with cored Upper Visean shales. Two sample sets have been selected. (1) For the investigation of the organic matter evolution due to thermal maturation, samples were chosen according to their vitrinite reflectance (measured following established procedures [17]). (2) A second sample set with comparable (marginal to oil window) maturity was obtained from different locations throughout the basin, to characterize lateral changes in the depositional environment and maceral composition. The location of sampled wells within the DDB is illustrated in figure 2a.

For conventional SEM investigations, freshly broken surfaces of rock samples have been coated with an Au layer to enhance conductivity. To prevent charging effects, relatively even surfaces were chosen for investigation. Investigations were conducted with a Zeiss Evo MA 10 SEM, equipped with a secondary electron (SE) and a backscattered electron (BSE) detector, as well as a Bruker Nano XFlash 430M energy-dispersive X-ray (EDS) detector.

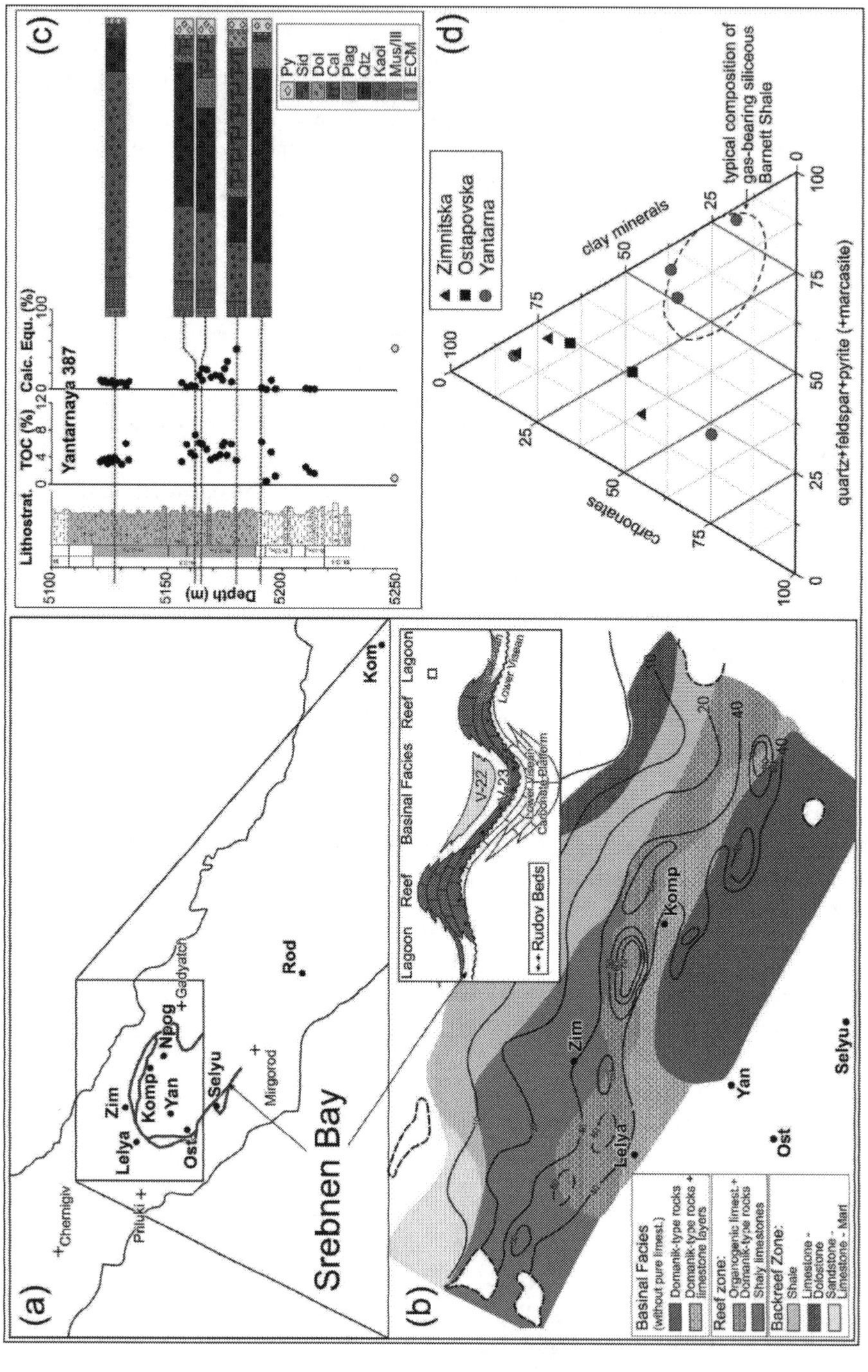

Figure 2. a) Position of studied wells in relation to the "Srebnen Bay". b) Lateral facies map (modified after [11]). Isopachs indicate thickness of V-23 interval. The schematic cross-section (insert; after [18]) shows the syncline surrounded by a reef belt. "Rudov Beds" (V-23) are deposited in the central depression, overlying a Lower Visean platform. c) Vertical change of mineralogy, total organic carbon and calcite equivalent content (calculated from total inorganic carbon) within basinal well Yantarna (Yan). d) Mineral assemblages of shale samples from X-ray diffractometry. The composition of gas-bearing, siliceous Barnett Shale is shown for comparison.

For FIB-SEM investigations, the specimen surfaces were prepared using a JEOL IB-09010 broad ion beam (BIB) cross-section polisher with an Ar ion beam (5 kV; 6 h), followed by an Au coating to make the specimen conductive for high resolution imaging. For the final polishing of the investigated surface, a FEI Versa 3D Dual focussed ion beam (SEM-FIB) microscope was used. The particular area of interest was polished with a Ga ion beam at 30 kV for 30 s at low current (10 pA or less). With the electron beam of the microscope, the image surface was investigated at 5 kV and 10 kV using SE and BSE detectors. The gradient distribution of pores to depth within the organic matter was obtained by progressive ion polishing and image capturing.

EDS analysis was done with an Octane silicon drift detector (SDD) from EDAX and a Genesis software package. In order to decrease the interacting volume, measurements were performed with 6 kV.

X-ray diffraction (XRD) measurements of texture-free and textured powder mounts were done following established procedures [19]. Measurement parameters were set at a goniometer speed rate of 0.5° 2θ/minute and a registration range from 2° to 66° 2θ. Particularly for the analyses of clay minerals, textured mounts were measured four times at a goniometer speed rate of 0.5° θ/minute with a registration range from 2° to 42° 2θ in the following order: (1) in untreated condition, (2) after solvation with ethylene glycol (12 h, 60 °C), after thermal treatment (2 h) at (3) 350 °C and (4) at 550 °C. The quantification followed the method of Schultz [20]. The results were cross-checked using the ADM software (Version 6.22) of Wassermann Röntgenanalytik (Germany) which is based on the Rietveld method. To verify the quantitative analysis, a comparison between the percentage-composition based on XRD-data of the shale samples, XRD-data of quartz standards and data of calcimeter measurements in accordance with the method of Scheibler [3] took place.

4. Results

4.1. Mineralogy

Facies zones were established by Gavrish *et al.* [10] and Machulina and Babko [11] for parts of the north-western DDB (Srebnen Bay; figure 2b). Within the frame of the present study, investigated samples cover basinal (siliceous), transitional (siliceous-carbonaceous-clayey), and clayey reef facies (for locations see figures 2a,b).

As a consequence of lateral facies zones, the mineralogical composition is strongly variable throughout the V-23 interval. Apart from lateral variations, highly variable vertical composition has been observed for well Yantarna (Yan, see figures 2c,d).

4.1.1. Basinal siliceous facies. Samples from the basin-centred well Yantarna show strong vertical variations in mineral assemblages, with a quartz-rich composition in the lowermost (~ 65 wt%) and central part (40 - 55 wt%), downwards decreasing contents of kaolinite as well as minor amounts of feldspar, pyrite, and detrital mica (< 5 - 10 wt%). Expandable clay minerals (ECM) are present in trace amounts in the upper part of the sampled interval. One sample exhibits a remarkably high calcite content (~ 50 wt%), corresponding to a significantly lower quartz content than observed in surrounding areas. A kaolinite-rich sample (~ 70 wt%) with total clay minerals exceeding 80 wt% was investigated in the uppermost part of the stratigraphic section. The cumulated amount of brittle mineral phases (quartz + feldspar + carbonates + pyrite + marcasite) exceeds 60 wt% in the Yantarna well, except of the uppermost sample.

SEM imaging shows that quartz is mainly microcrystalline with grain sizes < 5 μm (figures 3g,h), whereas rhombohedral carbonate minerals reach sizes of several tens of μm. Clay minerals are randomly arranged, therefore, a layering is not visible at the given magnification. Trace amounts of apatite have been detected from EDS measurements, occasionally occurring as fine layers or randomly distributed, isolated crystals. Organic matter mainly occurs as dispersed vitrinite grains (10 - 30 μm), sharp-edged, partly branched inertinite (up to 300 μm), or irregularly shaped, bituminous (?) organic matter (several tens of μm; figure 3h).

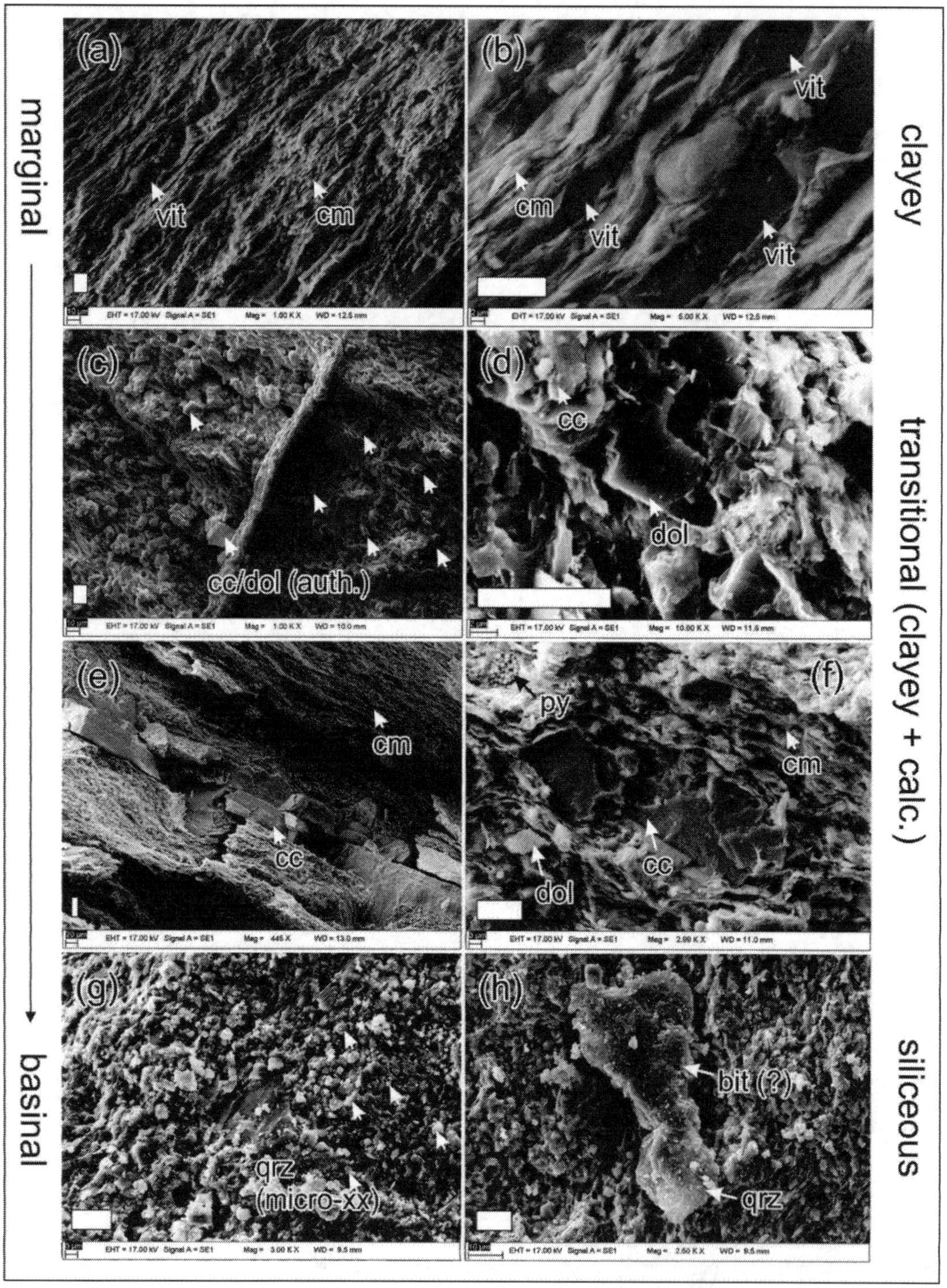

Figure 3. SEM (SE) images of selected Rudov (V-23) samples from marginal (Zimnistka; a-b), transitional (Selyukhov, Kompanska, Lelyakov; c-f) and basinal (Yantarna; g-h) wells, showing facies variations. White bar represents a width of 10 μm. bit – bitumen; cc – calcite; cm – clay minerals; dol – dolomite; qrz – quartz; py – pyrite; vit – vitrinite.

4.1.2. Marginal carbonate and clay rich facies. Within wells from the calcareous and clayey reef facies (Ostapovska; Zimnitska), the content of total clay minerals ranges between 45 and 85 wt%. The clay mineral fraction of both samples from Ostapovska consists of variable amounts of ECM, identified as an illite-dominated illite-smectite mixed layer mineral, detrital mica, illite, and kaolinite. In one sample, the dominant non-clay mineral is calcite. In the second, deeper sample, the content in kaolinite is ~ 45 wt%, whereas the quartz content is < 15 wt%. Samples from Zimnitska show high clay mineral and only minor quartz contents. The dominant mineral phase is kaolinite, although the percentages vary. Small amounts of chlorite are present in all samples. Two samples contain trace amounts of marcasite. One sample exhibits a total carbonate content > 30 wt%, including a significant proportion of siderite. Plagioclase occurs in trace amounts in all samples. Within the transitional and clayey facies, the content of brittle mineral phases generally ranges below 60 wt%.

A fine layering of clay minerals is visible within samples from the clayey facies zone in SEM images (figures 3a,b), elongated mineral grains as well as fine, carbonate-rich layers frequently occur within the clayey matrix. In contrast, within the transitional zone, samples exhibit both blocky, layered carbonates as well as finely dispersed partly rhombohedral crystals (figures 3c-f). In carbonate rich samples, the layering of clay minerals is obscured by partly authigenic, mainly microcrystalline (< 5 - 10 μm) carbonate.

Vitrinite is present as layers or elongated grains embedded into the clay matrix (see figures 3a,b). In samples with high content of inertinite macerals, irregularly branched particles, up to several 100 μm in size, have been detected. According to their habitus, they are suggested to represent higher plant material, which has been oxidized during syn-depositional wildfires.

4.2. Maturity related evolution of organic matter

Four samples with thermal maturity from 0.9 to 2.7 % Rr were picked for FIB-SEM and SEM imaging to create a maturity profile of organic matter evolution (for locations see figure 2a).

4.2.1. Oil window (N. Pogarshchinska). Organic matter in an oil mature sample (0.9 % Rr; ~ 5,100 m) from well N. Pogarshchinska (Npog) appears as primary (vitrinite) macerals as well as (secondary) solid bitumen, generated by thermal alteration of kerogen. Both types could be visualized by SEM imaging of freshly broken surfaces (figures 4a-d). Primary macerals occur as layers or isolated, equidimensional grains, while occasionally visible solid bitumen is preferentially found near larger rigid grains indicating precipitation in pre-existing pore space. Vitrinite macerals exhibit a homogeneous, pore-free surface. In contrast, solid bitumen hosts (sub-)micrometre open pores (figures 4c,d).

FIB-SEM imaging on primary macerals confirms their homogeneous composition without visible nanopores (figures 4e,f). Occasionally, pores in the sub-micrometre range occur at the border between maceral compounds and surrounding clay sheets, but not within the organic matter itself.

4.2.2. Wet gas window (Rodnikova). A sample from the wet gas window (1.35 % Rr) was obtained from well Rodnikova (Rod) at a depth of ~ 5,500 m, southeast of the Srebnen Bay. Maceral composition is dominated by primary vitrinite and inertinite fragments. According to SEM (figures 4g,h) and FIB-SEM measurements, no sub-micrometre pores are present in these macerals. Solid bitumen with secondary pores has not been detected (see figures 4i,j). Nevertheless, clay matrix-hosted nanopores, partly occurring at the borders between organic matter and clays, are clearly visible in the FIB-SEM image. Furthermore, nanometre-scale flakes of clay minerals (?), fully incorporated into solid bitumen, have been detected in few cases. However, EDS mapping could not be performed on such inhomogeneities due to very small particle sizes.

4.2.3. Dry gas window and overmature (Komyshovatska). Dry gas mature (2.1 % Rr; ~ 3,800 m) and overmature (2.7 % Rr; ~ 4,300 m) samples from well Komyshovatska (Kom), located in the inverted south-eastern part of the basin, host both non-porous, primary macerals (vitrinite; inertinite), and

IOP Conf. Series: Materials Science and Engineering **109** (2016) 012010 doi:10.1088/1757-899X/109/1/012010

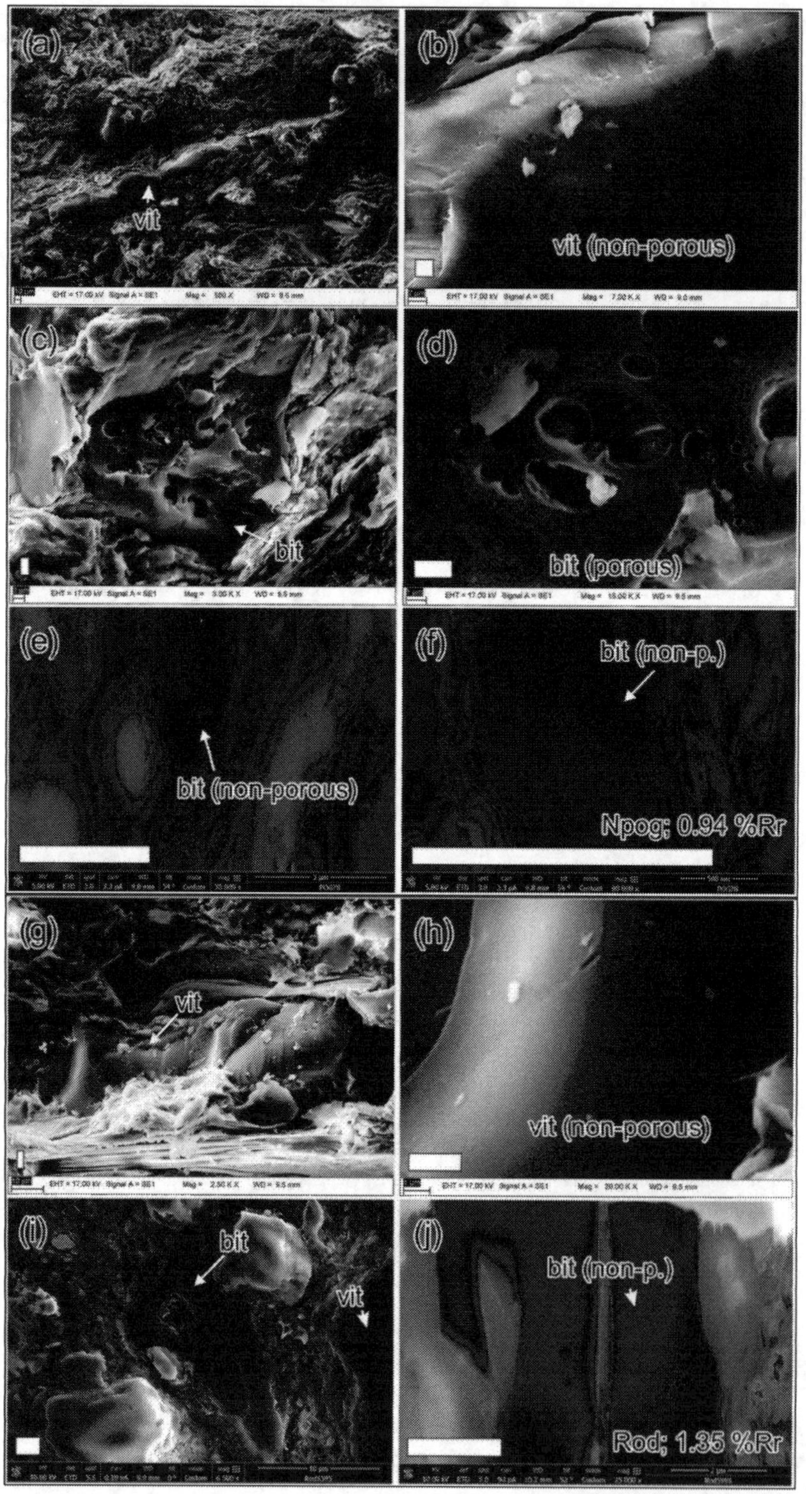

Figure 4. SEM (SE) and FIB-SEM (SE) images of non-porous, homogeneous vitrinite, non-porous as well as porous bitumen at a vitrinite reflectance of 0.9 - 1.35 % Rr. White bar represents a width of 2 μm.

nanoporous, secondary organic matter (pyrobitumen). While primary macerals mainly form isolated, randomly dispersed grains within the clay- and quartz-rich matrix (figures 5a-g), secondary bitumen often forms elongated or irregularly shaped bodies or fillings along large mineral grains (figures 5b,e,k). Two types of "inclusions" within bitumen have been detected, (1) pores down to 10 nm in size (figures 5c,d,i,j,l) and (2) clay flakes or calcite nodules in the nanometre-scale (figures 5e,f,k). In case of the latter, EDS measurements show Al, Si and low amounts of K, Mg or Fe (clay), or Ca (calcite) enrichments with well-defined borders within the organic carbon. Although a quantification of those minerals using the required small spot size is not possible, the qualitative identification within a large-scale organic phase without present inorganic elements is reliable. FIB milling was performed and confirmed that the inorganic phases are fully incorporated into the secondary organic matter and not part of the surrounding clay matrix. In contrast, primary vitrinite or inertinite macerals exhibit a homogeneous, pore-free fabric without included inorganic elements (see figure 5h). In case of the overmature sample, merging nanopores partly form micrometre-scale, irregularly shaped connected pores within pyrobitumen (figure 5l).

5. Discussion

5.1. Facies zones

The obtained results agree with the facies zones established by former studies [10, 11]. Nevertheless, XRD measurements revealed a highly variable mineralogical composition within Rudov Beds, both in lateral and vertical direction. Even within the basinal, siliceous facies, one sample exhibited a clay mineral portion of > 80 wt%. A good estimation of mechanical properties can be established according to the accumulated content of brittle mineral phases, which should exceed 60 wt% for unconventional HC production. This value is reached by all samples from the siliceous basinal facies except one, whereas no sample from transitional (carbonate-dominated) or clayey facies shows such a high amount of brittle phases.

Apart from that, a higher permeability can be suggested for the basinal samples rich in microcrystalline quartz grains, compared to their fine-layered, clay rich counterparts recovered from distal positions. The decrease of such microcrystalline, slightly elongated quartz particles towards the basin rim is most likely a result of decreasing contribution from deeper water radiolaria. The change from siliceous and clay dominated facies is marked by a transitional zone, often rich in carbonates. The amount of inertinite macerals formed by wildfires also increases towards the basin margin. Carbonates occur both as fine layers as well as authigenic rhombs of dolomite and calcite.

The presence of ECM to depths of > 5,000 m argues for a low heat flow during burial, consistent with the low gradient of vitrinite reflectance to depth within the north-western part of the basin.

Mineral pores of sub-micrometre-scale were frequently observed within Rudov Beds, especially in the oil and wet gas mature samples. However, illitic shales are commonly suggested to be strongly water wet, whereas kaolinite-rich clays might impart a higher possibility for shales to become oil/gas wet [21]. Therefore, the clay mineral ratio (illite/kaolinite) might be a useful parameter for production behaviour from shale reservoirs.

In general, pore geometries and distribution in the mineral matrix might be strongly related to clay mineral composition; therefore, assignment of pore classes to certain clay phases could reveal valuable insights.

5.2. Thermal evolution of organic matter

Organic matter within the investigated black shale samples can be separated into (1) primary macerals and (2) secondary (pyro-)bitumen, formed by thermal alteration of the primary kerogen. Jarvie [22] distinguishes between earlier solid bitumen (oil window) and later pyrobitumen generated at gas window maturities. Depending on the type of kerogen, porous bitumen might occur already at oil window maturity (>0.9 Rr; see also [23, 24]), but in most cases is not abundant below 1.4 % Rr [25, 26]. Porous pyrobitumen was frequently observed within samples showing thermal maturity

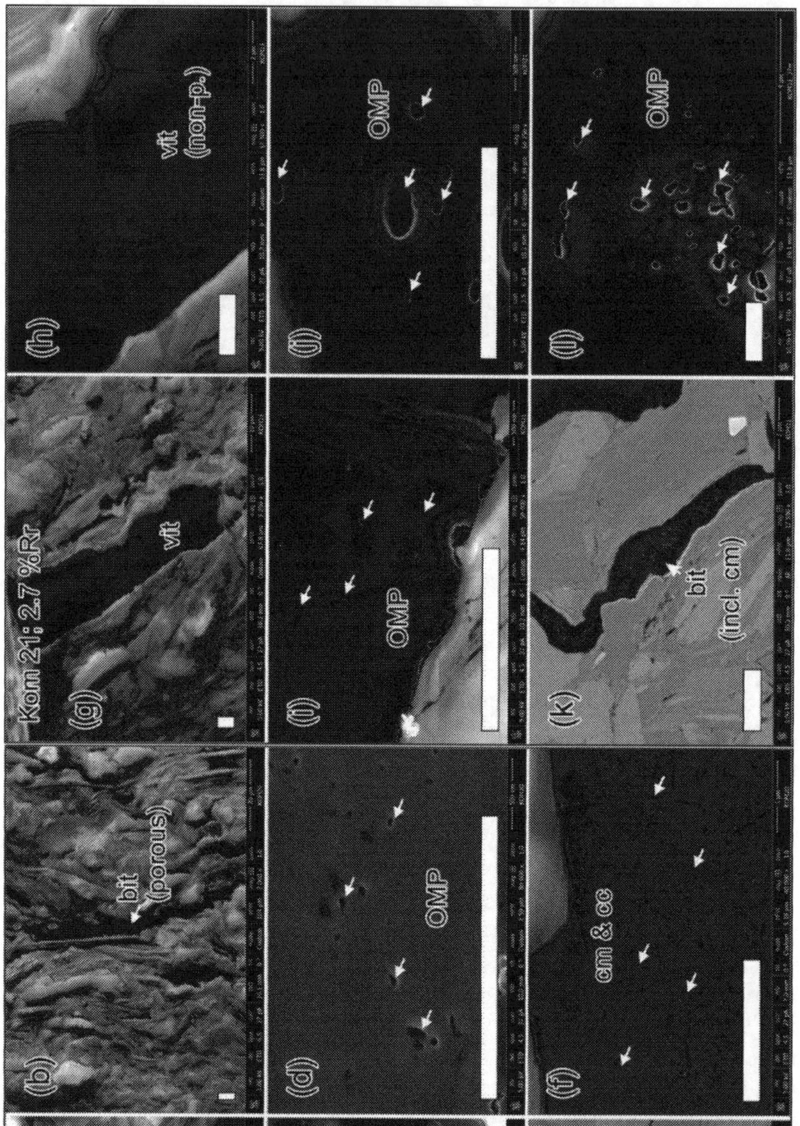

Figure 5. FIB-SEM images (SE and BSE), showing organic matter porosity and inhomogeneities within pyrobitumen as well as non-porous, homogeneous vitrinite at a vitrinite reflectance of 2.1 % Rr (a - f) and 2.7 % Rr (g - l). White bar represents a width of 2 μm. OMP – organic matter porosity.

corresponding to the dry gas window (2.1 % Rr). The mostly equidimensional nanopores show limited connectivity, whereas, in zones with enhanced porosity, connected (former) nanopores form irregularly shaped pores in the micrometre-scale. The size and abundance of connected micrometre-scale pores increases within the overmature (2.7 % Rr) sample. Oval pores up to micrometre-scale occur at the boundaries between organic matter and clay matrix, possibly also holding gas storage potential. Apart from that, elongated, sharp-edged pores occasionally have been visualized at the interface of organic particles with the surrounding matrix. Klaver *et al.* [27] interpreted such features as artificial micro-cracks formed probably during or after coring.

Milliken *et al.* [28] observed that variations in TOC have a stronger influence on organic matter-hosted pores than thermal maturity. However, the present study suggests that apart from kerogen type, the presence of organic matter-hosted porosity is strongly controlled by the maturity related generation of (pyro-)bitumen.

Another feature observed within the maturity profile was the development of authigenic clay flakes and calcite nodules, often in the scale of < 100 nm, within pyrobitumen. This was not observed within the least mature sample (0.9 % Rr), rarely observed within the wet gas mature sample (1.35 % Rr) and frequently observed within the dry gas (2.1 % Rr) and overmature (2.7 % Rr) samples. The growth mechanisms of such mineral phases during thermal evolution of organic matter need further investigation. However, similar findings have been described by Reed *et al.* [23], who observed pellets of mixed organic-clayey composition, even if the clay flakes in those pellets were an order of magnitude larger than in this case, and seemed not to be fully incorporated into bitumen.

The possible presence of nanopores in primary kerogen was discussed controversially and is subject to an ongoing debate [23, 29]. However, we did not identify nanoporous organic matter that likely can be referred to as kerogen.

6. Conclusions

SEM is used as a standard method for the characterisation of rock fabrics within conventional reservoir rocks. However, also within the field of unconventional reservoir rocks with small grain sizes, SEM imaging performed on freshly broken surfaces reveals valuable insights for the understanding of rock properties like fraccability and permeability. Large variations, especially in the content of brittle mineral phases, have been observed within the investigated Rudov samples in lateral extent. Furthermore, strongly variable XRD data also suggests high variations in vertical direction. Siliceous, transitional and clayey facies could be distinguished from SEM imaging. Therefore, mechanical properties of Rudov Beds and Upper Visean time equivalents outside of the Srebnen Depression have to be considered highly variable in both lateral and vertical extent.

As already reported by Wüst *et al.* [30], the characterisation of organic matter by SEM on freshly broken surfaces depends on its specific properties and also on the hosting mineral fabric. Primary macerals and bituminous pore filling appeared more often in rock samples with a higher content of brittle mineral phases (e.g., quartz, calcite), whereas they might be easily obscured in laminated, clay-rich shales. However, OM was identified in all rocks investigated with conventional SEM and bituminous organic matter hosting micrometre-scale pores could be easily discriminated from primary vitrinite within the oil mature sample from well N. Pogarshchinska.

FIB-SEM and BIB-SEM proved to be a valuable tool for the investigation of sub-micrometre porosity within the organic matter hosted by shales (see also [23, 25, 26, 29, 31-33]). OM-hosted nanopores are restricted to secondary bituminous organic matter within Rudov Beds. Their frequency strongly increases at a vitrinite reflectance of > 2.0 % Rr, corresponding to a dry gas window maturity. Furthermore, nanoscale features like authigenic clay and calcite particles within pyrobitumen at gas window maturity have been visualized. At a vitrinite reflectance of 0.9 % Rr, sub-micrometre pores to nanopores have been detected only in very few cases within bituminous pore fillings. Most investigated particles did neither host nanopores, nor did they host mineral inclusions. Therefore, in case of Rudov Beds, minor generation of nanopores within bitumen, due to hydrocarbon generation and release, might already start at oil window maturity, but the main stage of development occurs at maturities between 1.35 % Rr and 2.0 % Rr.

The performed investigations showed that non-invasive preparation techniques like BIB milling further improve processing due to fast coverage of larger areas within a single sample and lower possibility of creating artefacts caused by improper FIB milling conditions. Working on organic matter with a FIB-SEM, handling parameters like the acceleration voltage of the ion beam turned out to be a crucial influencing factor possibly even causing the concealment of pores. In contrast, clay minerals seem less sensitive to beam damage. Reduced working distance as well as the use of a microscope equipped with a field emission electron source enhanced the maximum resolution down to < 10 nm of equal pore diameter, helping to visualize even the smallest features within the studied kerogen.

Acknowledgments
The authors thank V. Makogon (Ukrainian State Geological Institute) for his support during the collection of samples from the Chernigiv Core Store and the staff members of Wuppermann AG for the preparation of BIB milled samples. The manuscript was improved considerably by the reviews of two anonymous reviewers.

References

[1] King H E, Eberle A P R, Walters C C, Kliewer C E, Ertas D and Huynh C 2015 *Energy Fuels* **29** 1375-1390
[2] Falk K, Coasne B, Pellenq R, Ulm FJ and Bocquet L 2015 *Nature Comm.* **6** 6949
[3] ON L 1084-99 1999 (Vienna, Austria: Austrian Standards Institute ÖNORM)
[4] Ivanyuta M M, Fedyshyn V O, Denega B I, Arsiriy Y O and Lazaruk Y G 1998 *Atlas of oil and gas fields of Ukraine (6 volumes).* (Lviv, Ukraine: Ukrainian Oil and Gas Academy)
[5] Shpak P F (ed.) 1989 *Geology and petroleum productivity of the Dniepr-Donetsk depression - petroleum productivity.* (Kiev, Ukraine: Naukova Dumka)
[6] Ulmishek G F 2001 Petroleum Geology and Resources of the Dnieper-Donets Basin, Ukraine and Russia. *US Geol. Survey Bull.* **2201-E**
[7] Sachsenhofer R F, Shymanovskyy V A, Bechtel A, Gratzer R, Horsfield B and Reischenbacher D 2010 *Petroleum Geosci.* **16** 377-399
[8] Gradstein F, Ogg J and Smith A 2004 *A Geological Time Scale.* (Cambridge: Cambridge University Press)
[9] Lukin A E, Shpak P F, Chepil P M and Machulina S A 1994 *Dokl. Akademii Nauk Ukrainy* **8** 101-104
[10] Gavrish V K, Machulina S A and Kurilenko V S 1994 *Dokl. Akademii Nauk Ukrainy* **7** 92-95
[11] Machulina S A and Babko I M 2004 *Naft. i gasova prom-st.* **5** 3-8
[12] Kabyshev B, Krivchenkov B, Stovba S and Ziegler P A 1998 *Marine Petroleum Geol.* **15** 177-190
[13] Dvorjanin E S, Samoilyuk A P, Egurnova M G, Zaykovsky N Y, Podladchikov Y Y, van den Belt F J G and de Boer P L 1996 *Tectonophysics* **268** 169-187
[14] Kabyshev B P, Kabyshev Y B, Krivosheev V T, Prigarina T M and Ulmishek G F 1999 *Rep. Nat. Acad. Sci. Ukraine* **12** 112-117
[15] Sachsenhofer R F, Privalov V A and Panova E A 2012 *Int. J. Coal Geol.* **89** 26-40
[16] Shymanovskyy V A, Sachsenhofer R F, Izart A and Li Y 2004 *Tectonophysics* **381** 61-79
[17] Taylor G H, Teichmüller M, Davis A, Diessel C F K, Littke R and Robert P 1998 *Organic petrology. A New Handbook incorporating some revised parts of Stach's Textbook of Coal Petrology.* (Berlin-Stuttgart: Gebrüder Borntraeger)
[18] Babko I M, Gontcharov V E and Kononenko L P 2003 *Mineral Resources of Ukraine* 20-25
[19] Brunton G 1955 *Am. Mineralogist* **40** 124-126
[20] Schultz L G 1964 *Geol. Survey Prof. Pap.* **391-C** 31pp

[21] Josh M, Esteban L, Delle Piane C, Sarout J, Dewhurst D N and Clennell M B 2012 *J. Petroleum Sci. Engng.* **88-89** 107-124
[22] Jarvie D M 2012 *AAPG Memoir* **97** 69-87
[23] Reed R M, Loucks R G and Ruppel S C 2014 *Int. J. Coal Geol.* **127** 111-113
[24] Lu J, Ruppel S C and Rowe H D 2015 *AAPG Bull.* **99** 333-357
[25] Bernard S, Wirth R, Schreiber A, Schulz H M and Horsfield B 2012 *Int. J. Coal Geol.* **103** 3-11
[26] Bernard S, Horsfield B, Schulz H M, Wirth R, Schreiber A and Sherwood N 2012 *Marine Petroleum Geol.* **31** 70-89
[27] Klaver J, Desbois G, Urai J L and Littke R 2012 *Int. J. Coal Geol.* **103** 12-25
[28] Milliken K L, Rudnicki M, Awwiller D N and Zhang T 2013 *AAPG Bull.* **97** 177-200
[29] Bernard S and Horsfield B 2014 *Int. J. Coal Geol.* **127** 114-115
[30] Wüst R A J, Nassichuk B R and Bustin R M 2014 *AAPG Memoir* **102** 81-100
[31] Bernard S, Brown L, Wirth R, Schreiber A, Schulz H M, Horsfield B, Aplin A C and Mathia E J 2014 *AAPG Memoir* **102** 53-66
[32] Curtis M E, Sondergeld C H, Ambrose R J and Rai C S 2012 *AAPG Bull.* **96** 665-677
[33] Curtis M E, Cardott B J, Sondergeld C H and Rai C S 2012 *Int. J. Coal Geol.* **103** 26-31

IOP Conf. Series: Materials Science and Engineering **109** (2016) 012011 doi:10.1088/1757-899X/109/1/012011

Effect of shape and thickness of asbestos bundles and fibres on EDS microanalysis: a Monte Carlo simulation

D Moro and G Valdrè

University of Bologna, Interdisciplinary Research Centre of Biomineralogy, Cristallography and Biomaterials, Department of Biological, Geological and Environmental Sciences, Geological Section, Piazza di Porta San Donato 1, 40126 Bologna, Italy

E-mail: giovanni.valdre@unibo.it

Abstract. Quantitative microanalysis of tiny asbestos mineral fibres by scanning electron microscopy equipped with energy-dispersive X-ray spectroscopy (SEM-EDS) still represents a complex analytical issue. This complexity arises from the variable fibre shape and small thickness (< 5 μm) compared with the penetration of the incident electron beam. Here, we present the results of Monte Carlo simulations of chrysotile, crocidolite and amosite fibres (and bundles of fibres) of circular and square section and thicknesses from 0.1 μm to 10 μm, to investigate the effect of shape and thickness on SEM-EDS microanalysis. The influence of shape and thickness on the simulated spectrum was investigated for electron beam energies of 5, 15 and 25 keV, respectively. A strong influence of the asbestos bundles and fibres shape and thickness on the detected EDS X-ray intensity was observed. The X-ray intensity trends as a function of fibre thickness showed a non-linear dependence for all the elements and minerals. In general, the X-ray intensities showed a considerable reduction for thicknesses below about 0.5 μm at 5 keV, 2 μm at 15 keV, and 5 μm at 25 keV. Correction parameters, *k*-ratios, for the asbestos fibre thickness effect, are reported.

1. Introduction

Precise chemical analysis of asbestos fibres is important for their identification and in depth mineral characterisation. Fibres composition is also an important criteria in understanding the minerals' environmental and health impacts, and determining the potential to have toxicological effects [1-3].

In vitro and *in vivo* studies revealed that asbestos fibres toxicity is strongly related to the chemical composition, particularly as regards the Fe content and its oxidation state [4]. In addition, asbestos-induced cytotoxicity and genotoxicity is thought to be also due to the generation of reactive oxygen species and other radicals that are catalysed by iron ions at the fibre surface [5]. A strong biochemical reaction is induced in biological systems by the mineral composition of fibres with bonded metal ions. Fundamental cell biology research on the interaction of cells with long fibres or fibres containing highly reactive chemical species will be facilitated by techniques able to accurately quantify the chemistry of mineral fibres in situ [6].

Scanning electron microscopy equipped with X-ray energy dispersive spectroscopy (SEM-EDS) is of great importance for minerals, materials and biomaterials characterisation, and in evaluating the phenomena and possible mechanism of interaction between materials and bio-entities [7-9].

Energy-dispersive X-ray spectroscopy (EDS) is by far the most routine method for asbestos chemical analysis [1]. However, SEM-EDS quantitative microanalysis of asbestos mineral fibres may suffer from systematic errors because of the variable fibre shape and small thickness (< 5 μm) compared with the penetration of the incident electron beam. In the classical correction methods used in quantitative microanalysis of bulk samples for the correction of matrix effects, e.g., ZAF or $\phi(\rho z)$ procedures, both standard and sample are assumed to have flat polished surfaces and be infinitely thick with respect to the penetration of the electron beam.

In the case of micro- and sub-micrometric particles their size and shape may cause large errors in the quantification due to particle effects on the generation and measurement of X-rays from the sample. These ones are related to the elastic scattering of electrons in the finite size (mass) of the fibre, strongly influenced by the average atomic number. For a given mean atomic number, the thickness of the particle is key, with a shape component affecting the absorption and fluorescence contributions to the correction routine. In the analysis of micro- and sub-micrometric particles, X-ray emergence angle and absorption path length cannot be determined as for flat polished samples [10, 11]. Furthermore we point out that EDS systems, because they are often set to normalize the results, are prone to hide poor analytical results, like high or low totals, which then hide any problem in the analytical protocol.

The empirical methods proposed to overcome these issues are, however, cumbersome and need characterized standards for thickness, geometry and composition [11, 12].

The Monte Carlo method is a valuable tool for investigating the basis, the reliability and the limitations of quantification procedures. In particular, Monte Carlo simulations can be of paramount importance for studying the electron transport and X-ray generation in samples with complex geometry or under unconventional measurement conditions [13-15].

Here, we report the results of Monte Carlo simulations of asbestos fibres and bundles of fibres to investigate the effect of shape and thickness on SEM-EDS microanalysis. NISTMONTE [16] was used to simulate electron transport, X-ray generation and detection in complex sample geometries. The investigated asbestos minerals were chrysotile, amosite and crocidolite.

Chrysotile, white asbestos, is a phyllosilicate of the serpentine group. About 95% of asbestos used in commercial products was chrysotile [17, 18].

Crocidolite, blue asbestos, and amosite, brown asbestos, are members of the amphiboles group and are the most commonly used amphibole asbestos. The term crocidolite is used for the asbestos variety of the amphibole mineral riebeckite, whereas amosite for the asbestos variety of grunerite, as indicated by Gunter and co-workers [1]. They contain high levels of iron as a normal constituent of the crystal structure. Amphibole exposure is greater than chrysotile in the natural environment, since amphiboles are more widespread than chrysotile [1, 17, 19-21].

2. Methods and simulation models

2.1. Methods

The Monte Carlo simulations were carried out using the programme NISTMONTE version 2 [16], which simulates electron transport, X-ray generation and transmission in complex sample geometries. Monte Carlo method applied to electron and X-ray transport is used to simulate electrons trajectories and X-rays through the mineral fibre and to the X-ray detector. Accurate quantitative analysis by X-ray spectroscopy requires the understanding of electron scattering, X-ray generation, absorption and fluorescence.

NISTMONTE 2 implements a single scattering model to track electrons trajectories as they interact with matter. It uses three different elastic scattering models: a basic screened Rutherford model, the Mott scattering cross-section of Czyzewski and co-workers [22], and the Mott cross-section of Jablonski and co-workers [23]. The energy loss is modelled using the Joy-Luo expression [24], which is an empirical modification of the Bethe energy loss equation [25]. The ionisation cross-section is modelled using the empirical expression of Casnati [26]. The mass absorption coefficients are those of Heinrich [27], and the fluorescence yields are tabulated experimental values [28].

The electron source is defined to be a Gaussian beam. The emitted x-ray events are convolved with a response function that mimics the energy resolution of an EDS detector. The simulated EDS spectra of single asbestos and bundles of fibres were used to evaluate the effect of shape and thickness by integrating and comparing peaks intensity as a function of thickness, shape and beam energy for each mineral, after background subtraction.

2.2. Simulation models

Bundles and fibres of variable shape and thickness on a pure carbon holder were simulated taking into account realistic experimental conditions, such as sample geometry, SEM set-up and detector physics.

We investigated chrysotile, crocidolite and amosite with chemical formulae as specified in Table 1.

Table 1. Chemical formulae of the simulated asbestos.

Mineral name	Chemical formulae
Chrysotile	$Mg_3Si_2O_5(OH)_4$
Crocidolite	$NaFe^{2+}{}_{2.25}Mg_{0.25}Fe^{3+}{}_2Si_8O_{22}(OH)_2$
Amosite	$Fe^{2+}{}_{5.25}Mg_{1.75}Si_8O_{22}(OH)_2$

Two models for each mineral were taken into account: (i) a three-dimensional fibre of circular section, 100 μm long and with thicknesses ranging from to 0.1 μm to 5 μm, deposited on a flat graphite support, and (ii) a three-dimensional bundle of square section, 100 μm long and with thicknesses ranging from to 0.1 μm to 10 μm, deposited on a flat graphite support.

An electron probe of 40 nm in diameter focussed in parallel illumination onto the surface of the fibre or bundle, in a mid position with respect to the edges, was taken into account. The modelled EDS detector used to generate the spectrum had a gold layer of 7 nm, a dead layer of 10 nm, a detector diode thickness of 3 mm, solid angle of $5 \cdot 10^{-3}/4\pi$, 4,096 channels each of 10 eV and a resolution of 130 eV (at Mn Kα). The detector elevation angle was set to 40°, with an azimuthal angle of 0°. The fibre or bundle was oriented with its long axis in the same direction as the detector.

The influence of shape and thickness on the simulated spectrum was investigated for electron beam energies of 5, 15 and 25 keV.

3. Results and discussion

In conventional SEM-EDS microanalysis (with beam energy 10 - 30 keV) the sizes of the electron beam interaction volume and X-ray generation volume are on the order of micrometres in most materials. When particle size is smaller than the primary electron beam penetration depth, a portion of the beam escapes the particle before exciting X-rays (finite size effect). Thus, a fraction of the electron beam scatters out of the particle volume and can excite X-rays from the substrate or adjacent particles. These phenomena can affect the analytical results.

Figure 1 shows, as an example, a section image of the simulated trajectories of electrons for a 25 keV electron beam focussed onto the surface of a chrysotile circular fibre, in a mid position with respect to the edges. The fibre has a diameter of 2 μm and is deposited on a graphite substrate. The primary electrons (green vertical beam) interact with the fibre as visualized by the fuchsia trajectories. At this energy level a fraction of the electrons penetrate into the graphite substrate (grey trajectories) for several micrometres. Furthermore, a fraction of the electrons scatters out of the fibre volume from its sides, both toward the substrate, green trajectories, and leaving the fibre and the substrate, blue lines.

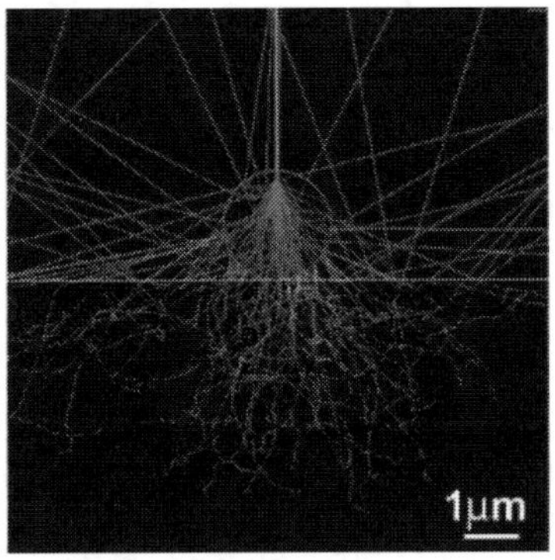

Figure 1. Simulated trajectory of electrons as they interact with the sample (chrysotile) and the substrate (graphite). 2D view of the 3D simulated trajectory of electrons (fuchsia: interacting with the sample; green and grey: towards and penetrating the substrate, respectively; blue: exiting) as they interact with a chrysotile cylindrical fibre (marked with a fuchsia circular line) deposited on a bulk graphite substrate (marked with a grey horizontal line). The view is from the direction parallel to the long axis of the cylindrical fibre.

To help the reader in the analysis of the results we report in Table 2 the nominal range value (for a bulk sample) of electron penetration depth, and X-rays intensity (generated and emitted) for the three mineral samples and the three beam energies considered in this work. The calculation was performed with the same SEM set-up reported in Section 2.2 considering 1,000 electron trajectories. The table report the absolute magnitude of generated and emitted X-ray intensity for each X-ray line calculated on bulk samples, in arbitrary units as reported by Ritchie [29].

Table 2. Electron penetration depth (e.p.d., in μm) and absolute magnitude (arbitrary units) of generated (g) and emitted (e) X-ray intensity for each X-ray line calculated on bulk samples (as reported by Ritchie [29].

	Chrysotile			Crocidolite			Amosite		
	5keV	15keV	25keV	5keV	15keV	25keV	5keV	15keV	25keV
e.p.d. [μm]	0.38	2.20	5.09	0.26	1.76	4.12	0.27	1.47	3.76
Na K_α	—	—	—	(g) 7.0 (e) 5.9	(g) 38 (e) 14	(g) 74 (e) 11	—	—	—
Fe K_α	—	—	—	0	(g) 21 (e) 21	(g) 81 (e) 78	0	(g) 24 (e) 24	(g) 97 (e) 93
Fe K_β	—	—	—	0	(g) 2.6 (e) 2.5	(g) 10 (e) 10	0	(g) 3.0 (e) 3.0	(g) 12 (e) 12
Fe L	—	—	—	(g) 29 (e) 27	(g) 130 (e) 30	(g) 240 (e) 20	(g) 34 (e) 26	(g) 150 (e) 40	(g) 290 (e) 30
Mg K_α	(g) 27 (e) 26	(g) 170 (e) 110	(g) 340 (e) 130	(g) 2.2 (e) 2.0	(g) 14 (e) 7	(g) 27 (e) 6	(g) 5.0 (e) 4.5	(g) 30 (e) 15	(g) 61 (e) 15
Si K_α	(g) 12 (e) 12	(g) 100 (e) 70	(g) 220 (e) 100	(g) 16 (e) 15	(g) 140 (e) 100	(g) 280 (e) 130	(g) 16 (e) 15	(g) 130 (e) 90	(g) 270 (e) 130
O K_α	(g) 130 (e) 120	(g) 550 (e) 220	(g) 1,000 (e) 170	(g) 120 (e) 100	(g) 470 (e) 190	(g) 860 (e) 150	(g) 110 (e) 100	(g) 440 (e) 180	(g) 820 (e) 150

3.1. Chrysotile

To investigate the significance of the thickness and shape effects on chrysotile, a fibre of circular section and a bundle of square section were taken into account with a composition as reported in Table 1. See section 2.2 for details on the simulation model. Figure 2 shows the trend of the integrated intensities (counts), obtained from the simulated EDS spectra, as a function of thickness (fibre diameter or bundle thickness) for each element (Si, Mg, O) of the mineral at beam energies of 5 keV, 15 keV and 25 keV (see the legend in figure 2e). Figures 2a, 2c and 2e shows the simulation results of the circular fibre for thickness form 0.1 μm to 5 μm, whereas figures 2b, 2d and 2f those of the bundle of square section for thickness form 0.1 μm to 10 μm.

A strong dependence of the simulated X-ray intensity on the fibre thickness was observed for all the elements of both models. Note that the intensity values reported for thickness of 10 μm can be considered as a reference value valid for a massive sample. By comparing the intensity values for a 10 μm thick sample with those calculated for a bulk one, we verified that a thickness of 10 μm represents, at the simulated experimental conditions, a bulk sample for all minerals and energies here considered. We report in Table 3 the reference intensity values for bulk samples calculated at an electron beam energy of 25 keV. Generally speaking, the integrated intensities start to have a reduction for thickness below 5 μm. In particular, for all the considered elements, Si, Mg and O, and both fibre and bundle, the integrated intensities strongly decrease starting from thickness of 5 μm at 25 keV, 2 μm at 15 keV and 0.5 μm at 5 keV. Notably, the square section bundle model presents a more marked effect of intensity enhancement before the reduction starting point. In terms of the previously cited finite size and reduced absorption effects, for the bundle model the reduced absorption effect initially prevails on the finite size one, which in turn becomes predominant for smaller thicknesses.

Table 3. Reference intensity values for a bulk sample at 25 keV in our experimental set up (§ 2.2).

	$Na_{K\alpha}$	$Fe_{K\alpha}$	$Fe_{K\beta}$	Fe_{L}	$Mg_{K\alpha}$	$Si_{K\alpha}$	$O_{K\alpha}$
Chrysotile	—	—	—	—	$1.78\cdot10^{6}$	$1.40\cdot10^{6}$	$1.32\cdot10^{6}$
Crocidolite	$1.43\cdot10^{5}$	$1.14\cdot10^{6}$	$1.42\cdot10^{5}$	$1.86\cdot10^{5}$	$8.54\cdot10^{4}$	$1.90\cdot10^{6}$	$1.12\cdot10^{6}$
Amosite	—	$1.36\cdot10^{6}$	$1.70\cdot10^{5}$	$2.25\cdot10^{5}$	$2.00\cdot10^{5}$	$1.80\cdot10^{6}$	$1.09\cdot10^{6}$

3.2. Crocidolite

A fibre of circular section and a bundle of square section were simulated also for crocidolite, with a composition as reported in Table 1. See section 2.2 for details on the simulation model. Figure 3 shows the results for the circular section fibre. It reports the trend of the integrated intensities (counts), obtained from the simulated EDS spectra, as a function of fibre diameter for each element of the mineral at beam energies of 5 keV, 15 keV and 25 keV (see legend in figure 3g). Fibres with a diameter from 0.1 μm to 5 μm were simulated. Bundles with a thickness from 0.1 μm to 10 μm were also simulated (data not shown, but available on request).

A strong dependence of the simulated X-ray intensity on the fibre thickness was observed for all the elements of both the models. The intensity values calculated for thicknesses of 10 μm were found very close to those calculated for a massive sample. Table 3 reports the reference intensity values for bulk samples calculated at an electron beam energy of 25 keV. Generally speaking, the integrated intensities start to have a reduction for thickness below 5 μm, with the exception of the Fe L-line which starts decreasing from 2 μm in the case of square section bundle. In particular, for almost all the considered elements, Si, Mg, O, Na, Fe (Kα), Fe (Kβ), Fe (L), and both fibre and bundle, the integrated intensities strongly decrease starting from thickness of 5 μm at 25 keV, 2 μm at 15 keV and 0.5 μm at 5 keV.

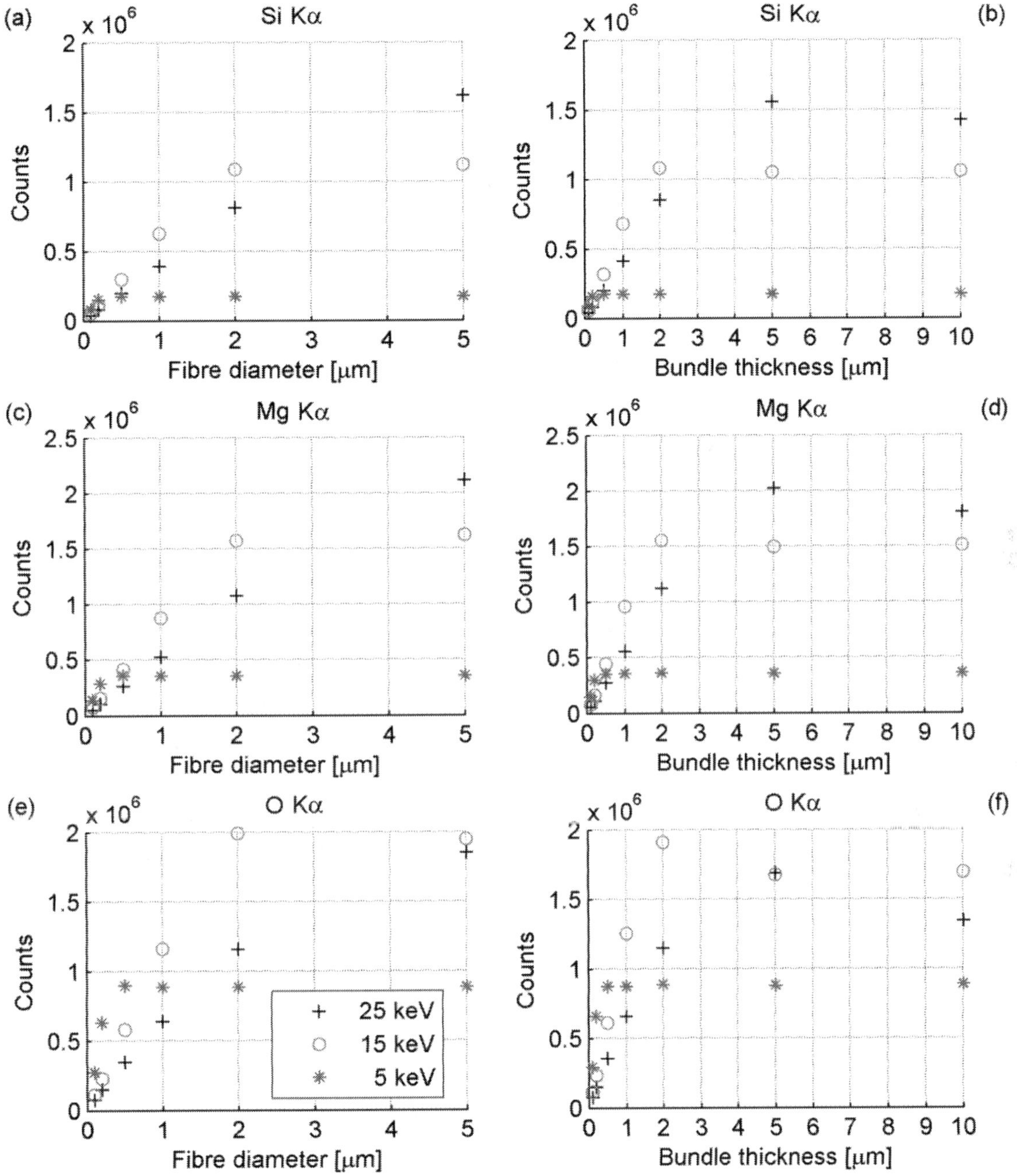

Figure 2. Simulated SEM-EDS Kα-line X-ray intensity of Si, Mg and O as a function of thickness for a single chrysotile cylindrical fibre, (a), (c), (e), and for a bundle of fibres, (b), (d), (f), with beam energy of 5 keV, 15 keV and 25 keV; see legend in (e).

IOP Conf. Series: Materials Science and Engineering **109** (2016) 012011 doi:10.1088/1757-899X/109/1/012011

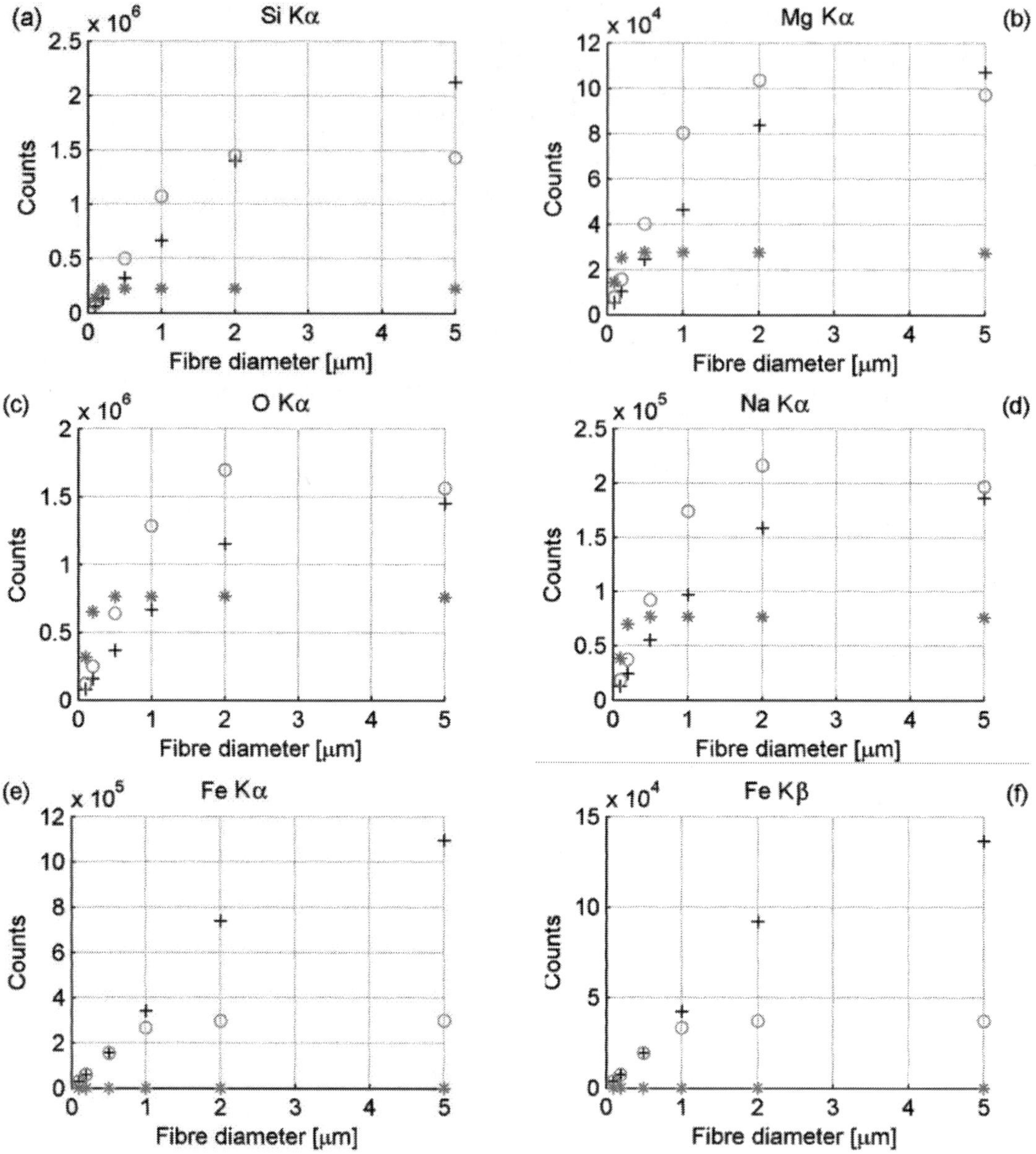

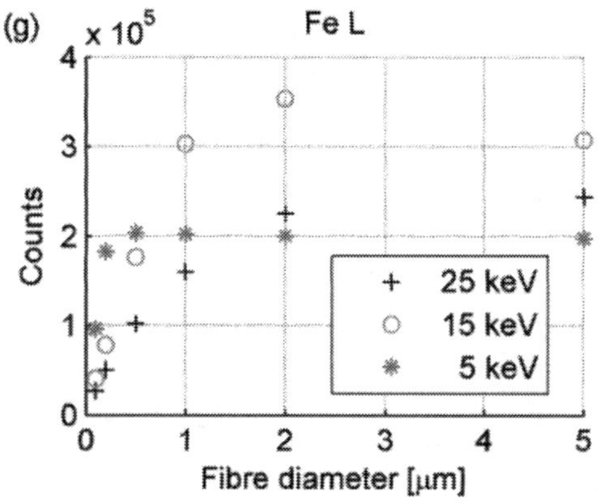

Figure 3. Simulated SEM-EDS X-ray intensity as a function of thickness for a single crocidolite cylindrical fibre with beam energy of 5 keV, 15 keV and 25 keV.

The two models present a similar effect of intensity enhancement before the reduction starting point. Again, in terms of the previously cited finite size and reduced absorption effects, the reduced absorption effect initially prevails on the finite size one, that in turn become predominant for smaller thicknesses. The magnitude of the reduced X-ray absorption effect is generally largest in the case for "soft" X-rays, when there is high absorption, in agreement with the experimental results of Paoletti and co-workers, and Small [10, 11].

3.3. Amosite

An amosite fibre of circular section and a bundle of square section were simulated, with a composition as reported in Table 1. See section 2.2 for details on the simulation model. Figure 4 shows the results for the circular section fibre. It reports the trend of the integrated intensities (counts), obtained from the simulated EDS spectra, as a function of fibre diameter for each element of the mineral at beam energies of 5 keV, 15 keV and 25 keV (see the legend in figure 4f). Fibres with a diameter from 0.1 μm to 5 μm were simulated. Bundles with a thickness from 0.1 μm to 10 μm were also simulated (data not reported, but available on request).

A strong dependence of the simulated X-ray intensity on the fibre thickness was observed for all the elements of both the models. The intensity values calculated for thicknesses of 10 μm were found very close to those calculated for a massive sample. Table 3 reports the reference intensity values for bulk samples calculated at an electron beam energy of 25 keV. Generally speaking, the integrated intensities start to have a reduction for thickness below 5 μm, with the exception of the Fe L-line which starts decreasing from 2 μm in the case of square section bundle. In particular, for almost all the considered elements, Si, Mg, O, Fe (Kα), Fe (Kβ), Fe (L), and both fibre and bundle, the integrated intensities strongly decrease starting from thickness of 5 μm at 25 keV, 2 μm at 15 keV and 0.5 μm at 5 keV. The two models present a similar effect of intensity enhancement before the reduction starting point. Again, in terms of the previously cited finite size and reduced absorption effects, the reduced absorption effect initially prevails on the finite size one, that in turn becomes predominant for smaller thicknesses.

Our investigation may be used to develop specific thickness correction factors for SEM-EDS quantitative analysis. For the sake of example we report in Table 4 the *k*-ratios (intensity ratio of X-rays emitted by the mineral fibre to those produced by a massive sample) calculated for cylindrical fibres of different thickness at an electron beam energy of 15 keV.

4. Conclusions

A strong influence of the asbestos bundles and fibres thickness and shape on the detected EDS X-ray intensity was observed. In particular, the variation of the simulated X-ray intensity was significantly correlated with the mineral fibre thickness. The X-ray intensity trends as a function of thickness showed a non-linear dependence for all the elements and minerals. Generally speaking, the X-ray intensities showed a considerable reduction for thicknesses below about 0.5 μm at 5 keV, 2 μm at 15 keV, and 5 μm at 25 keV. Low voltages greatly reduce the excitation volume. Notably, an X-ray intensity enhancement was often observed before the reduction starting point, with a magnitude that depends on the mineral type, shape, specific element and beam energy. Monte Carlo simulation methods proved to be effective to face the thickness effects in SEM-EDS microanalysis with approaches that appear less complex and cumbersome than experimental ones.

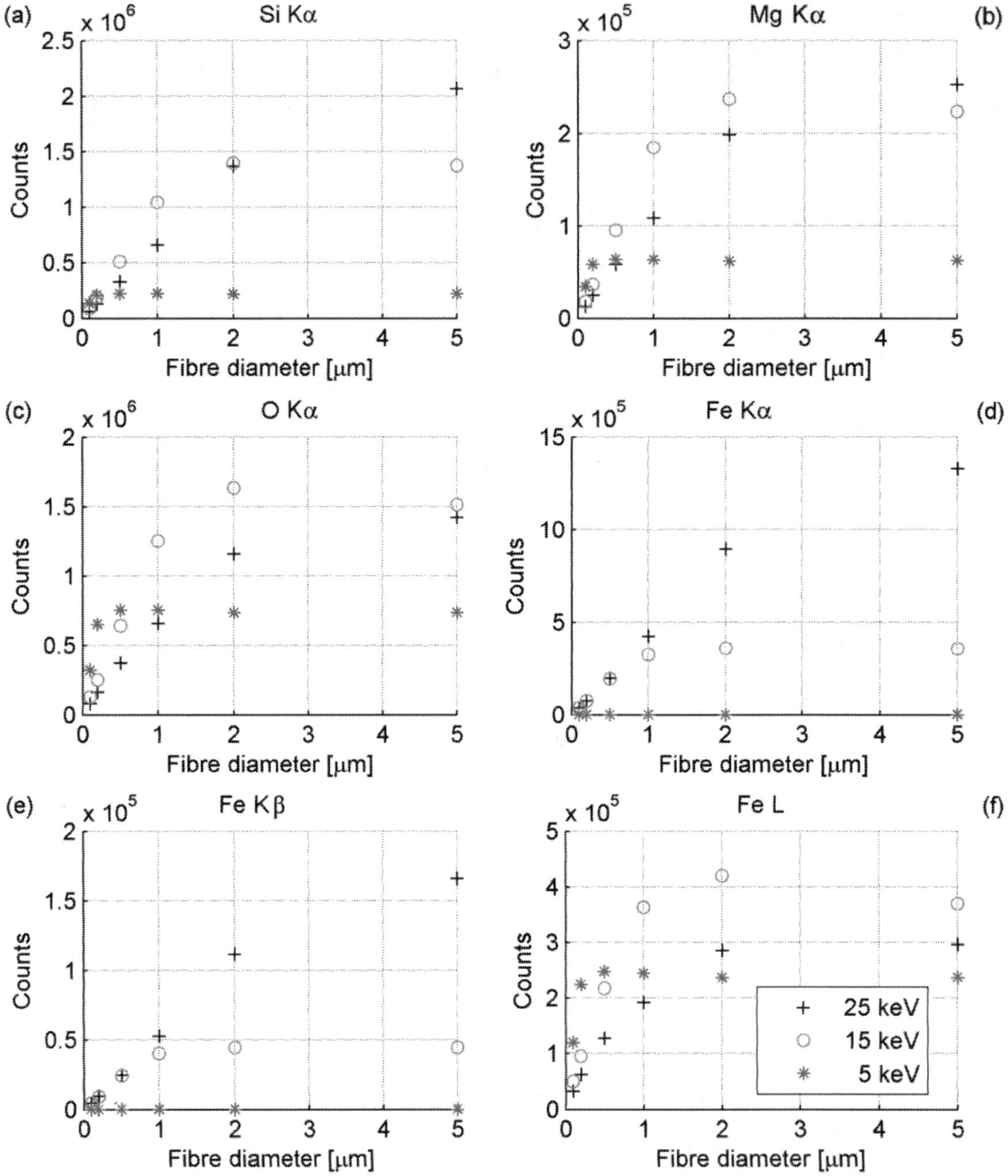

Figure 4. Simulated SEM-EDS X-ray intensity as a function of thickness for a single amosite cylindrical fibre with beam energy of 5 keV, 15 keV and 25 keV.

Table 4. *k*-ratios calculated at 15 keV for cylindrical fibres of different thickness.

Mineral	Thickness	X-ray lines						
		$Na_{K\alpha}$	$Fe_{K\alpha}$	$Fe_{K\beta}$	Fe_L	$Mg_{K\alpha}$	$Si_{K\alpha}$	$O_{K\alpha}$
Chrysotile	2.0 μm	—	—	—	—	1.04	1.03	1.18
	1.0 μm	—	—	—	—	0.58	0.59	0.69
	0.5 μm	—	—	—	—	0.28	0.28	0.35
	0.1 μm	—	—	—	—	0.05	0.05	0.07
Crocidolite	2.0 μm	1.20	0.99	0.99	1.29	1.14	1.04	1.18
	1.0 μm	0.97	0.88	0.88	1.10	0.88	0.77	0.90
	0.5 μm	0.51	0.51	0.51	0.64	0.44	0.36	0.44
	0.1 μm	0.10	0.10	0.10	0.15	0.08	0.06	0.09
Amosite	2.0 μm	—	1.00	1.00	1.31	1.16	1.06	1.21
	1.0 μm	—	0.91	0.90	1.13	0.91	0.79	0.93
	0.5 μm	—	0.54	0.54	0.68	0.46	0.39	0.47
	0.1 μm	—	0.10	0.10	0.16	0.09	0.07	0.09

References

[1] Gunter M E, Belluso E and Mottana A 2007 Amphiboles: environmental and health concerns. in: *Amphiboles: Crystal Chemistry, Occurrence, and Health Issues* (*Reviews in Mineralogy & Geochemistry* vol 67) (Hawthorne F C, Oberti R, Della Ventura G and Mottana A; Eds.) (Chantilly, VA: Mineralogical Society of America) chapter 12 pp 453-516
[2] Berstein D M and Hoskin J A 2006 *Regul. Toxicol. Pharmacol.* **45** 252
[3] Berstein D M 2014 Serpentine and amphibole asbestos. in: *Inhalation Toxicology* (3rd edition) (Salem H and Katz S A; Eds.) (Boca Raton, FL: CRC Press) 295-326
[4] Hardy J A and Aust A E 1995 *Chem. Rev.* **95** 97
[5] Gazzano E, Turci F, Foresti E, Putzu M G, Aldieri E, Silvagno F, Lesci I G, Tomatis M, Riganti C, Romano C, Fubini B, Roveri N and Ghigo D 2007 *Chem. Res. Toxicol.* **20** 380
[6] Yao S, Della Ventura G and Petibois C 2010 *Anal. Bioanal. Chem.* **397** 2079
[7] Bocchi G and Valdrè G 1993 *J. Inorg. Biochem.* **49** 209
[8] Borgia G C, Brown R J S, Fantazzini P, Mesini E and Valdrè G 1992 *Nuovo Cimento* **14** 745
[9] Gatti A M, Valdrè G and Tombesi A 1996 *J. Biomed. Mater. Res.* **31** 475
[10] Small J A 2002 *J. Res. Nat. Inst. Stand. Technol.* **107** 555
[11] Paoletti L, Bruni B M, Gianfagna A, Mazziotti-Tagliani S and Pacella A 2011 *Microsc. Microanal.* **17** 710
[12] Paoletti L, Bruni B M, Arrizza L, Mazziotti-Tagliani S and Pacella A 2008 *Per. Mineral.* **77** 63
[13] Merlet C and Llovet X 2012 *IOP Conf. Series: Mater. Sci. Engng.* **32** 012016
[14] Merlet C and Llovet X 2011 *X-ray Spectrom.* **40** 47
[15] Salvat F, Llovet X, Fernández-Varea J M and Sempau J 2006 *Microchim. Acta* **155** 67
[16] Ritchie N W M 2005 *Surf. Interface Anal.* **37** 1006
[17] Skinner H C W 2003 *Indoor Built Environ.* **12** 385
[18] Wiewióra A 1990 *Clay Min.* **25** 93

[19] Hawthorne F C and Oberti R 2007 Amphiboles: crystal chemistry. in: *Amphiboles: Crystal Chemistry, Occurrence, and Health Issues.* (*Reviews in Mineralogy & Geochemistry* vol 67) (Hawthorne F C, Oberti R, Della Ventura G and Mottana A; Eds.) (Chantilly, VA: Mineralogical Society of America) chapter 1 pp 1-54

[20] Hawthorne F C and Oberti R 2007 Classification of the amphiboles. in: *Amphiboles: Crystal Chemistry, Occurrence, and Health Issues* (*Reviews in Mineralogy & Geochemistry* vol. 67) (Hawthorne F C, Oberti R, Della Ventura G and Mottana A; Eds.) (Chantilly, VA: Mineralogical Society of America) chapter 2 pp 55-88

[21] Hawthorne F C, Oberti R, Harlow G E, Maresch W V, Martin R F, Schumacher J C and Welch M D 2012 *Am. Mineralogist* **97** 2031

[22] Czyzewski Z, MacCalium D O'N, Romig A and Joy D C 1990 *J. Appl. Phys.* **68** 3066

[23] Jablonski A, Salvat F and Powell C J 2010 *NIST electron elastic-scattering cross-section database.* (Gaithersburg, MD: National Institute of Standards and Technology)

[24] Joy D C and Luo S 1989 *Scanning* **11** 176

[25] Bethe H A and Ashkin J 1953 Passage of radiations through matter. in: *Experimental Nuclear Physics* vol 1. (Segré E; Ed.) (New York, NY: Wiley)

[26] Casnati E, Tartari A and Baraldi C 1982 *J. Phys. B: At. Mol. Opt. Phys.* **15** 155

[27] Heinrich K F J 1986 in: *Proc. 11th Int. Congr. of X-ray Optics and Microanalysis (London, Canada)* (Brown J D and Packwood R H; Eds.) (Ontario: University of Western Ontario) p 67

[28] Bambynek W *et al.* 1972 *Rev. Mod. Phys.* **44** 716

[29] Ritchie N W M 2010 *Microsc. Microanal.* **16** 248

IOP Conf. Series: Materials Science and Engineering **109** (2016) 012012 doi:10.1088/1757-899X/109/1/012012

Effects of cold rolling deformation on microstructure in 18/8 grade stainless steel

A Núñez Galindo and J F Almagro bello

ACERINOX EUROPA SAU, Technical Department, Pol. Industrial Palmones s/n, 11379 Los Barrios, Spain

E-mail: andres.nunez@acerinox.com

Abstract. 18-8 stainless steel is one of the most versatile materials. Some of its uses are related to structural applications; in this case, the high mechanical properties required are obtained by controlled lamination after re-crystallisation annealing. In this work, detailed analyses by FEG-SEM and EBSD have been done to correlate the mechanical properties, the amount of martensite and the structural misorientation generated. A linear relation between yield strength and local misorientation has been obtained.

1. Introduction

AISI 304 (EN 1.4301) is the most versatile and widely used stainless steel. It is characterized by an austenitic fcc crystallographic structure with a nominal composition of 18 wt% of chromium and 8 wt% of nickel as alloying elements. This grade is used for structural applications where a stainless steel is required and high strength is needed, which is obtained by cold rolling after an intermediate annealing. In addition, Type 304 stainless steel is an austenitic grade that can be severely deep drawn. This property has resulted in AISI 304 being the dominant grade used in architectural applications.

The deformation-induced martensite transformations have been studied for many years and it is known that the austenite structure or phase is unstable during cold deformation and breaks down to the much stronger, less ductile martensite phase. So, the stainless steel SS304 is a metastable austenite material and its crystallographic structures can transform to martensite phase under plastic deformation. The presence of martensite phase produces significantly different properties compared to the austenitic steel, both in plastic deformation and failure. The phase transformation changes and disorders the microstructure of grains, resulting in resistance to dislocation motions. The material with martensite phase has higher strength and less ductility [1-2].

In this context, it is necessary to improve the basic knowledge about cold rolling for fcc structure, as well as the effect on microstructure. This required the use of techniques such as electron channelling contrast imaging, ECCI (backscattered electron Mott detection [3, 4]), which allows the analysis of variations in sample structures, and electron backscatter diffraction (EBSD) to study textures and secondary phases [5].

In this study, cold rolled AISI 304 (EN 1.4301) steel samples, subjected to different degrees of deformation controlled by thickness reduction. An undeformed sample, with final conventional annealing treatment, was used as a reference. The resulting samples were first analysed by using conventional techniques available in stainless steel work in order to determine bulk composition, mechanical properties, structure and grain size. In a second step, textures and secondary phases were characterized by EBSD [6-9], and crystallographic orientation and defects were studied by ECCI

IOP Conf. Series: Materials Science and Engineering **109** (2016) 012012 doi:10.1088/1757-899X/109/1/012012

(backscattered electron Mott detection [10, 11]). The correlation between strain level obtained, mechanical properties, structure and EBSD results are discussed.

2. Experimental procedure

In this study, EN 1.4301 steel samples of thickness 0.70 - 0.75 mm, taken from the daily production of ACERINOX EUROPA SAU (www.acerinox.es), with the same chemistry, were submitted to different degrees of deformation as determined by thickness. The rolling reduction was 22.0 % (R1) and 37.5 % (R2). As a reference, another sample without deformation and with final conventional annealing treatment is taken. The summary of the samples is listed in table 1.

Table 1. Summary of the samples.

Sample Id	Final thickness (mm)	Rolling reduction (%)	Content of martensite (%)	Percentage of not indexed EBSD patterns
A	0.70	0	0.04	1.1
R1	0.70	22.0	2.4	16.1
R2	0.75	37.5	4.3	26.7

The chemical composition of the samples was analysed depending on the element, specifically Si, P, Ti, Cr, Mn, Co, Ni, Cu, Nb, Mo and Sn were analysed by X-ray fluorescence (XRF) using a PANALYTICAL PW 2606 XRF spectrometer (www.panalytical.com); finally, C and S, and N were analysed using LECO CS 600 and TC 600 analysers, respectively (www.leco.com).

The mechanical tests were carried out using different equipment, depending on the parameters to be measured. For the study of high-tensile strength, yield strength, elongation, coefficient of plastic anisotropy and hardening coefficient, a RK100 ROELL+KORTHAUS Universal Testing Machine was used; to obtain the Erichssen, KWI and LDR parameters, a ZWICK ROELL BUP600 drawing machine (www.zwick.com) was used.

Volume fraction of ferromagnetic phase along the surface was measured with a Feritscope Fischerscope MMS (www.helmut-fischer.es). The measured values were converted to martensite contents with a correlation factor of 1.7 defined by Papula *et al.* [12]; the results are shown in table 1.

Grain size analysis was performed after polishing and etching the samples (Vilella's reagent), for 60 to 75 s. The observation and measurement of grain size were performed with an OLYMPUS GX71 optical microscope equipped with an "AnalySIS Inclusion Inspector" image analysis system (www.olympus.co.uk).

The characterisation by EBSD and BSE Mott detectors was performed on a ZEISS ULTRA 55 FEG-SEM (www.zeiss.es), using polished longitudinal sections of the samples. The sample sections were prepared by conventional metallographic procedures, with a final polish with colloidal silica suspension, OPS (www.struers.com). The FEG-SEM used is equipped with an HKL CHANNEL 5 EBSD system, from Oxford Instruments (www.oxinst.com). Backscattered electron (BSE) images were obtained at an accelerating voltage of 20 kV and a working distance of 1-2 mm; EBSD maps were acquired at 20 kV and a working distance of 10 mm.

3. Results and discussion

3.1. Composition, structure and mechanical properties

The materials selected for this study are summarized in Table 2, together with their global chemical composition and type of annealing treatment applied. The composition of the samples is typical for this grade, with lower contents of interstitial elements such as C and N.

Table 2. Composition, annealing treatment and metallurgical properties of the materials.

Sample Id	Chemical composition (wt%)*	Treatment	Mechanical tests			Grain size	
			Rm (Mpa)	Rp (Mpa)	A (%)	OM (ASTM)	EBSD (ASTM)
A	C: 0.049; N: 0.051; Si: 0.35; Cr: 17.7; Ni: 8.0; Mo: 0.30; Cu: 0.32	Annealing	638	274	59	8.0	9.9
R1	C: 0.046; N: 0.048; Si: 0.33; Cr: 17.7; Ni: 8.1; Mo: 0.26; Cu: 0.32	Rolling	1,004	857	14	8.7	15.3
R2	C: 0.049; N: 0.050; Si: 0.37; Cr: 17.7; Ni: 8.1; Mo: 0.31; Cu: 0.37	Rolling	1,108	929	11	8.7	15.3

*Other elements in their residual conventional levels.

Optical microscope images showing the structure of samples A, R1 and R2 are displayed in figures 1, 2 and 3 respectively. The structure of material A shows complete re-crystallisation, with pits at grain boundaries (distribution of carbides within the grains well differentiated); material R1 can be seen to show a partial re-crystallisation with texture marked and distortion preserved; and in material R2 the texture is very marked and there is more evident deformation.

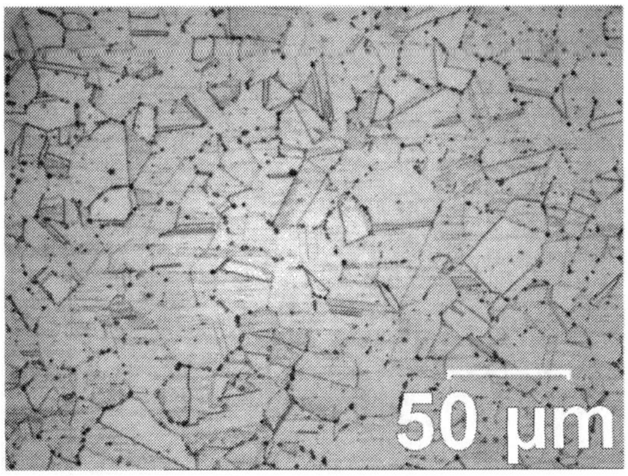

Figure 1. Optical microscope image showing the structure of material A.

Figure 2. Optical microscope image showing the structure of material R1.

The results of the mechanical tests performed on the A, R1 and R2 samples are listed in table 2, which include the tensile strength, Rm, the yield strength, $Rp_{0.2}$ [13], and the elongation test, (A). The results obtained for parameters Rm and $Rp_{0.2}$ indicate that these properties are directly proportional with rolling reduction. In addition, as it is known, the A parameter, which gives valuable information on plastic deformation, is correlated with rolling reduction (inversely proportional) in these materials.

Figure 3. Optical microscope image showing the structure of material R2.

The results of grain size measurements obtained by conventional optical microscopy (OM) and EBSD are listed in table 2 [14, 15]. EBSD measurements respond only to variations in the relative misorientation between adjacent areas, and they are not affected by the action of chemical reagents as in conventional analysis. Grain sizes obtained by OM and EBSD are correlated [16] but not closely agree with each other.

3.2. High angle BSE information

Backscattered electron Mott detection provides information on crystallographic orientation and defects in the material under study [17]. ECCI images are usually obtained by tilting the sample, which increases the BSE yield and enhances electron channelling. High-angle BSE information is used to locate areas of tension, dislocations, or other crystalline defects [18, 19].

Here it should be noted that the high accelerating voltage (20 kV) and the short working distance (1 - 2 mm) used allow the observation of fine structures in grain boundaries, which largely coincide with the dislocations that can be observed by transmission electron microscopy (TEM). In these conditions, tilting the sample is not required to get the ECCI effect.

Figures 4 and 5 show general BSE micrographs corresponding to the reference sample, A. The contrast observed is the result of the different crystallographic orientation of the grains with respect to the impinging electron beam (different zone axis). For all the samples, a crystalline structure of equiaxed grains randomly oriented can be observed. Twins are clearly observed.

Figure 4. ECCI (backscattered electron Mott detection) of material A.

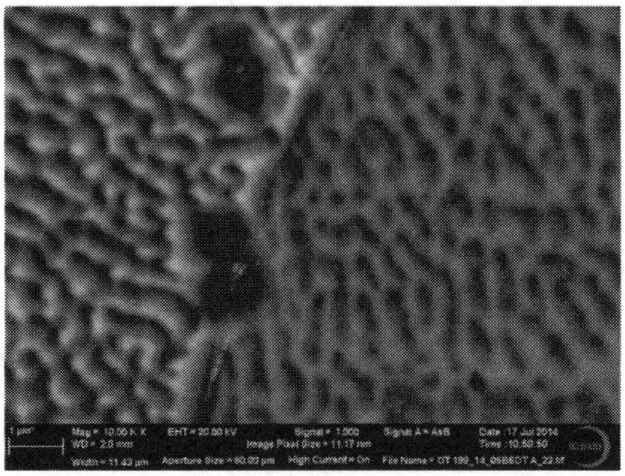

Figure 5. Detail image of material A.

Figures 6 and 7 show general and detailed BSE micrographs corresponding to the rolling sample R1. The grains present changing contrast due to stress associated with the presence of dislocation networks by accumulated tension in grain boundaries, which present complex structures of shear and slip bands [20-23].

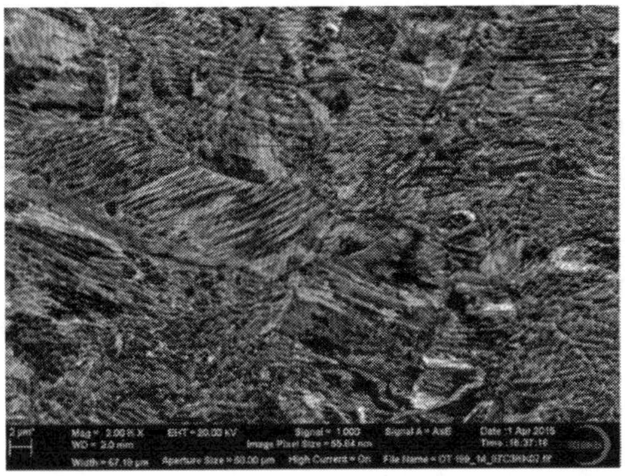

Figure 6. ECCI (backscattered electron Mott detection) of material R1.

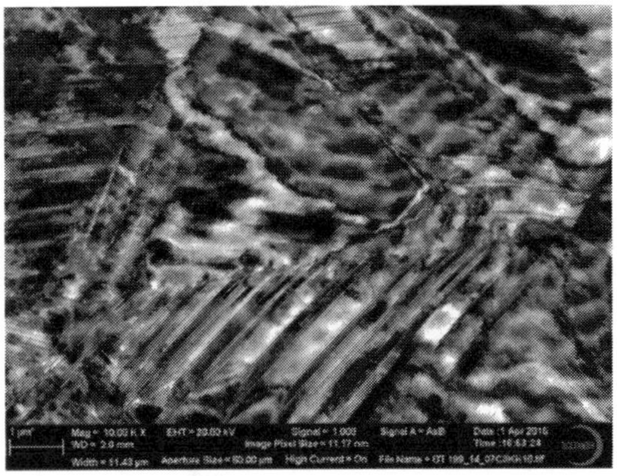

Figure 7. Detail image of material R1.

Figures 8 and 9 show general and detailed BSE micrographs corresponding to the rolling sample R2. In addition to the grain stress contrast, shear bands and other line defects can be observed. These line defects have resulted in dislocation networks defining sub-grains. In comparison to the reference sample, twin content has been considerably reduced, probably by deformation twin [24].

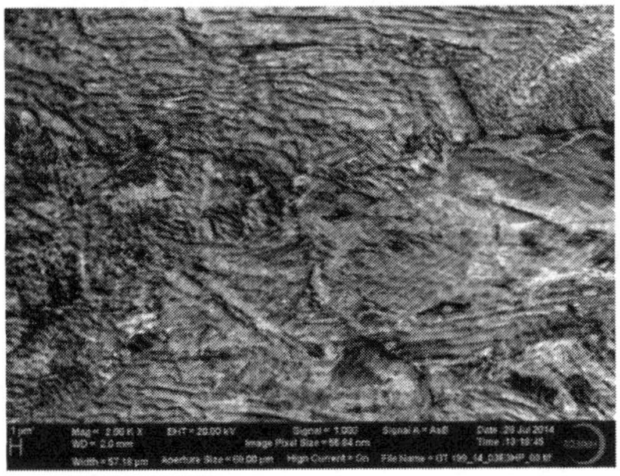

Figure 8. ECCI (backscattered electron Mott detection) of material R2.

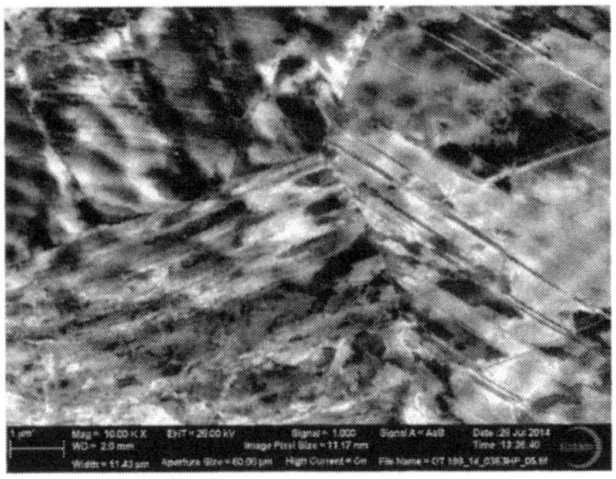

Figure 9. Detail image of material R2.

Therefore, the density of crystalline defects increases with plastic deformation in the grains, especially dislocations. This results in a complex sub-grain structure, which is generated in response to the deformation.

IOP Conf. Series: Materials Science and Engineering **109** (2016) 012012 doi:10.1088/1757-899X/109/1/012012

3.3. EBSD

EBSD analysis allowed obtaining information of the whole sample thickness for the main phase, as the observation plane was ND-LD (ND and LD stands for normal direction and line, or rolling, direction). The FEG-SEM characterisation was carried out using a polished sample in longitudinal section (ND-LD plane). After the analysis, a cleaning procedure was applied to the data, which consisted of removing EBSPs that could not be indexed and isolating points that were incorrectly indexed and appeared as Wild Spikes. The removed points were filled in using copies of their neighbouring points (i.e., if a given point had more than the minimum number of indexed neighbours (5-6), then it was replaced by the most common neighbouring orientation).

Figures 10 to 12 show grain, sub-grain, and coincidence site lattice (CSL) maps corresponding to the A, R1 and R2 materials. The grain and sub-grain maps are obtained from misorientation angles (sub-grain, 1° - 10°, in aqua and grain boundaries, > 10°, in black). The CSL maps are obtained from a specific coincidence of the crystal lattices across the boundary. This information is needed to define more accurately the grain size by EBSD (table 2) because CSL are not considered as grain boundaries but special boundaries (i.e., Σ3 is the well-known twin boundary in fcc materials). The lines with different colours correspond to CSLs with different Σ-values (Σ is the reciprocal number density of lattice points that are coincident in two adjacent grains), axis and angle.

Figure 10. Grain (black colour), sub-grain (aqua colour and CSL maps (CSL colour code: Σ3, red; Σ5, green; Σ7, blue; Σ9, pink; Σ11 yellow; Σ41c, light blue) of material A. BC+GB+CSL; Grid 1127 x 845 pixels; Step 0.5 µm.

Figure 11. Grain (black colour), sub-grain (aqua colour and CSL maps (CSL colour code: Σ3, red; Σ5, green; Σ7, blue; Σ9, pink; Σ11 yellow; Σ41c, light blue) of material R1. BC+GB+CSL; Grid 1127 x 845 pixels; Step 0.5 µm.

In the reference material, sample A (figure 10), areas with a high density of sub-grain boundaries are not observed (except due to polishing artefacts in the sample). However, in rolling materials, R1 and R2 (figures 11 and 12), these areas represent almost the entire analysed zone (aqua colour). In addition, these areas with high densities of sub-grains correlate with deformed grains but not with grain areas with preferred orientation.

Table 3 shows the results of the measurements of sub-grains and twin areas by EBSD in all the studied materials, which are the proportion of grain boundaries analysed that correspond to sub-grain and twin boundaries

Figure 12. Grain (black colour), sub-grain (aqua colour and CSL maps (CSL colour code: Σ3, red; Σ5, green; Σ7, blue; Σ9, pink; Σ11 yellow; Σ41c, light blue) of material R2. BC+GB+CSL; Grid 1127 x 845 pixels; Step 0.5 µm.

Table 3. Sub-grain and twin area.

Sample Id	Sub-grain area (%)	Twin area (%)
A	11.97	41.39
R1	85.91	2.89
R2	87.47	2.73

The sub-grain area increases from material A to R2, being much higher in the rolled samples than in the annealed one. The twin area decreases from material A to R2, being much higher in the annealed sample that in the rolled ones.

Figures 13 to 15 show maximum misorientation per grain, or strain contouring, for each material. This is a component that provides an estimate of the extent of deformation, or strain, in individual grains in an area. The component measures the maximum misorientation between any two points in a grain and then weights this grain according to this misorientation value. The maps show variations in the amount and colour (from green to red) of area with misorientation per grain, which increases from sample A to R2. Red areas, observed mainly in sample R2, indicate higher values of strain in the samples.

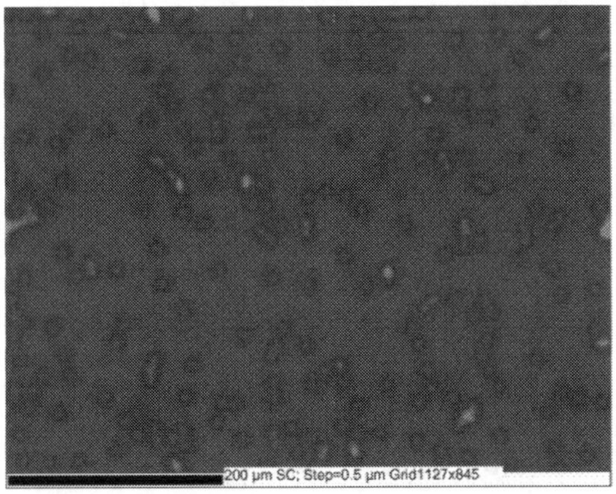

Figure 13. Strain contouring map of material A. SC; Grid 1127 x 845 pixels; Step 0.5 µm.

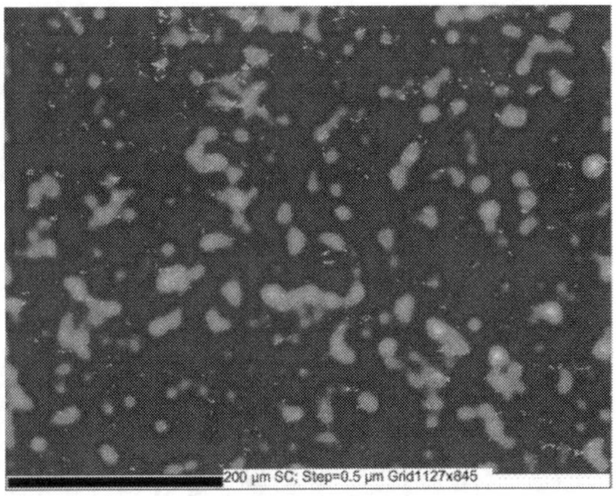

Figure 14. Strain contouring map of material R1. SC; Grid 1127 x 845 pixels; Step 0.5 µm.

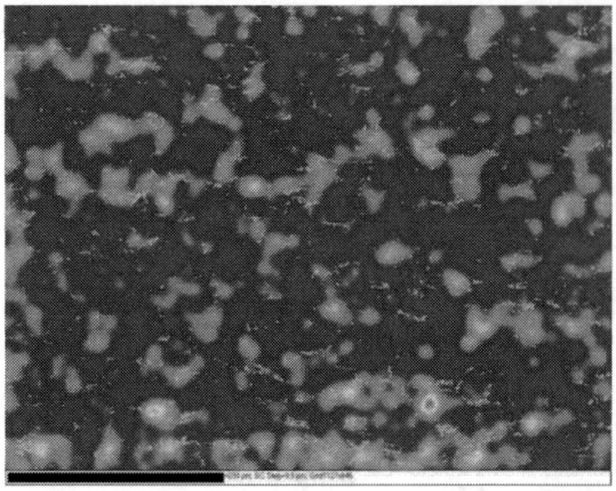

Figure 15. Strain contouring map of material R2. SC; Grid 1127 x 845 pixels; Step 0.5 μm.

Table 4 shows the results of the measurements of maximum misorientation per grain by EBSD in all the studied materials. The strain contouring increases from material A to R2, according to the higher deformation in the samples.

Table 4. Maximum misorientation per grain.

Sample Id	Max. misorientation (deg)
A	5
R1	11
R2	16

Figures 16 to 18 show local misorientation average angle histograms for each material. This component is for displaying small orientation changes in the map, highlighting regions of higher deformation. It calculates the average misorientation between each pixel and its surrounding pixels, and assigns the mean value to that pixel. Misorientations over a certain value are discarded, so that the misorientations associated with discrete sub-grain and grain boundaries are excluded.

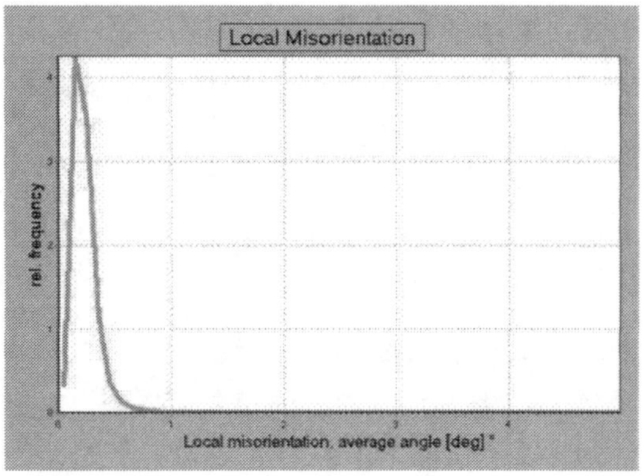

Figure 16. Local misorientation average angle histogram of material A.

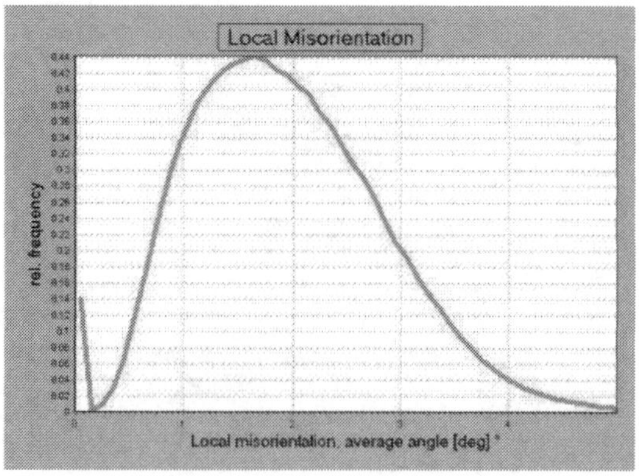

Figure 17. Local misorientation average angle histogram of material R1.

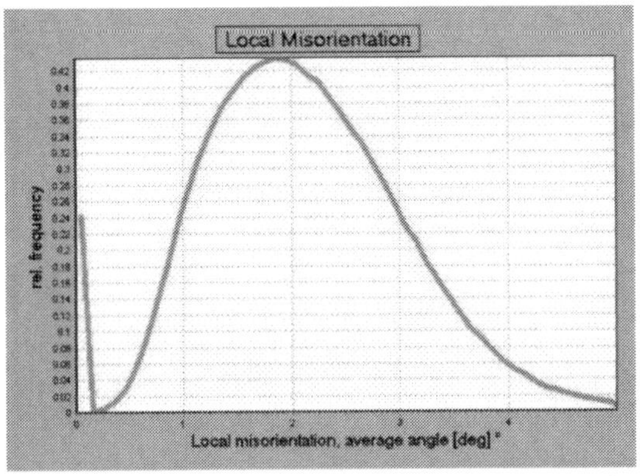

Figure 18. Local misorientation average angle histogram of material R2.

Table 5 shows the results of the measurements of local misorientation average angle by EBSD in all the studied materials. The local misorientation increases from material A to R2, corresponding to the higher deformation in the samples.

Table 5. Local misorientation.

Sample Id	**Local misorientation (deg)**
A	0.15
R1	1.65
R2	1.85

3.4. Correlations

Figures 19 to 22 show correlations between properties of each material studied by conventional techniques (mechanical tests and ferritoscope) and EBSD measurements. Figure 19 shows that the percentage of non-indexed EBSD patterns (see the data cleaning procedure outlined in 3.3.) and the content of martensite calculated for each material are directly proportional with rolling reduction. Also, the elongation calculated in each material is inversely proportional to rolling reduction. Figure 20 shows the correlation between rolling reduction and the mechanical properties tensile (Rm) and yield strength (Rp). Figure 21 illustrates that the content of martensite is directly proportional with the percentage of non-indexed EBSD patterns. Figure 22 demonstrates that the yield strength is directly proportional with the average local misorientation (as measured by EBSD).

4. Conclusions

The analyses performed by means of the EBSD technique as well as the results from conventional tests, have shown that AISI 304 austenitic stainless steel after different rolling reduction present an increasing degree of deformation. This is characterized by slip bands, deformation twins (parallel plates characteristic for martensite), strain hardening and the formation of martensite phase grains, which divide the elongated grains of austenite.

Plastic deformation during the rolling reduction of the samples induces a martensitic transformation ($\gamma \rightarrow \alpha'$) over the whole range of applied deformations.

The deformation structure is characterized by elongated austenite grains with α'-phase plates and by deformation twins.

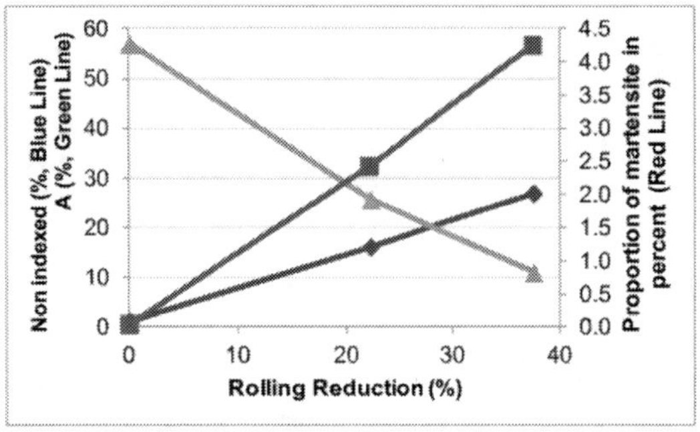

Figure 19. Correlation between rolling reduction % and percentage of non-indexed EBSD patterns, elongation (A) and proportion of martensite in percent.

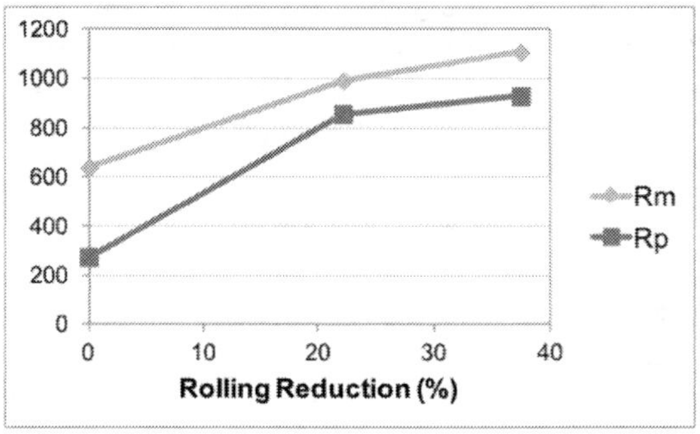

Figure 20. Correlation between rolling reduction % and tensile (Rm) and yield strength (Rp).

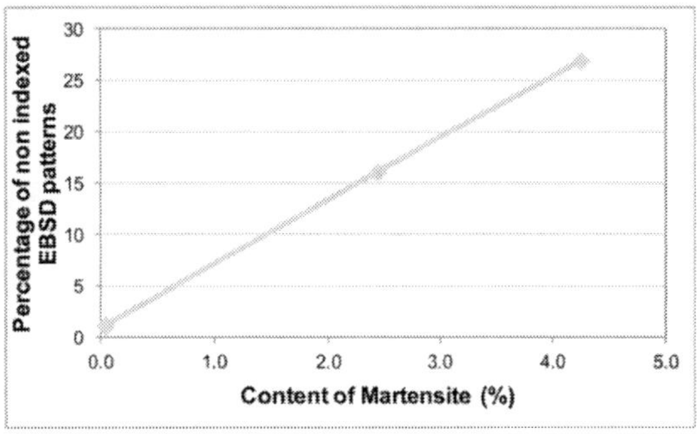

Figure 21. Correlation between proportion of martensite in percent and percentage of non-indexed EBSD patterns.

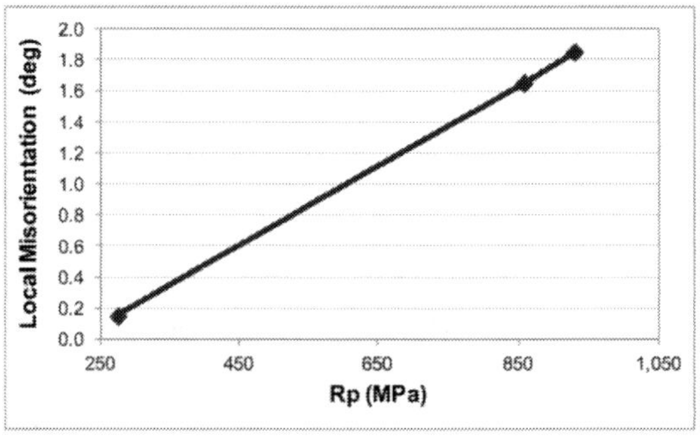

Figure 22. Correlation between yield strength (Rp) and average local misorientation.

Comparison between the materials subjected to rolling reduction and the reference sample, with conventional annealing and without deformation, show correlations between the strain level obtained, mechanical properties, structure and EBSD (linear relation between yield strength and local misorientation angle) as follows:

- There is a correlation of the rolling reduction with mechanical properties, amount of martensite (in percent) and percentage of non-indexed EBSD patterns. The increase in proportion of martensite in percent (A to R2 material) is linearly related to percentage of non-indexed EBSD patterns with a correlation very close to 1, so martensite proportionally increases with deformation.
- There is a correlation between the measurements by means of OM and SEM, both methods evidence very marked texture and lattice rotation in the rolling materials.
- By OM and SEM, the density of martensite needles increases with strain by plastic deformation, therefore, it is confirmed that the volume fraction of martensite formed is a function of the amount of deformation.
- By EBSD, the local misorientation angle presents a linear relationship with the yield strength (Rp). In addition, the strain contouring maps show that the maximum misorientation per grain increases from sample A to R2, while keeping constant the grain size in the samples R1 and R2.

References

[1] Technical article information 2015 British Stainless Steel Association (www.bssa.org.uk)
[2] Gang H and Huang Y 2013 in: *13th International Conference on Fracture.* (June 16-21) (Beijing, P.R. China)
[3] Jaksch H 2008 in: *Eur. Microscopy Congress 2008* **1** 553
[4] Bakaleinikov L, Berner A and Kalina M 2012 *Mika Group Meeting,* lecture 6
[5] Nuñez A, Llovet X and Almagro J F 2013 *Microsc. Microanal.* **19** 959
[6] Ferreira H, Santana S, Neto P, Pires R, Nogueira V, de Oliveira P and Souto S 2007 *Mater. Res.* **10** 359
[7] da Costa C S, Candido F S and Pinto A L 2005 *Mater. Sci. Forum* **495-497** 173
[8] Sato Y S, Nelson T W and Sterling C J 2005 *Acta Mater.* **53** 637
[9] Shimada M, Kokawa H, Wang Z J, Sato Y S and Karibe I 2002 *Acta Mater.* **50** 2331
[10] Trager-Cowan C, Sweeney F, Trimby P W, Day A P, Gholinia A, Schmidt N H, Parbrook P J, Wilkinson A J and Watson I M 2007 *Phys. Rev. B* **75** 085301
[11] Shimizu K and Mitani T 2010 *New Horizons of Applied Scanning Electron Microscopy.* (Berlin: Springer Verlag)
[12] Papula S, Talonen J and Hänninen H 2013 *The Minerals, Metals and Materials Society and ASM International.* (DOI: 10.1007/s11661-013-2090-3)
[13] *ASTM International* 2007 Designation: A 370-07a. V. 01.03, ISBN 978-0-8031-4284-8
[14] *ASTM International* 2011 Designation: E 112. V. 03.01, DOI 1520/E0112-10
[15] *International Standard. Measurement of average grain size.* ISO 13067, 2011 (E)
[16] Gao N, Wang S C, Ubhi H S and Starink M J 2005 *J. Mater. Sci. Lett.* **40** 4971
[17] Jaksch H, Steigerwald M and Drexel V 2005 *Microsc. Microanal.* **11** (Suppl. 2) 758
[18] Yasuda M, Kawata H and Murata K 1996 *Phys. Stat. Sol. (a)* **153** 133
[19] Karato S 1987 *Phys. Chem. Miner.* **14** 245
[20] Kamaladasa R J and Picard Y N 2010 in: *Microscopy: Science, Technology, Applications and Education.* (Méndez-Vilas A and Díaz J; Eds.) (Bajadoz, Spain: Formatex) 1583-1590
[21] Kuwano N, Itakura M, Nagatomo Y and Tachibana S 2010 *J. Electron Microsc.* **3** 1
[22] Epishin A, Link T and Nolze G 2007 *J. Microsc.* **228** 110
[23] Morikawa T and Higashida K 2010 *Mater. Trans.* **51** 620
[24] Abuzaid W, Sangid M, Carrol J, Sehitoglu H and Lambros J 2012 *J. Mech. Phys. Solids.* **60** 1201

IOP Conf. Series: Materials Science and Engineering **109** (2016) 012013 doi:10.1088/1757-899X/109/1/012013

Quantification of low concentration elements using soft X-rays at high spatial resolution

P T Pinard and S Richter

RWTH Aachen University, Gemeinschaftslabor für Elektronenmikroskopie (GFE), Ahornstrasse 55, 52074 Aachen, Germany

E-mail: pinard@gfe.rwth-aachen.de

Abstract. A review of the remaining challenges in electron probe microanalysis to quantify sub-micrometre features containing elements with a concentration less than 1 wt% is presented. These challenges derive from the equivocal influence of the main experimental parameters (accelerating voltage, beam current, measuring time and X-ray lines) on the achievable spatial resolution and the accuracy and precision of the quantification. This inter-dependence is demonstrated by examples from the literature and from new experimental results. From this survey, five key areas are identified as future research and development topics in order to solve for the discussed challenges and achieve the objective of quantification of low concentration elements using soft X-rays at high spatial resolution.

1. Introduction

This work elaborates on what could arguably be referred to as the two biggest remaining challenges of quantification using electron probe microanalysis (EPMA); as the title suggests, the quantification of elements with (1) a concentration less than 1 wt% at (2) a spatial resolution better than 1 µm. If taken individually, these challenges are not necessarily unheard-of and, for most cases, some solutions already exist for them. It is however more and more common to be confronted with measurements requiring, for instance, the quantification of small phases containing low concentration elements. This situation originates on the one hand from the increased complexity of modern materials and more advanced characterisation needs, but also on the other hand from the improved instrumental and computing capabilities.

Quantification in an electron microprobe or scanning electron microscope is the determination of the composition within a certain volume of a flat polished sample. A model is required to correct for matrix effects within this volume in order to transform measured intensities to concentrations. Up to now, most models assume that the composition within the analysed volume is homogeneous (ZAF) or layered ($\phi(\rho z)$). Although other models could be developed [1], the following argumentation will imply the use of conventional quantification models, or in other words that the quantified volume is equal to the X-ray excitation volume.

Quantification is influenced by several experimental parameters that can be roughly classified in three categories. First, the fixed parameters are defined by the instrumental setup, such as the type of electron emitter, the type of detector, the take-off angle and solid angle of the detectors, etc. As mentioned in the introduction, changes in these parameters can lead to important improvements of the quantification capabilities. For example, the introduction of field emission guns led to smaller probe size at low accelerating voltage or larger detectors, to improved sensitivity. For the subsequent discussion, these parameters will however be considered as constant as they cannot be routinely modified for a given instrument. The second category contains the parameters that are part of the calibration. The pulse height analyser for wavelength-dispersive X-ray spectrometers (WDS) or the energy calibration for energy-dispersive X-ray spectrometers (EDS) are two examples. The assumption that these parameters are properly adjusted will be implied in the following paragraphs. Finally, the third category includes the accelerating voltage, the beam current, the measuring time and the X-ray lines chosen for the analysis. The inter-dependence between the quantification of low concentration elements and the quantification at high spatial resolution results from the fact that variations of these experimental parameters will differently influence the accuracy and precision of the quantification.

These quantification challenges will be explained using examples from the literature and experimental results in the following sections. The inter-dependence will be demonstrated based on two different starting points, the quantification of low concentration elements (Section 2) and the quantification at high spatial resolution (Section 3).

2. Quantification of low concentration elements

Independently of the spatial resolution, the quantification of low concentration elements necessitates high counting statistics and proper determination of the background [2, 3]. High counting statistics translates into high beam currents and/or longer measuring times. In other words, the dose must be increased to increase the number of detected counts and therefore reduce the random uncertainties. Since the peak to background ratio will be close to unity for low concentration, counting statistics applies for both the peak and background evaluation. Furthermore, interferences of other X-ray lines, absorption edges near or below the peak as well as background curvatures have a large influence for the quantification of low concentration elements. Reference samples without the element of interest can be used to assess the effectiveness of the measuring strategy (blank test) [2].

In terms of spatial resolution, increasing the beam current has an adverse effect as the beam diameter increases with the beam current. In a previous work [4], experimental measurements of the beam diameter as a function of the beam current (figure 1) were performed for a field emission microprobe (JEOL JXA8530F).

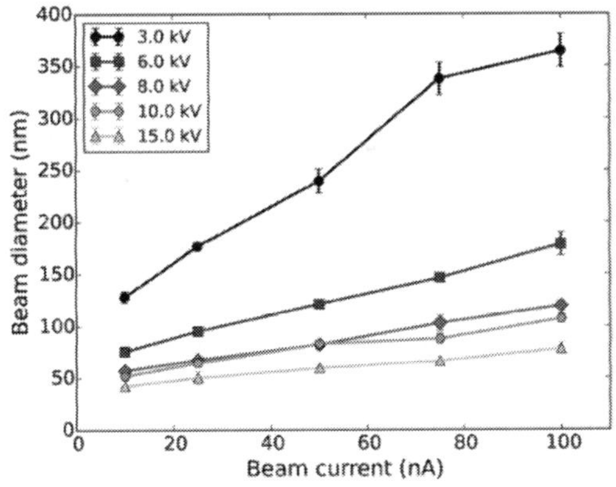

Figure 1. Variation of the beam diameter as a function of the beam current for five accelerating voltages (reproduced from [4]).

Based on these measurements, optimal parameters to measure carbide inclusions in a SAE52100 (100Cr6) steel were found using Monte Carlo simulations where the change in the precision as well as the depth and lateral intensity distributions of all X-ray lines were evaluated as a function of the accelerating voltage and beam current. The quantification of carbon in the carbides is a good example where the background determination is crucial due to the absorption edge and interferences of high order lines from Mn, Cr and Fe. Despite these inherent difficulties, the inclusions measured in [4] were relatively large (≈ 1 μm diameter) and the concentration of all elements was higher than 5 wt%. Finding optimal acquisition parameters would therefore be more challenging for smaller inclusions containing low concentration elements and would inevitably lead to longer measuring times. Nevertheless, the method described in [4] would still be applicable.

Unfortunately, longer measuring times are not only synonym to more costly measurements, but phenomena of contamination and beam damage become predominant. The formation of a carbonaceous deposit under and around the electron beam decreases the landing energy and increases the absorption path of the outgoing X-rays. Especially for soft X-rays, contamination can lead to significant artefacts in the quantification. Anti-contamination devices (e.g., plasma cleaning, cold trap, gas jet) can reduce the effects of contamination, but cannot completely eliminate the deposition of cracked hydrocarbons on the surface of the sample [5]. For high spatial resolution measurements where measurement locations are closely spaced, it is important to realize that the contamination leftover by previous neighbouring measurements has a large influence on the measurement of the following locations.

One way to evaluate this influence is to look at the precision of carbon measurements on steel reference samples. The precision on the carbon concentration can be evaluated using the standard deviation calculated from several measurement points. As the samples are homogeneous, the observed fluctuations originate from the combination of random uncertainties due to counting statistics (Poisson distributed) and systematic uncertainties, which strongly depends on the contamination behaviour. By keeping the counting time constant, similar random uncertainties are obtained and the influence of contamination can be observed from the variations in the systematic uncertainties [6]. This experiment was performed on pure iron and four low alloyed steel reference materials containing up to 1 wt% C. Line scans with a step size of 40 nm and 20 μm were run at 15 kV and 100 nA with a counting time of 5 s per point. An oxygen gas jet was used as an anti-contamination device throughout the measurements. Under these conditions a ring-shaped carbonaceous deposit is formed with a diameter of approximately 4 μm [5]. There is, therefore, a strong overlap between the deposits for the 40 nm line scan. Measured intensities were converted to concentrations using the quantification procedure described in [6]. Figure 2 shows the variations of the random and systematic uncertainties for the different samples and step sizes. The full height of each column corresponds to the total measured uncertainty. In all cases, the systematic uncertainties dominate the random uncertainties. This confirms the influence of contamination on the measurements of carbon. Fluctuations in the systematic uncertainties for both line scans indicate that the contamination varies from one measurement to another. This instability is higher for closely spaced measurements as illustrated by the stronger variations of the systematic uncertainties for the overlapping-spot line scan. Nevertheless, these measurements demonstrates that high resolution carbon measurements can be performed up to a detection limit of 0.05 wt% C in low alloyed steels despite the influence of contamination. Measuring lower carbon concentrations would require a better control of the contamination behaviour.

Apart from the anti-contamination devices, the influence of contamination on the measurements can be reduced by other methods. If the carbonaceous deposit is assumed to be a simple carbon film and there is no carbon in the sample of interest, a thin-film correction algorithm (e.g., STRATAGEM [7]) can be used to correct for the absorption in the carbonaceous film by measuring the carbon intensity [8]. When WDS are used, one method to minimize the measuring time is to skip the measurement of the background positions, either by using a calibration curve [9, 10], a pre-determined background [6] or a calculated background (e.g., mean atomic number [11]). For all these strategies, interferences and changes of the background curvature due to total reflection [12] must be carefully considered.

Beam damage comes in different flavours: ablation of the conductive coating layer of non-conductive specimens [13], ion migration [2, 14], volatilisation, sample degradation, etc. Solutions have been proposed to minimize the amount of beam damage: replacing the conductive carbon coating

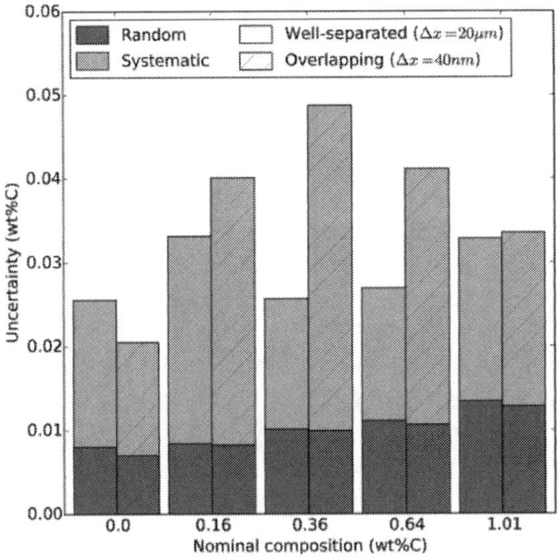

Figure 2. Random and systematic uncertainties on carbon concentrations for line scans performed with two different step sizes.

with metallic coatings [14-16], using lower beam currents [17], using higher accelerating voltages [14], or cooling the sample using a cryo-stage [18]. All except perhaps the last solution are detrimental to the objective of quantification of low concentration elements at high spatial resolution. Metallic coatings and lower beam currents worsen the detection limits, whereas higher accelerating voltages increase the emission volume.

One last consideration regarding the quantification of low concentration elements is the phenomenon of fluorescence. The emission of fluorescent X-rays far from the electron beam essentially expands the X-ray emission volume and thus worsens the spatial resolution. Although the fluorescence probability is low in most cases, this effect becomes less negligible when quantifying low concentration elements [19, 20]. Lowering the accelerating voltage can only provide a small improvement [21]. Correction based on Monte Carlo simulations [19, 20] or semi-analytical models [22] is a promising solution to improve the spatial resolution.

3. Quantification at high spatial resolution

Approaching the initial challenges from the perspective of the high spatial resolution, the most influential parameter is the accelerating voltage. Lowering the accelerating voltage has the combined effects of decreasing the electron interaction volume and additionally, due to the lower overvoltage, the X-ray emission volume. As for the beam current, the beam diameter is not constant as the accelerating voltage is decreased. The experimental measurements from [4] show a large increase of the beam diameter below 6 kV (figure 3). Practically this puts a limit on the spatial resolution improvement achievable by the current generation of field emission microprobes. Using Monte Carlo simulations, the authors have estimated that there is no advantage to use accelerating voltages below 5 kV in most cases.

Inevitably at lower accelerating voltages X-ray lines with lower excitation energy must be used to quantify the elements inside a sample. This has several implications, which also apply to all soft X-rays, including the Kα (K-L3) lines of the ultra-light elements (e.g., B, C, N and O). First X-rays are produced from transitions involving electrons taking part in chemical bonding [23-25]. This effect was extensively studied by [25] for borides, carbides, nitrides and oxides. Using standards with similar chemistry as the unknown is the typical solution, since chemical effects are not considered by any model used in the quantification procedure. The second implication is that low energy X-rays are more affected by surface effects such as oxidation and contamination due to increased absorption

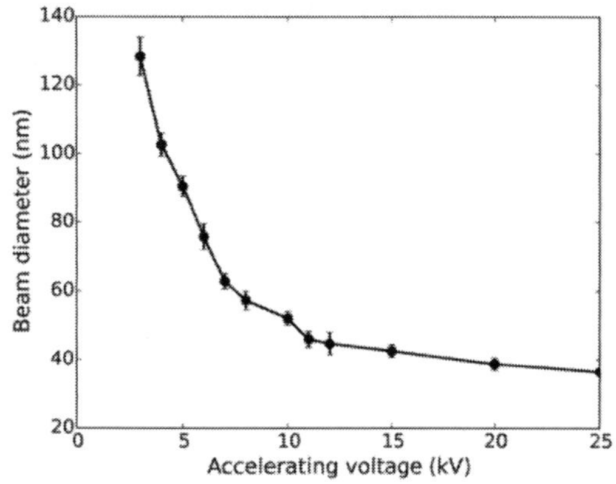

Figure 3. Variation of the beam diameter as a function of the beam energy for a 10 nA beam current (reproduction from [4]).

[5, 25, 26]. Although these effects can be minimized by proper sample preparation and anti-contamination devices, they can never be completely avoided. Finally, inaccuracies in the matrix correction models lead to errors in the quantification. Large errors were observed using the Lα (L3-M5) lines for alloyed steels [27–29], iron silicides [13], and Ni-base metallic glass [30]. Two examples are given in figure 4 where the quantification of alloyed steels (Acerinox, S.A.) using the Lα-line results in relative errors up to 32 % at an accelerating voltage of 5 kV.

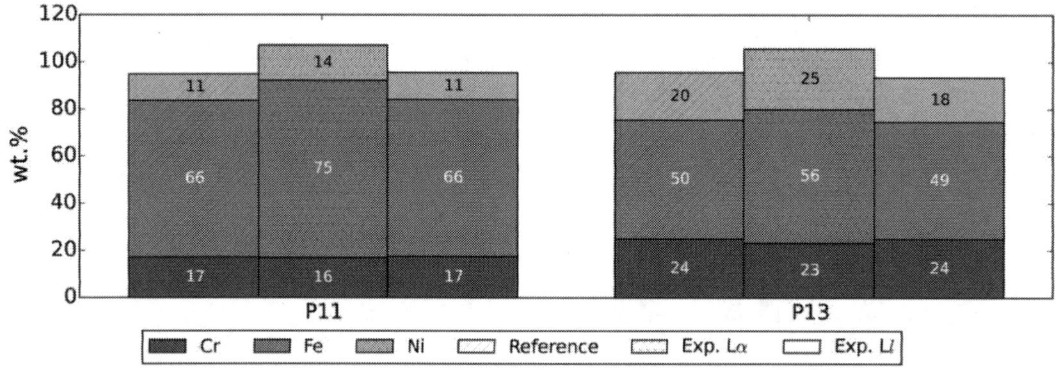

Figure 4. Quantification of two alloyed steels at an accelerating voltage of 5 kV using the Lα- and Ll-lines. Cr and Fe were measured on a LDE1 crystal, whereas a TAP crystal was used for Ni. Pure standards, DTSA-II [31] and mass absorption coefficients from Chantler *et al.* [32] were used for the quantification.

It should however be pointed out the anomalies related the Lα-lines seem to only influence the quantification of first-row transition metals. Measurements on binary alloys of Au-Cu and Pt-Rh at 5 kV gave accurate results even if the Mα-lines were used for Au and Pt and Lα-lines for Cu and Rh (figure 5).

One solution to the Lα-quantification problem is to use the Ll (L3-M1) X-ray line. Authors [13, 28, 29] showed satisfactory results for the major elements using this line. This improvement was attributed to the absence of absorption edges near the X-ray line and valence electrons involved in the X-ray transition [13, 28]. The improved quantification results for the major elements (Cr, Fe and Ni) can be seen in figure 4.

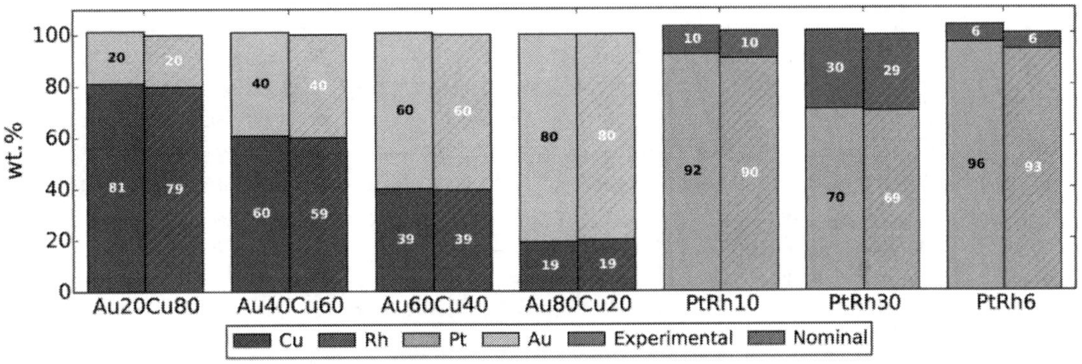

Figure 5. Quantification of 7 binary alloys at 5 kV using the Cu-Lα, Rh-Lα, Pt-Mα and Au-Mα. Measurements on Au-Cu alloys were performed at 300 nA using TAP (Cu) and PETJ (Au) crystals. Measurements on Pt-Rh were performed at 250 nA using PETJ crystals (Pt and Rh). Nominal concentrations for Au-Cu were taken from the NIST SRM482 whereas for Pt-Rh concentrations were measured at 15 kV. Pure standards, DTSA-II [31] and mass absorption coefficients from Chantler *et al.* [32] were used for the quantification.

Using soft X-ray lines has also an impact on the quantification of low concentration elements, and unfortunately an adverse impact affecting both the needs for high counting statistics and proper background determination. First, the emission probability of soft X-ray lines is normally smaller than the one of high energy lines, due to lower fluorescence yield, lower ionisation probability and higher absorption. To illustrate this point, intensities for different X-ray lines of Ti, Fe and Au from 3 to 30 kV were calculated using Monte Carlo simulations (DTSA-II/NISTMONTE [33]). The results are plotted in figure 6 as the factors required obtaining the same precision as the one at 20 kV using the conventional X-ray line (Kα for Ti and Fe, Lα for Au). The measuring time or beam current would need to be multiplied by these factors to get the same precision. At 5 kV, this would be equivalent to 54 times for Ti-Lα, 21 for Ti-Ll, 7 for Fe-Lα, and 20 for Fe-Ll, whereas the same precision can be achieved by the Au Mα than the Au Lα at 20 kV. For Ti and Fe, it then becomes obvious that low concentration quantification can only be done using the most intense low energy X-ray line, in occurrence the Ti-Ll and Fe-Lα.

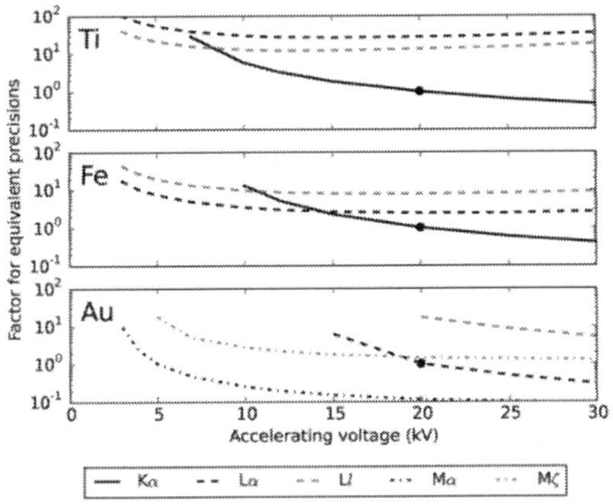

Figure 6. Variation of the equivalent precision factor as a function of the accelerating voltages for different X-ray lines of Ti, Fe and Au. The factors are defined as the multiplication factor in order to obtain the same precision than Ti-Kα, Fe-Kα and Au-Lα at 20 kV. Intensities used in the calculations are obtained by Monte Carlo simulations.

Secondly, the use of soft X-ray lines complicates the determination of the background. For WDS, synthetic layered dispersive crystals become necessary for X-ray lines with an energy lower than 0.7 keV (≈ Fe-Lα). The commercially available crystals are typically optimized for high reflectivity (high intensity) but at the expense of the spectral resolution. Combined with the fact that the energy separation between low energy X-ray lines is significantly reduced, overlapping and interference problems between adjacent X-ray lines are therefore common. Figure 7 shows an overlay of five WD spectra acquired on pure standards of Cr, Mn, Fe, Co and Ni on a LDE1 crystal. These elements which are very common in different steel grades would all require interference corrections. The latter would be substantial for the Ll-lines as the interfering lines (mainly Lα-lines) have a higher intensity than the interfered lines. Pinard *et al.* [29] obtained large errors (3 to 4 wt%) on the Mn-concentration in high Cr containing steels using the Mn Ll-line.

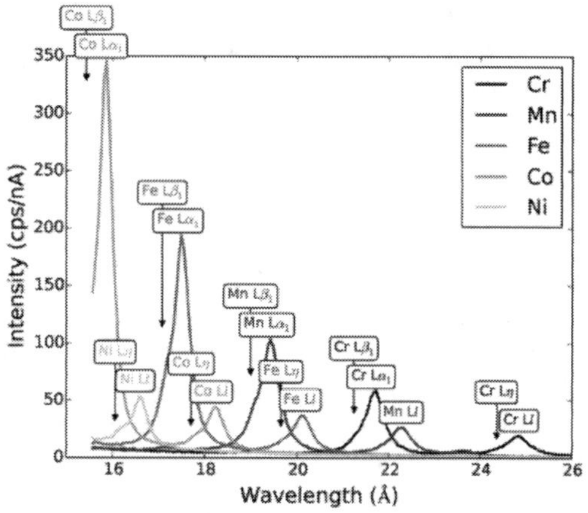

Figure 7. Five WD spectra acquired on pure standards of Cr, Mn, Fe, Co and Ni at 5 kV, 50 nA using a LDE1 crystal.

This interference problem is even more challenging for EDS, although this technique has the advantage that deconvolution is performed on the whole spectrum [34]. In the case of the first- row transition metals, it is however unlikely, due to the energy resolution, that the intensity of Ll-lines could be correctly extracted [28]. In most cases, the *k*-ratios obtained from EDS spectra are constructed using all lines within a family.

For completeness, another approach to reduce the emission volume and therefore increase the spatial resolution is to adjust the accelerating voltage in order to have a small overvoltage ratio for each X-ray line measured [35, 36]. In principal, this strategy circumvents many of the previously mentioned problems with the low energy X-ray lines as conventional, well-established X-ray lines can be used. Limitations for low concentration quantification still exist since the emitted intensity decreases with lower overvoltage, thus worsening the counting statistics. As shown by [36], instrumental stability and other experimental considerations also complicate the practicality of this approach.

4. Conclusion

The quantification of elements with a concentration less than 1 wt% at a spatial resolution better than 1 µm pushes the frontiers of EPMA. Review from the literature and new experiments revealed limited improvements to this objective when the accelerating voltage, beam current, measuring time and/or X-ray lines of analysis are changed. Each parameter will either have a positive effect on the quantification of low concentration elements or on the spatial resolution, but never for both. Five problems or areas of future research were identified to overcome these shortcomings: (1) better focussing optics to reduce the broadening of the beam diameter at high beam currents and low accelerating voltages, better understanding and minimisation of the (2) contamination and (3) beam damage phenomena, (4) higher spectral resolution crystals with good reflectivity, less total reflection for low energy X-ray lines and low intensity of high order X-ray lines, and (5) better matrix correction models and/or fundamental physical parameters to quantify using soft X-rays.

To leave the reader on a hopeful note, solutions to the presented challenges are under development. To sidestep the quantification problems of soft X-rays, standards with similar composition as the unknown could be used, if such reference samples exist. Moving away from soft X-rays, quantification at high spatial resolution using conventional X-ray lines is possible if the sample is a thin lamella [37] or if multiple accelerating voltages are used to reduce the overvoltage [36]. Another idea is to eliminate the constraint for chemical homogeneity inside the X-ray excitation volume by modelling X-ray emission in arbitrary geometries and iteratively reconstructing the composition of each constituent [1]. Needless to say all these solutions have advantages and disadvantages as well as their own challenges.

Acknowledgments
The authors would like to acknowledge the contribution of several colleagues to this work: Anke Aretz (RWTH Aachen University, Aachen, Germany), Oliver Schön (Aixtron, Herzogenrath, Germany), John Villarrubia (NIST, Gaithersburg, USA), Xavier Llovet (Universitat de Barcelona, Barcelona, Spain), and Erkki Heikinheimo (Aalto University, Espoo, Finland).

References
[1] Richter S, Pinard P, Mevenkamp N and Torrilhon M 2013 *Microsc. Microanal.* **19** (Suppl. S2) 1296-1297
[2] Jercinovic M J, Williams M L, Allaz J and Donovan J J 2012 *IOP Conf. Ser.: Mater. Sci. Engng.* **32** 012012
[3] Donovan J J, Lowers H A and Rusk B G 2011 *Am. Mineral.* **96** 274-282
[4] Pinard P T and Richter S 2014 *IOP Conf. Ser.: Mater. Sci. Engng.* **55** 012016
[5] Pinard P, Aretz A, Börder J, Donovan J and Richter S 2013 *Microsc. Microanal.* **19** (Suppl. S2) 1242-1243
[6] Pinard P T, Schwedt A, Ramazani A, Prahl U and Richter S 2013 *Microsc. Microanal.* **19** 996-1006
[7] Pouchou J L 1993 *Anal. Chim. Acta* **283** 81-97
[8] Buse B and Kearns S 2014 *Microsc. Microanal.* **20** (Suppl. S3) 704-705
[9] Robaut F, Crisci A, Durand-Charre M and Jouanne D 2006 *Microsc. Microanal.* **12** 331-334
[10] Technical Committee 202 2006 ISO 16592:2006(e) *Microbeam analysis - electron probe microanalysis - guidelines for determining the carbon content of steels using a calibration curve method.*
[11] Donovan J J and Tingle T N 1996 *J. Microsc. Soc. Am.* **2** 1-7
[12] Rehbach W and Karduck P 1992 in: *Electron Microbeam Analysis.* (Boekestein A and Pavicevic M K; Eds.) *Mikrochim. Acta* **Suppl. 12** 153-160
[13] Gopon P, Fournelle J, Sobol P E and Llovet X 2013 *Microsc. Microanal.* **19** 1-11
[14] Kearns S, Buse B and Wade J 2014 *Microsc. Microanal.* **20** (Suppl. S3) 740-741
[15] Smith M P 1986 *J. Sediment. Res.* **56** 560-561
[16] Jercinovic M J and Williams M L 2005 *Am. Mineral.* **90** 526-546
[17] Morgan G B I and London D 1996 *Am. Mineral.* **81** 1176-1185
[18] Nielsen C H and Sigurdsson H 1981 *Am. Mineral.* **66** 547-552
[19] Llovet X and Galan G 2003 *Am. Mineral.* **88** 121-130
[20] Fournelle J 2007 *Microsc. Microanal.* **13** (Suppl. S02) 1390-1391
[21] Wade J, Buse B and Kearns S 2014 *Microsc. Microanal.* **20** (Suppl. S3) 728-729
[22] Llovet X, Pinard P T, Donovan J J and Salvat F 2012 *J. Phys. D: Appl. Phys.* **45** 225301

[23] Willich P and Bethke R 1996 in: *Microbeam and nanobeam analysis.* (Benoit D, Bresse J-F, Van't dack L, Werner H and Wernisch J; Eds.) *Mikrochim. Acta* **Suppl. 13** 631-638
[24] Pouchou J L 1996 in: *Microbeam and nanobeam analysis.* (Benoit D, Bresse J-F, Van't dack L, Werner H and Wernisch J; Eds.) *Mikrochim. Acta* **Suppl. 13** 39-60
[25] Bastin G F and Heijligers H J 2013 *Quantitative electron probe microanalysis of boron, carbon, nitrogen and oxygen.* ISBN 9789462282223
[26] Wendt M 1980 *Krist. Tech.* **15** 1367-1375
[27] Llovet X, Heikinheimo E, Galindo A N, Merlet C, Bello J F A, Richter S, Fournelle J and van Hoek C J G 2012 *IOP Conf. Ser.: Mater. Sci. Engng.* **32** 012014
[28] Statham P and Holland J 2014 *IOP Conf. Ser.: Mater. Sci. Engng.* **55** 012017
[29] Pinard P T, Heikinheimo E, Llovet X and Richter S 2014 *Microsc. Microanal.* **20** (Suppl. S3) 700-701
[30] Ruste J 2013 *Circuit d'intercomparison 2012-2013.* Tech. Rep. Groupement National de Microscopie Electronique à` Balayage et de Microanalyses (Paris: GN-MEBA)
[31] Ritchie N W 2009 *Microsc. Microanal.* **15** 454-468
[32] Chantler C, Olsen K, Dragoset R, Kishore A, Kotochigova S and Zucker D 2005 X-ray form factor, attenuation and scattering tables (version 2.1). Online http://physics.nist.gov/ffast
[33] Ritchie N W M 2005 *Surf. Interface Anal.* **37** 1006-1011
[34] Statham P J 2014 *Microsc. Microanal.* **20** (Suppl. S3) 690-691
[35] Armstrong J T 2011 in: *AGU Fall Meeting Abstracts* C2538
[36] Richter S and Pinard P T 2014 *Microsc. Microanal.* **20** (Suppl. S3) 680-681
[37] Galbert F and Berger D 2008 in: *European Microscopy Congress - Instrumentation and methods.* (Luysberg M, Tillmann K and Weirich T; Eds.) (Vienna: Springer)

IOP Conf. Series: Materials Science and Engineering **109** (2016) 012014 doi:10.1088/1757-899X/109/1/012014

Combined EPMA, FIB and Monte Carlo simulation: a versatile tool for quantitative analysis of multilayered structures

S Richter and P T Pinard

RWTH Aachen University, Central Facility for Electron Microscopy (GFE), Ahornstrasse 55, 52074 Aachen, Germany

E-mail: richter@gfe.rwth-aachen.de

Abstract. Electron probe microanalysis and focussed ion beam milling are combined to improve the sensitivity and applicability of depth profiling quantification. With the nanoscale milling capabilities of the ion beam, very shallow bevels are milled by using a special preparation procedure to reduce any curtaining effect and minimize Ga ions implantation. A Ni/Cr multilayered specimen is used to evaluate the depth resolution. The best results are obtained by a well-focussed electron beam offered by a field-emission microprobe. A new evaluation algorithm is presented to quantify the structure in terms of mass thicknesses or if the density is known in terms of real thicknesses. The quantification procedure is based on Monte Carlo simulations where calculated *k*-ratios (calibrated X-ray intensities) are compared to the experimental ones to find the optimal structure. In comparison with an ion milled cross-section, the proposed bevel technique is more sensitive and provides more information about the material's structure.

1. Introduction

Today, wavelength-dispersive (WD) and energy-dispersive (ED) detectors show excellent properties for highly sensitive measurements due to high peak to background ratios or high output count rates. By moving the beam step by step across small structures variations in the X-ray intensities are detectable although the dimensions of the structures are much smaller than the X-ray excitation volume. With the help of Monte Carlo simulations the local ionisation probability can be calculated in a virtual specimen, leading to the description of a three dimensional function. Laterally averaging this function gives the $\Phi(\rho z)$ function used in well-established quantification procedures for bulk, thin films and even multilayered structures [1]. However, conventional thin film analysis is mainly limited to layers containing elements, which are not present in the surrounding matrix, and limited to depths smaller than the mean free path of the characteristic X-rays [2]. To improve the sensitivity and applicability of depth profile analysis electron probe microanalysis (EPMA) can be combined with a destructive preparation technique, for instance, using a sputtering procedure. One possibility is an in-situ technique similar to Auger depth profile analysis, where X-ray intensity measurements and thin surface film removal by inert gas ions are alternated [2-4]. Another possibility is to scan the beam across the wedge of a sputtered crater. This requires the assumption of a laterally homogeneous sample. First attempts of the crater edge profiling EPMA were successfully applied to heat insulation coatings on glass [5]. The technique presented in this paper uses EPMA in combination with the focussed ion beam (FIB) milling. The first development steps have been presented earlier [6]. It has been shown that the FIB preparation technique delivers excellent properties due to the capability of its highly focussed ion beam to mill a precise, well-defined geometry [6]. This property makes the FIB preparation superior to all other sputtering based depth profile techniques and allows the determination of the depth scale by two independent ways. The aim of this work is to improve the depth resolution of this technique by reducing FIB artefacts and by using microprobes with better focussing optics [10] in order to extend the limits of depth profiling in EPMA. Finally, using a well-defined multilayered structure, the sensitivity and resolution of the proposed bevel technique are compared to those obtained from the measurements of a cross-section. In summary, the possibilities and limitations of the new method are evaluated based on the following points: (1) improvements of the FIB milling procedure, (2) theoretical considerations regarding the X-ray intensity profiles and depth resolution for the bevel geometry, (3) influence of the type of electron emitter and spectrometer on the depth resolution, (4) quantification of the multilayered structure, and finally (5) comparison to cross-section analysis.

2. Experiment

To test the technique a NIST certified multilayered sample was used [7]. It consists of nine alternating metal thin film layers, five layers of pure chromium with a thickness of 57 nm interlaced with four pure nickel layers with a thickness of 56 nm. The substrate consists of polished silicon. The thickness of each layer is defined with an uncertainty of 3 - 4 nm.

For the focussed ion beam (FEI Strata 205) preparation of the bevel, the coated wafer was cleaved and fixed on a holder in the vertical direction. The Ga ion beam was focussed on the cleavage plane with angles smaller than 1° relative to the surface of the coating. The bevels were milled by scanning the beam over an area of about 0.5 x 20 μm^2. The operating conditions were an ion accelerating voltage of 30 kV, an ion current between 300 and 1,000 pA, a dwell time of 1 µs, an overlap of 50 % and a total milling time between 10 and 60 min. The FIB prepared cross-section was performed in a similar manner except that the ion beam was positioned perpendicular to the coating's surface and parallel to the cleavage plane. All the other conditions were kept the same as described above. For studying the curtaining effect a gas injection system for W deposition was used.

To evaluate the depth resolution of the method the same sample was inserted into three different devices: a W-filament electron microprobe (Cameca SX50), a field emission (FE) microprobe (JEOL JXA8500F) both equipped with WD spectrometers, and a FE scanning electron microscope (SEM) (Zeiss Supra 40) equipped with an ED spectrometer (Thermo Fisher Scientific NSS 302). For all X-ray measurements an electron beam energy of 10 keV was taken. The other measuring conditions like the beam current and the measuring time per analysed point are shown in table 1. For calibration of the X-ray intensities pure standards were chosen. Line scans were performed along the bevel starting from the intact multilayered coating towards the uncovered substrate.

Table 1. Measuring conditions used for the different instruments and detectors.

	Beam current (nA)	Measuring time (s)	Type of spectrometer
Cameca SX50	100	10	WDS: PET for Si-Kα and Cr-Kα, LIF for Ni-Kα
JEOL JXA-8500F	50	10	WDS: PETJ for Si-Kα, LiF for Cr-Kα and LIFH for Ni-Kα
Zeiss Supra 40	1	15	EDS

3. Advanced preparation of the FIB-bevel

To get the best depth resolution the angle of the bevel relative to the sample surface must be as small as possible. The interdependence between the angle and the depth resolution can be demonstrated by the following experiment. Bevels were milled under the same conditions with decreasing angle (step width was 0.5°). As can be seen from figure 1 decreasing the angle widens the uncovered layers. It is therefore obvious, that a better depth resolution will be achieved by a line scan across the bevel milled with a smaller angle. However, a limiting factor of the depth resolution is the curtaining effect. At very low bevel angles the milling process causes a pronounced smearing, which looks like a curtain (see figure 1). As a result, the sharpness of the interfaces is lost and the depth resolution worsens.

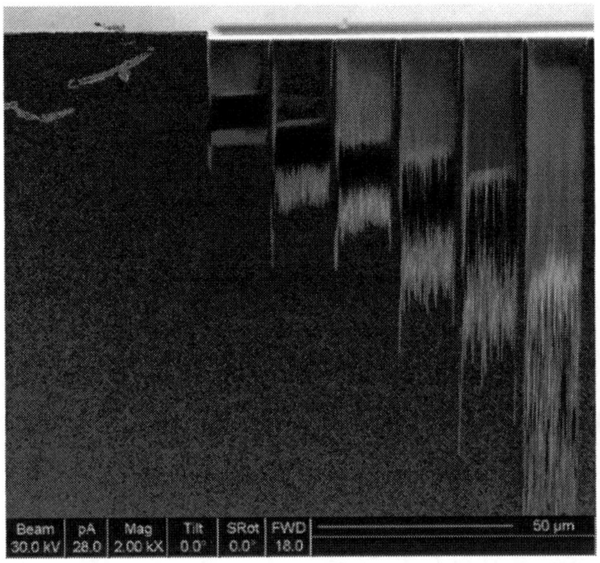

Figure 1. FIB induced secondary electron imaging: top view of the milled bevel with decreasing angle from left to right. The specimen is a DRAM structure ($STO/Pt/TiO_2/SiO_2$ on a silicon substrate).

Below a critical bevel angle the curtaining effect is unavoidable. The reason is the interdependence between the critical bevel angle αc and the divergence angle β of the ion beam. If the bevel angle approaches β, a part of the ions, indicated by the red arrows in figure 2, digs into the surface of the bevel and causes an irregular removal of the material, which causes the curtaining effect. The question arises whether it is possible to decrease the critical bevel angle and thus reduce the curtaining effect. One possibility is decreasing the ion beam current but this means longer milling time and stability problems of the ion beam positioning.

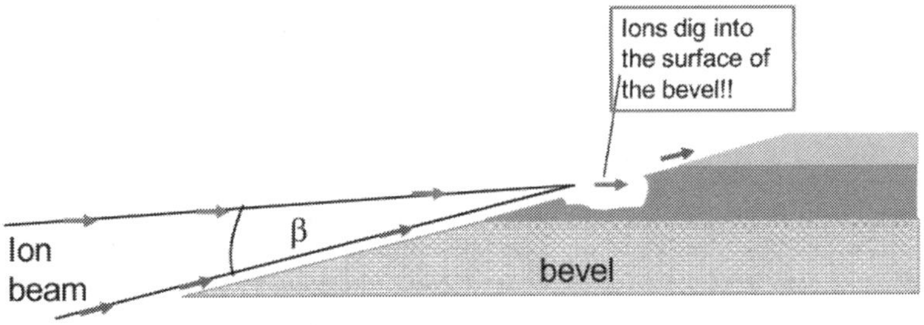

Figure 2. Schematic presentation illustrating the curtaining phenomenon. Due to the small bevel angle some ions are digging into the surface and causing an irregular removal of the surface material.

Alternatively, a tungsten mask, a thin film deposited in-situ by the FIB, can be used. By shadowing a part of the ion beam, it decreases the divergence angle of the beam (see figure 3). The efficiency of the tungsten mask is illustrated in figure 4a where a bevel was milled under the same operation parameters with and without a tungsten mask. As a result, the tungsten mask minimizes the curtaining effect.

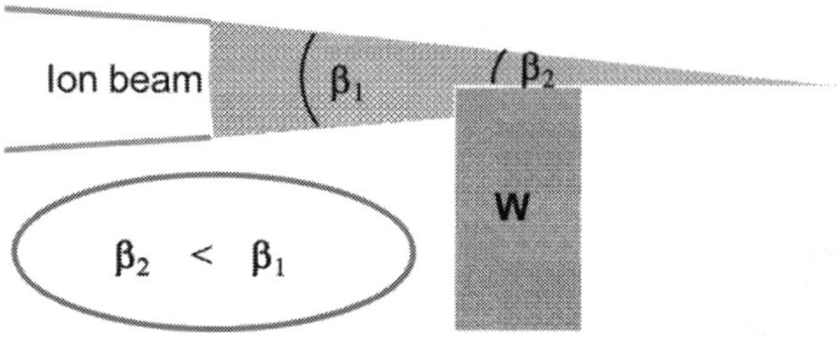

Figure 3. Schematic presentation illustrating the influence of a tungsten mask to reduce the divergence angle of the ion.

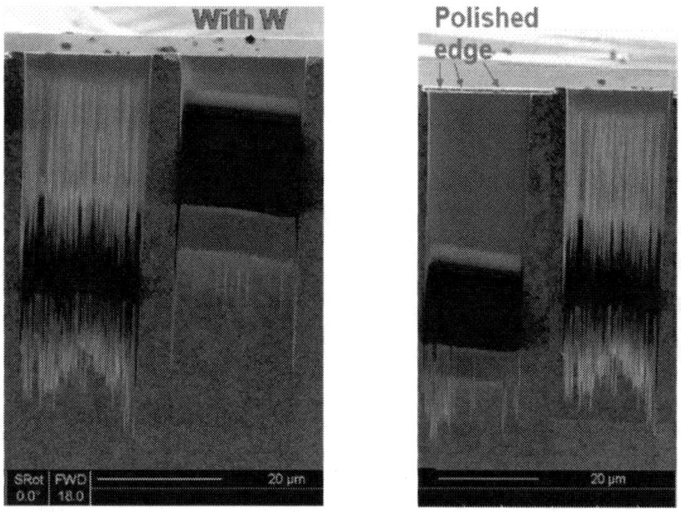

Figure 4. FIB induced SE images of bevels milled under different conditions to demonstrate the efficiency a) of a tungsten mask, and b) of pre-polishing the edge of the cleavage plane to reduce the curtaining effect.

Furthermore, the curtaining effect can also be reduced by polishing the edge of the cleavage plane using the FIB before milling the bevel. In the polishing mode the ion beam is parallel to the cleavage plane and a low current of less than 100 pA is used. Polishing removes topographical irregularities and small oxide particles, which may interfere with the milling process. The advantage of polishing is clearly seen in figure 4b. The curtaining effect has almost disappeared from the bevel milled from a polished edge (left).

4. Theoretical considerations

4.1. Modelling of the experiment

Due to the relatively large information depth of EPMA the measured X-ray intensity profiles give no direct information on the desired concentration depth profiles. It was shown in a previous work [9] that, in principle, the X-ray intensity profiles can be calculated using Monte Carlo simulations [8]. For the calculations an experiment, which describes an in-situ depth profile technique, was simulated as follows: small amounts of the material are step by step removed from the specimen's surface by ions and the X-rays of the interesting elements are measured, iteratively. This results in a distribution of emitted X-ray intensities as a function of sputtered depth (in terms of mass coverage). After each sputtering step the ionisation depth distribution, $\Phi(z)$, is updated based on the structure of the residual material. This is necessary since changes in the elements present in the structure at each sputtering step may lead to large differences in interaction volume. To remove dependencies on fundamental and instrumental parameters intensities are calculated for bulk standards, i.e., the same calibration procedure as in an experiment. These calibrated intensities are described by the *k*-ratios ($I_{unknown}$ / I*standard*). Further details have been described elsewhere [9].

A typical simulation result is presented in figure 5. Kinks are visible at each interface. This is a characteristic feature of the method caused by the steep increase or decrease of the concentration at the interfaces. Consequently, the position of the interfaces can be easily identified. The amplitude of the step between two kinks in the *k*-ratio profiles is proportional to the mass coverage of each element. For instance, the increase of the Ni-Kα *k*-ratio is due to a successive excitation of the Ni layer with a simultaneous removal of the upper Cr layer (see figure 6). These results can be also used for the simulation of the bevel technique. In this configuration, instead of measuring X-ray intensities as a function of the sputtered depth, intensities are acquired along the distance of the bevel starting from the surface of the coating towards the substrate. In other words, each sputtering step of the in-situ depth profile technique is equivalent to a lateral step along the bevel. The increase of the Ni-Kα *k*-ratios on the bevel is equivalently due to a successive excitation of the Ni layers, but it is now caused by the movement of the beam along the uncovered structure from the upper Cr/Ni interface to the lower one. Kinks at each interface are also a characteristic feature of the bevel profile. Finally, it should be also pointed out that the similarities between the two techniques are only valid if a very shallow bevel, i.e., very small angles of about 1° or less, is considered.

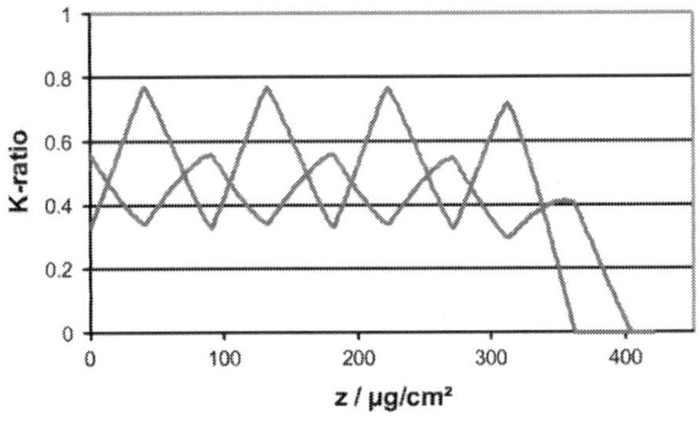

Figure 5. *k*-ratio profiles of Ni- and Cr-Kα versus sputter depth (proportional to the distance along a bevel) calculated by means of a Monte Carlo simulation.

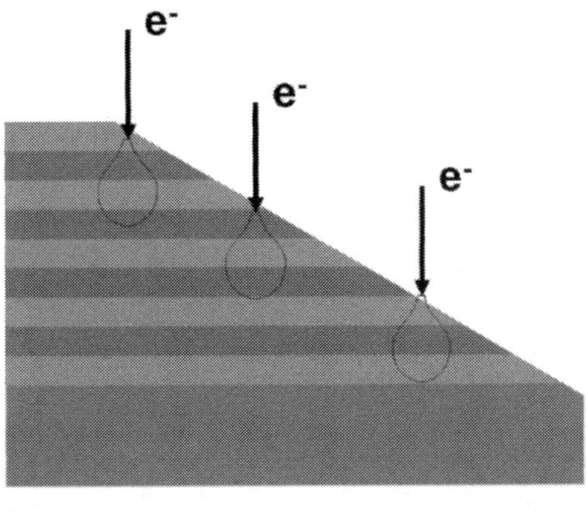

Figure 6. Schematic presentation of the bevel technique.

4.2. Definition of the depth resolution

To a first approximation, the *k*-ratio profiles can be considered as a convolution of the material's structure with the X-ray excitation volume where the kinks correspond to the steep increase of the concentration at each interface. The differentiation of the simulated curves (figure 7) gives a better understanding of the actual concentration profiles and allows the determination of the depth resolution of the method. The differentiated k-ratio profiles show a similar behaviour as the concentration profiles, steep increases are visible at the interfaces. In a real experiment these steep increases are broadened due to the finite depth resolution of the method. Thus, the depth resolution can be quantified by the transition width, over which the measured signal drops from 84 % to 16 % of its original value.

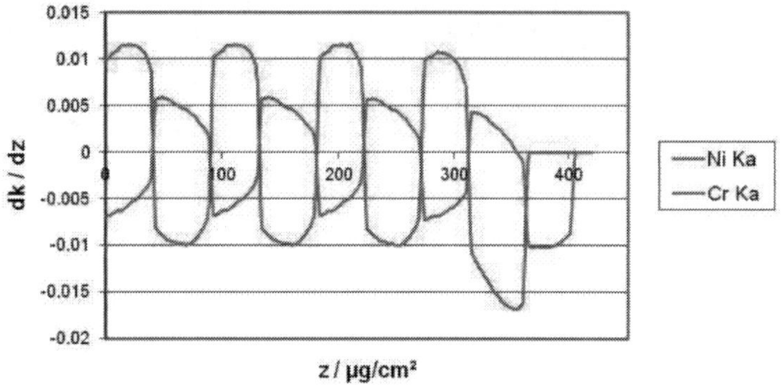

Figure 7. Profiles of the differentiated *k*-ratios (shown in figure 5) by the sputter depth.

4.3. Quantification procedure

In the previous section the characteristics of the *k*-ratio profiles obtained from the combined FIB-EPMA method were studied using modelling. The next step is to extract from the data the chemical structure of the material. Two possibilities exist for the quantification of the chemical structure. First, a manual step by step procedure based on a thin film algorithm can be applied, where the multilayered structure is successively reconstructed starting with the determination of the composition of the substrate using the *k*-ratios at the lowest interface (kink). Then, using this composition and the *k*-ratios from the next interface (next kink) the thickness and composition of the next layer are calculated. This step is repeated until the composition and thickness of all layers are calculated. Further details are described in [6].

The second possibility is to use an automated evaluation procedure, where all the unknown parameters are simultaneously determined from the whole set of measured data. To be more precise in terms of a mathematical description let us define a parameter vector **P** consisting of the partial mass coverage $\mathbf{m}_j$ (j = 1…N) and a data vector **K** consisting of the experimental *k*-ratios $\mathbf{k}_i$ (I = 1...M). The data vector **K** is linked with the parameter vector **P** by a set of non-linear equations

$$k_i = f_i(m_1, m_2, \ldots m_j, \ldots, m_N) \tag{1}$$

The aim is to find a solution **P***, for which the Euclidian norm

$$\left\| k(P^*) - K^{\exp} \right\|_2 \Rightarrow \min! \tag{2}$$

or the deviation between **k(P*)** and $\mathbf{K}^{exp}$ is a minimum. Using a non-linear solver such as the Levenberg-Marquardt algorithm an optimized solution **P*** can be iteratively determined from an initial value for **P**.

5. Results

5.1. Intensity profiles

X-ray intensity measurements of Ni-Kα, Cr-Kα and Si-Kα were acquired along the milled bevels of the NIST Ni/Cr multilayered structure using three different devices and detectors: 1) Cameca SX50, 2) JEOL JXA-8500F, and 3) Zeiss Supra 40 (see figure 8).

In general, all the results show the periodic structure of the specimen as expected from the modelling (figure 5). The maxima and minima in each profile have approximately the same values, indicating the uniform thickness of the Ni and Cr layers. A different behaviour is observed for the first Cr maximum after the Si substrate (see figures 8b/8c). This corresponds to the situation where the electron beam is crossing the first interface of Cr/Ni multilayered structure. In other words, at this location the specimen consists in a thin Cr film on top of a Si substrate. Due to the lower atomic number of Si the intensity of backscattered electrons is lower, resulting in a lower emitted Cr-Kα intensity in comparison to the intensities emitted at the other interfaces.

The first Cr maximum is almost disappearing in the profile measured by the tungsten filament microprobe (figure 8a). Furthermore, in comparison to the predicted theory the oscillations are generally smaller and no kinks are visible at the interfaces.

Results from the FE microprobe (figure 8c) show stronger oscillations and kinks are clearly visible at the interfaces. Similarly, the ED *k*-ratio profiles measured by the FE-SEM (figure 8b) also exhibit pronounced oscillations, but the signal is noisier due to the lower intensity.

Both FE instruments deliver well-focussed beams with diameters smaller than 100 nm [10], whereas the tungsten microprobe offers larger beam sizes at the chosen operating conditions. These experimental results demonstrate the influence of beam size on the depth resolution of the combined FIB-EPMA method.

5.2. Determination of the depth resolution

The depth resolution of each instrument was determined using the aforementioned method of measuring the transition width from differentiated *k*-ratio profiles (figure 9). The depth resolution was determined at each interface. The averaged values are listed in table 2, in terms of mass coverages (first column), mass coverages corrected for the uncertainty of the standard (second column) and thicknesses in nm (third column). These values correspond to the depth resolution of FIB-EPMA under the chosen devices and operating conditions. As expected, the smaller beam diameter, produced by field emission instruments, improves the depth resolution of the FIB-EPMA method up to about a few nm.

Table 2. Depth resolution determined by different instruments in terms of mass coverages (first column), mass coverages corrected for the uncertainty of the certified standard (second column) and thicknesses in nm assuming the density of pure Cr and Ni (third column).

	Δz /μg/cm²	**Δz* /μg/cm²**	**Δz' /nm**
Cameca SX50	16.8	14.2	16.0
JEOL JXA-8500F	8.6	6.0	6.7
Zeiss Supra 40	9.4	6.8	7.6

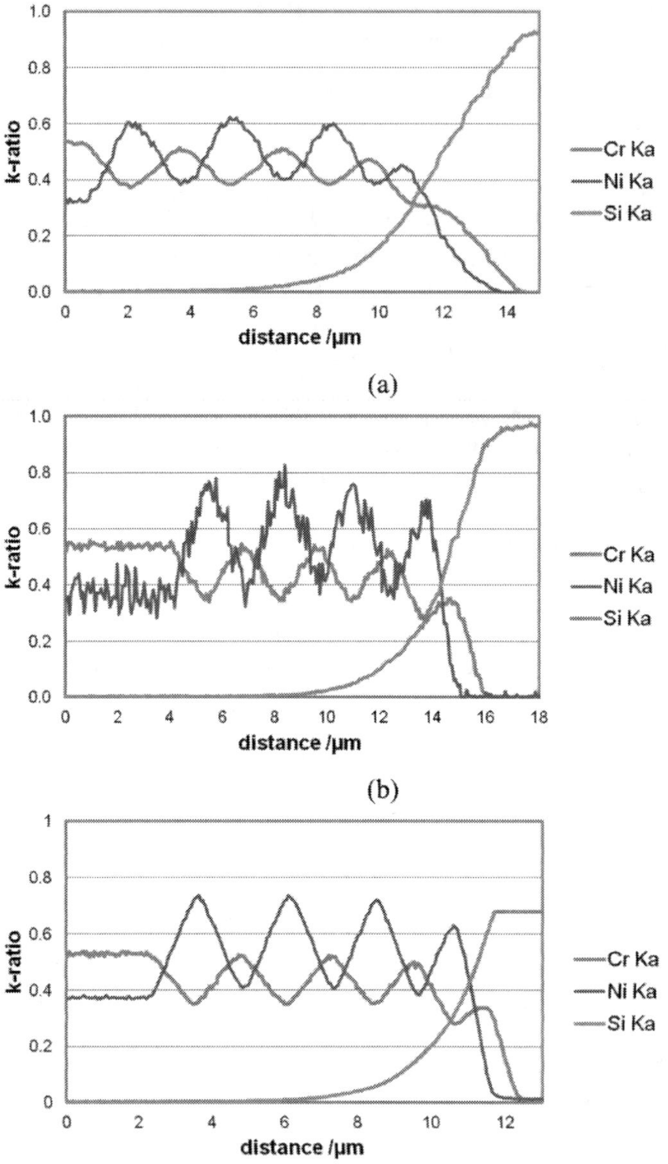

(a)

(b)

(c)

Figure 8. Line scan measurements of calibrated Ni-Kα, Cr-Kα and Si-Kα intensities across a bevel at 10 keV performed by different instruments: a) Cameca SX50, b) Zeiss Supra 40, and c) JEOL JXA 8500F.

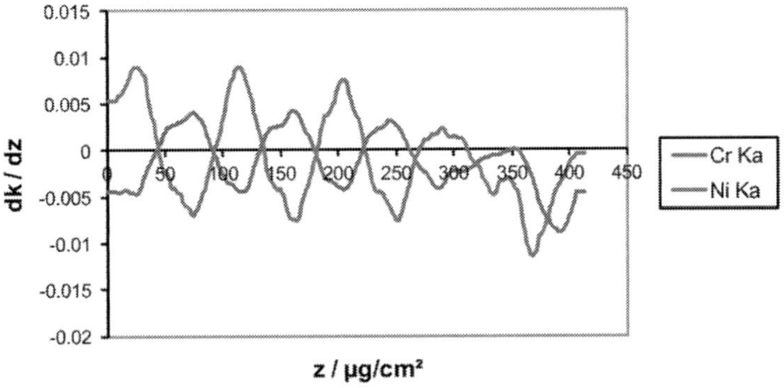

(a)

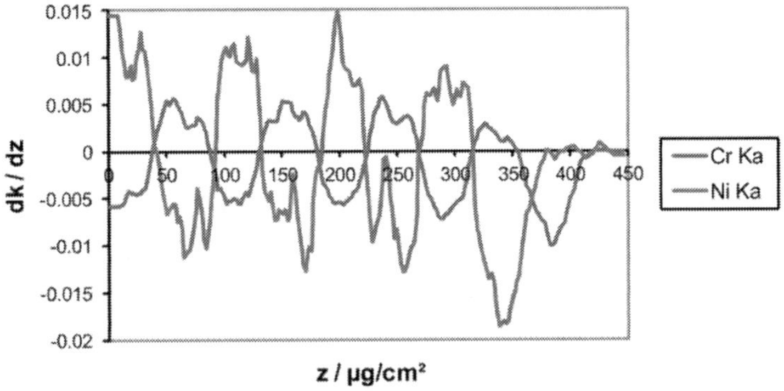

(b)

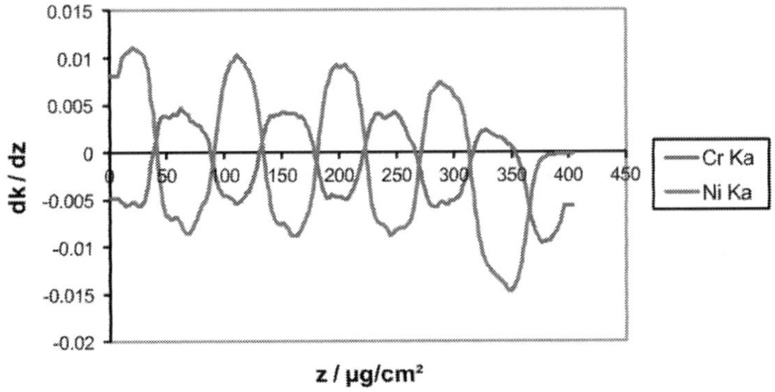

(c)

Figure 9. Measured *k*-ratio profiles (see figure 8) differentiated by sputter depth for different instruments: a) Cameca SX50, b) Zeiss Supra 40, and c) JEOL JXA 8500F.

5.3. Quantification

For the quantification of the multilayered structure the measurements from the FE microprobe were taken. All the Cr-Kα and Ni-Kα *k*-ratios acquired along the bevel were used to create the data vector **K** (eq. 1). The Si-Kα *k*-ratios were not used, since the only valuable information about the mass thicknesses is contained in the Cr-Kα and Ni-Kα profiles. The unknown parameters described by the structure vector **P** are the mass thicknesses. The calculated *k*-ratios for the different X-ray lines and sputter depths (**K(P)**) were obtained by means of Monte Carlo simulations. A non-linear solver was used to solve eq. 2 and obtained the optimized solution **P***. The optimized thicknesses of each layer are presented in table 3 together with the relative errors. The relative errors are defined as the relative deviation between the nominal and calculated thicknesses. On average the errors correspond to the certification limits of the specimen which are in the order of 7 % for Ni and Cr.

Table 3. Thickness of each layer and relative deviation with respect to the certificated values.

Layer n°	Type of layer	Thickness / nm	Rel. dev. /%
1	Cr	60.0	+5.2
2	Ni	54.5	-2.6
3	Cr	56.1	-1.6
4	Ni	56.9	+1.6
5	Cr	56.3	-1.1
6	Ni	54.2	-3.2
7	Cr	50.3	-11.7
8	Ni	59.8	+6.8
9	Cr	52.5	-7.9

To evaluate the potential influence of Ga implantation during the FIB preparation, the Ga-Lα intensity was acquired in the case of the tungsten microprobe measurements. For the calibration a GaAs standard was used. As a result, Ga-Lα *k*-ratios of less than 0.001 were detected at the chosen conditions (see figure 10). This corresponds to approximately less than 0.01 wt% Ga. Thus, FIB-induced Ga implantation can be neglected for the quantification of the above described multilayered structure.

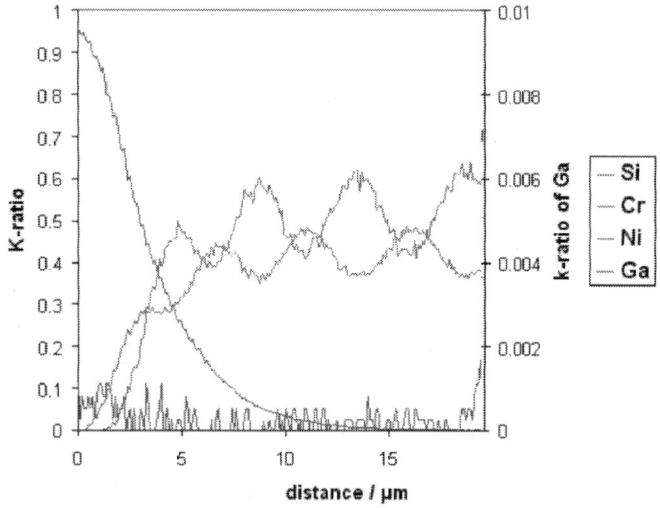

Figure 10. *k*-ratio profiles measured along the bevel including the acquisition of Ga-Lα.

6. Discussion

6.1. Cross-section versus bevel technique

Another method to analyse a sample as a function of depth is to prepare a cross-section and performs a line scan across the prepared surface, ideally using a FE electron microprobe or FE-SEM to benefit from the smaller beam diameters (figure 11).

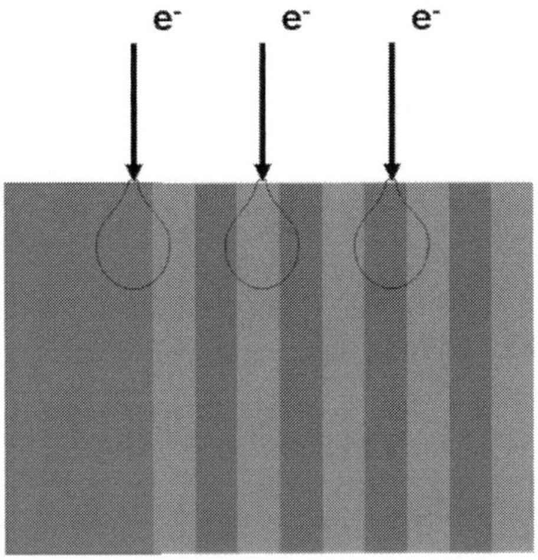

Figure 11. Schematic presentation of the measurement across the cross-section.

In comparison to the bevel, the FIB preparation of a cross-section is generally easier, since the curtaining effect is less pronounced (figure 12). To directly compare the depth sensitivity of the two techniques, a cross-section was prepared from the same certified multilayered system from NIST (see section 2). A gold layer was sputtered on top of the structure for protection against ion etching during the FIB preparation. Intensity measurements of the Ni-Kα and Cr-Kα X-ray lines were acquired across the structure starting from the Si substrate (figure 13). The same operating conditions on the JEOL JXA-8500F FE microprobe were chosen as the ones used for the bevel technique, a beam energy of 10 keV and a beam current of 10 nA. A beam drift correction procedure was used. The normalized X-ray intensities are presented in figure 13. The comparison of these profiles with those on the bevel (figure 8c) reveals lower amplitudes in the X-ray profiles along the cross-section. In addition, the interfaces are more diffuse and their exact position is more ambiguous.

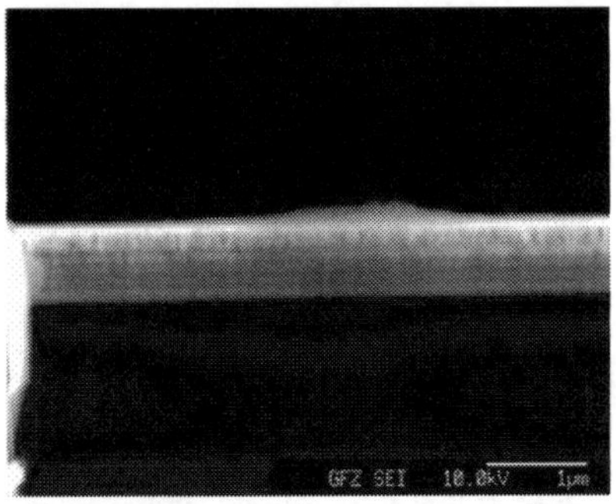

Figure 12. FIB-induced SE image of the milled cross-section of the multilayered structure.

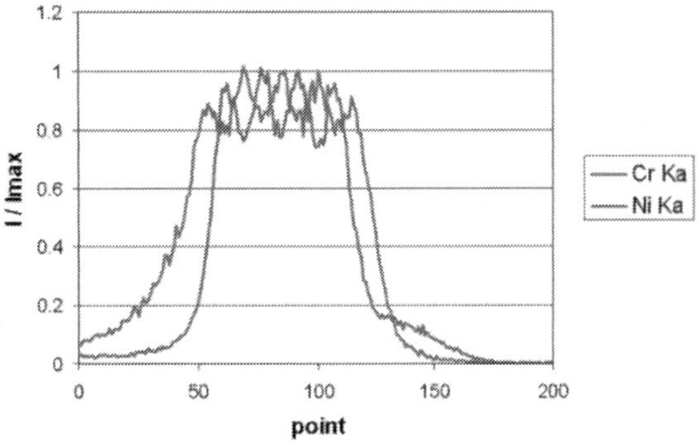

Figure 13. Normalized X-ray intensity profiles measured along the cross-section.

The quantification of multilayered structures based on X-ray intensity profiles across a cross-section is not established. Apart from the lack of analytical models to calculate the lateral X-ray ionisation distribution, the size and profile of the electron beam become crucial parameters. Assuming a Gaussian beam with a standard deviation of 50 nm and using the nominal thicknesses, simulations were performed using the Monte Carlo programme DTSA-II [11] to better understand the experimental profiles. The simulated profiles are shown in figure 14. To quantify a multilayered structure with unknown compositions and thicknesses, Monte Carlo simulations and a non-linear solver could be combined as presented in [12], albeit the very long calculation times.

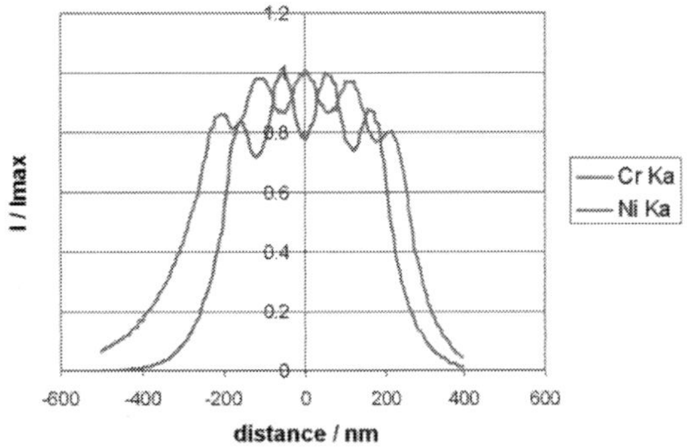

Figure 14. Normalized X-ray intensity profiles calculated by Monte-Carlo simulations.

Some conclusions can be drawn from the comparison between the bevel and cross-section techniques. It is more challenging to achieve an artefact-free FIB preparation of a bevel, a large milled area with an appropriate surface finish and a small bevel angle. However, this technique has the advantage to be less sensitive to the beam size and beam profile. No further measurement or assumption is needed for the beam diameter. Furthermore, the interfaces, layers and elements present in each layer can be directly determined from the experimental profiles. Finally, the quantification of an unknown multilayered structure based only on X-ray intensities acquired along the bevel can be done by a special thin film algorithm using Monte Carlo simulations and an optimisation algorithm [6].

7. Conclusion

This work shows the great potential of combining EPMA with the FIB specimen preparation. By improving the milling process to reduce curtaining effects (edge polishing or tungsten mask), low bevel angles were achieved. With the focussing capabilities of FE microscopes, a depth resolution of a few nanometres was obtained using the bevel technique. Using an optimisation algorithm based on

Monte Carlo simulations, the multilayered structure of a reference test material was quantified with an accuracy for the layer thicknesses comparable to the certified values.

Acknowledgements
The authors would like to acknowledge Dieter Rhede (GFZ, Postdam), Mathias Proccop (IFG, Berlin), Martina Schiffers (GFE, Aachen), Alexander Schwedt (GFE, Aachen), and Niklas Mevenkamp (GFE, Aachen).

References

[1] Pouchou J L and Pichoir F 1984 *Rech. Aerospaciale* **5** 47
[2] Karduck P 1998 *Mikrochim. Acta* **Suppl. 15** 109
[3] Lesch N, Richter S and Karduck P 1998 *Mikrochim. Acta* ***Suppl.*** **15** 133
[4] Briggs D and Seah M P 1983 *Practical Surface Analysis Volume 1: Auger and X-ray Photoelectron Spectroscopy*. (Chichester: John Wiley & Sons)
[5] Richter S, Lesch N and Karduck P 2000 *Surf. Interface Anal.* **30** 260
[6] Richter S, Kyrsta S, Aretz A, Bückins M, Spähn M and Mayer J 2004 *Mikrochim. Acta* **145** 187
[7] Fine J and Navinsek B 1988 *Surf. Interface Anal.* **11** 542
[8] Ammann N and Karduck P 1990 in: *Microbeam Analysis.* (Michael J R and Ingram P; Eds.) (San Francisco, CA: San Francisco Press) 150
[9] Richter S, Lesch N and Karduck P 1998 *Mikrochim. Acta* **Suppl. 15** 125
[10] Richter S and Pinard P T 2014 *Microsc. Microanal.* **20** Suppl. 3 680
[11] Ritchie N W M 2005 *Surf. Interface Anal.* **37** 1006
[12] Richter S, Pinard P T, Mevenkamp N and Torrilhon M 2013 *Microsc. Microanal.* **19** Suppl. 2 1296

Electron probe microanalysis of the dopant concentrations in complex perovskite ferroelectrics

Z Samardžija

Jožef Stefan Institute, Department for Nanostructured Materials, Jamova cesta 39, 1000 Ljubljana, Slovenia

E-mail: zoran.samardzija@ijs.si

Abstract. Quantitative EPMA-WDS microanalyses were applied for the compositional characterisation of complex perovskite ferroelectrics based on cerium-doped barium titanate, a solid solution between lead-magnesium-niobate/lead-titanate and niobium-doped barium-bismuth-titanate. The analyses were optimized for high analytical sensitivity, precision and an ultimate accuracy of $\approx \pm 1$ % relative. The inherent problem with the WDS peak overlap of the Ce-Lα_1 and Ba-L$\beta_{1,4}$ spectral lines was solved by introducing overlap-correction methods in order to obtain consistent quantitative results for the Ce-doped $BaTiO_3$. The quantitative results made it possible to obtain accurate chemical formulae for these materials, to determine the solubility of the dopants as well as to define the mode of the dopant incorporation and the charge-compensation mechanisms.

1. Introduction

Complex perovskite ferroelectrics belong to the group of oxide ceramic materials with a ternary ABO_3 crystal structure where the ions on the A and/or B sites are partially substituted by doping with other ions, thus forming numerous solid-solution compounds [1]. Among the large number of such compounds, many electroceramic dielectrics are based on a modified barium-titanate ($BaTiO_3$) ceramic host, which was one of the first synthetic perovskites [2]. Following the crystal chemistry of perovskites, a multiple ion substitution on the same structural site was also involved [3]. With an appropriate change in the composition, using various types and amounts of dopants, the properties of these materials can be tailored and modified for a wide range of electroceramic devices with different dielectric, piezoelectric, electro-optic, microwave and related applications [4]. Since small quantities of dopants can significantly change the ferroelectric properties, the *structure-composition-property* relations have to be accurately defined. Therefore, an important part of the studies of complex ferroelectrics is associated with an accurate analysis of their chemical composition. For this reason, quantitative electron probe microanalysis (EPMA) has a key role and can be applied for the compositional analysis of ferroelectric ceramics in forms of multi-phase or single-phase polycrystalline materials and single crystals.

In this work, selected case studies of EPMA for complex perovskite ferroelectrics based on Ce-doped $BaTiO_3$ (BTC), the solid solution between lead-magnesium-niobate and lead-titanate $Pb(Mg_{1/3}Nb_{2/3})O_3$-$PbTiO_3$ (PMN-PT) and the layer-structure (Aurivillius type) perovskite barium-bismuth-titanate ($BaBi_4Ti_4O_{15}$) doped with Nb (BBTN) were presented, demonstrating the usability and power of a quantitative EPMA in these cases. The aim of the work was to achieve reliable, accurate, quantitative analyses of the elemental concentrations in these materials in order to study the solubility and the incorporation of the dopants in the host perovskites. In order to achieve this, the ultimate quantitative microanalysis was carried out using wavelength-dispersive X-ray spectroscopy (WDS) that was optimized to achieve a high analytical sensitivity, precision and accuracy.

2. Experimental

The microstructures of selected polished ceramic samples are shown in figure 1, revealing that the BTC and BBTN materials are polycrystalline, multiphase samples, while the PMN-PT material is single phase with a polycrystalline and a single-crystal part. For a consistent microanalysis of the doped-perovskite matrix phases in multiphase microstructures it is necessary to perform precise positioning of the focussed electron beam in order to avoid the influence of adjacent secondary phases and/or pores.

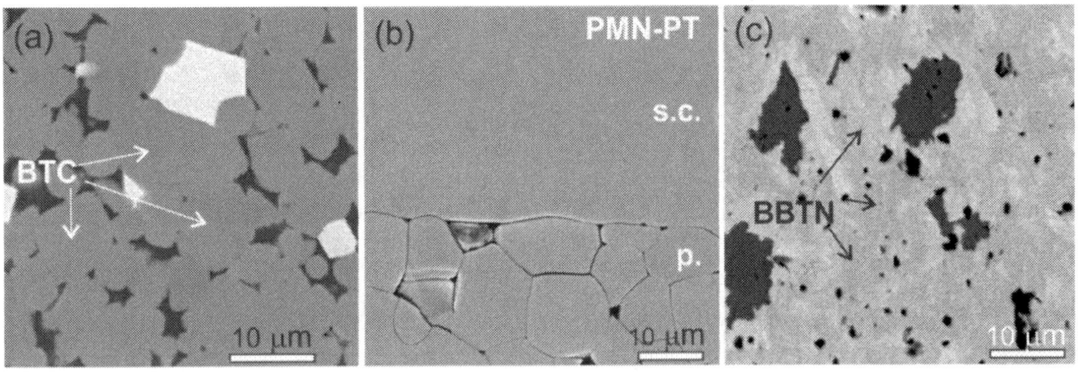

Figure 1. SEM compositional contrast micrographs of the microstructures of the analysed samples: (a) BTC; (b) PMN-PT with polycrystalline part (p.) and single-crystal part (s.c.); (c) BBTN.

Preliminary analyses using energy-dispersive X-ray spectroscopy (EDS) revealed several problems that show why quantitative EDS analyses in these cases cannot be sufficiently accurate. The problems are related to the poor detection limits of the method and to several peak overlaps, i.e., between the Ba-L and Ti-K spectral lines, the more complicated overlap of both the Ba-L and Ti-K with the Ce-L line, and the partial overlap of the Pb-M and Bi-M lines with the Nb-L line. These overlaps are especially difficult to deconvolute properly when the dopant concentrations are low/minor, as is the situation in the selected materials. The parts of the EDS spectra from the BTC and BBTN samples showing the current peak overlaps are presented in figure 2.

Consequently, the EPMA-WDS method was employed for the spectroscopic measurements of the elemental analytical spectral lines in the selected samples, taking advantage of the high spectral resolution and the high peak-to-background (P/B) ratios that can be achieved using properly optimized WDS. A JEOL JXA-840A electron-probe microanalyser with two wavelength spectrometers and a TN5600 system with TASK programme were used for the X-ray data acquisition and processing. For all the WDS measurements the designated diffracting crystals were tuned to peak and background positions that were accurately determined from precise, slow wavelength scans performed on standards and additionally re-checked on the samples. In the case of the PMN-PT and BBTN samples, the Nb-Lα, Pb-Mα and Bi-Mα spectral lines were easily separated and measured without any interference using the PET crystal. For the determination of the relatively low (1 - 2 m.f.%) dopant concentration of Mg in the PMN-PT, the Mg-Kα line was measured using a TAP crystal with the background positions (B1, B2) that were carefully determined, taking into account that the multiple-λ ($n = 2$) peak from the Pb-Mβ line emerges close to the Mg-Kα peak, as illustrated in figure 3a. When analysing the samples containing Ba and Ti (BTC, BBTN), the Ba-Lα and Ti-Kα spectral lines were well separated and correctly measured using the LiF crystal, as shown in figure 3b.

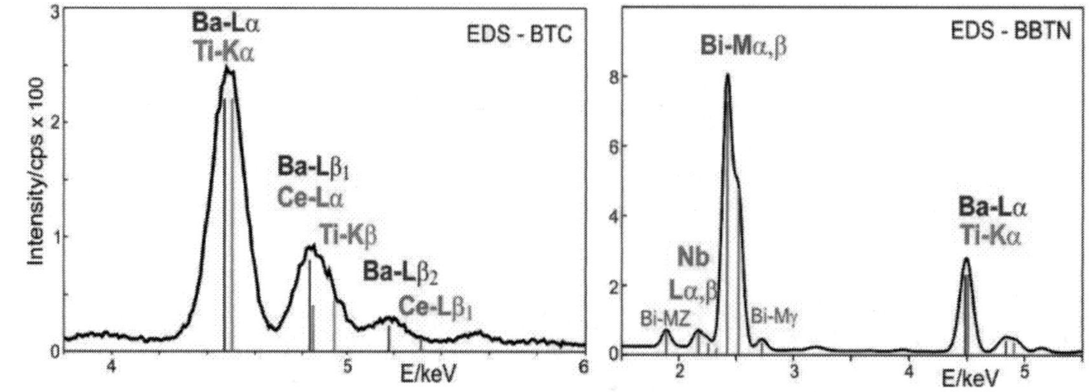

Figure 2. Characteristic EDS peak overlaps of the spectral lines in the spectra acquired from the doped-perovskite phases in the BTC and BBTN materials.

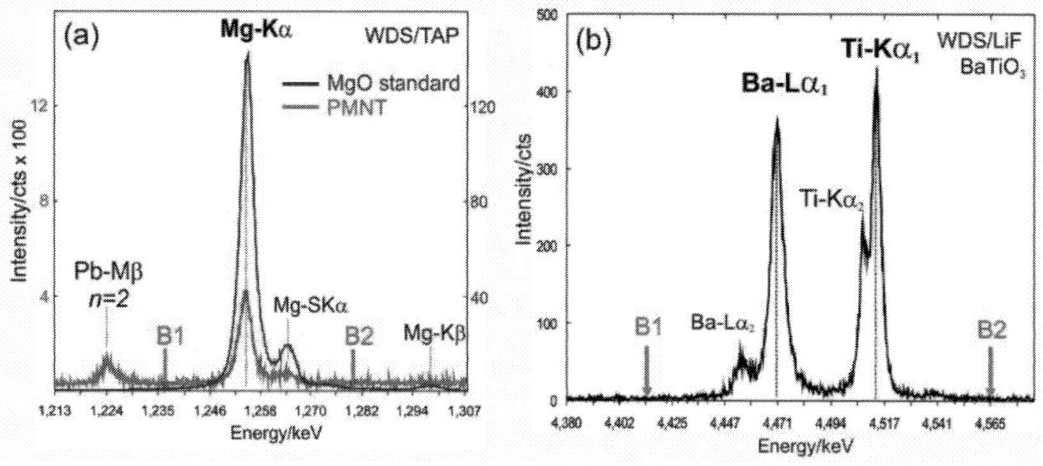

Figure 3. (a) TAP wavelength scans for the Mg-Kα line on the MgO standard and PMN-PT sample; (b) LiF wavelength scan for the Ba-Kα and Ti-Kα lines on the $BaTiO_3$ standard.

However, a specific situation occurs when the $BaTiO_3$ is doped with cerium, and the spectral interference between the Ce-Lα_1 and the Ba-L$\beta_{1,4}$ lines also causes a peak overlap when using the WDS and a high-resolution LiF crystal, as shown in figure 4a. This overlap introduces a systematic error in the quantification of the concentrations of the Ce dopant in the $BaTiO_3$. The evaluation of the Ba-L$\beta_{1,4}$-Ce-Lα_1 peak overlap for the BTC material that was obtained by deconvolution using pseudo-Voigt peak-profiles is shown in figure 4b. Here, it is clear that the measured Ce-Lα_1 intensity (I) is always greater than its true intensity (I^*), i.e., the *k*-ratios obtained for the Ce are too high, leading to an overestimated and erroneous quantification.

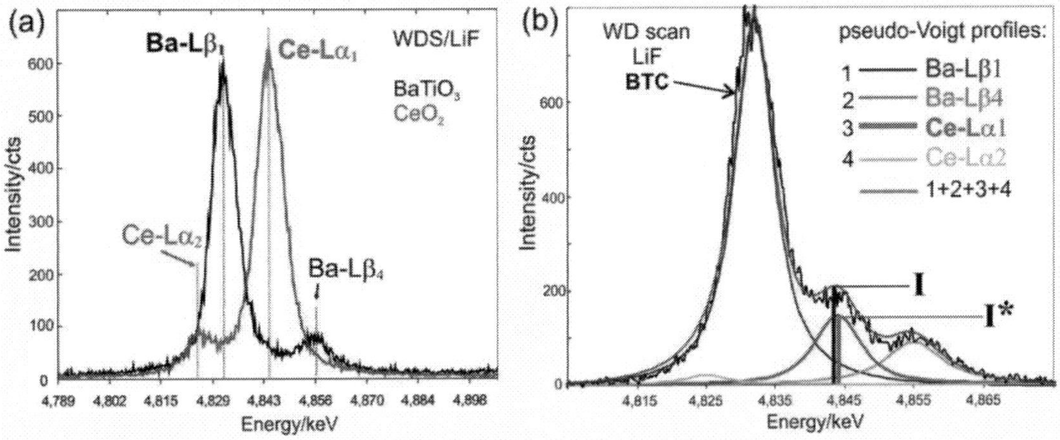

Figure 4. (a) Merged wavelength scans for the Ba-L$\beta_{1,4}$ and Ce-L$\alpha_{1,2}$ lines, showing their mutual overlap that exists in the wavelength spectrum; (b) deconvolution of the wavelength spectrum acquired from the BTC matrix using pseudo-Voigt profiles, revealing the difference between the measured (I) and true (I*) Ce-Lα_1 intensity.

In order to obtain the true Ce-Lα_1 intensity (I^*) a correction of the existing peak overlap has to be performed. Basically, the contribution of the Ba-L$\beta_{1,4}$ at the peak position for Ce-Lα_1 has to be estimated and then subtracted from the measured Ce intensity. In order to achieve this, three approaches were proposed to provide a reliable quantitative analysis of the BTC material:

A) Correction of the Ce-Lα_1-Ba-L$\beta_{1,4}$ peak overlap using the method based on relative spectral line intensities. Here we assume that the ratio between the Ba-Lα_1 intensity and the intensity of the Ba-L$\beta_{1,4}$ measured at the Ce-Lα_1 peak position is the same in the $BaTiO_3$ standard (BT) and in the Ce-doped $BaTiO_3$ sample (BTC). Then the true Ce-Lα_1 intensity (I^*) corrected for the contribution of the Ba-L$\beta_{1,4}$ is calculated using equation

$$I^{*\,BTC}_{CeL\alpha 1} = I^{BTC/measured}_{CeL\alpha 1} - \left[I^{BT}_{BaL\beta 1,4\text{ at }CeL\alpha 1\text{ position}} \times \left(I^{BTC}_{BaL\alpha 1} / I^{BT}_{BaL\alpha 1}\right)\right] \qquad (1)$$

B) Correction of the Ce-Ba overlap using the calibration-curve method [5]. Here we assume that the Ba-L$\beta_{1,4}$ intensity measured at the Ce-Lα_1 peak position increases linearly with an increase of the Ba concentration. This intensity was then measured versus the Ba-Lα_1 peak intensity in three Ba-Ti-O compounds (without Ce) and the linear calibration curve presented in figure 5a was obtained. The slope of this curve gave a correction factor that was used to estimate the contribution of the Ba-L$\beta_{1,4}$ to the Ce-Lα_1 in the Ce-doped $BaTiO_3$ matrix (BTC). The true Ce-L$\alpha\alpha_1$ intensity is then calculated using

$$I^{*\,BTC}_{CeL\alpha 1} = I^{BTC/measured}_{CeL\alpha 1} - 0.066 \times I^{BTC}_{BaL\alpha 1} \qquad (2)$$

C) An analysis using an alternative spectral line for Ce that is not interfered with by other lines, i.e., the Ce-Lβ_1 line, as illustrated in figure 5b. Here, two background positions (B1, B2) have to be precisely determined because of the presence of the Ba-L$\beta_{2,15}$ and Ce-Lβ_3 peaks in the vicinity of the Ce-Lβ_1.

For the WDS measurements the elemental P/B ratios were maximized to obtain a high analytical sensitivity (i.e., low detection limits) using optimized conditions for the analysing crystals (P and B positions, slit size and position) and the detector's electronics (pulse-height analyser settings: bias, baseline, window). The BTC and BBTN were analysed at 20 kV / 40 nA and the PMN-PT at 15 kV / 60 nA. A high analytical precision was achieved, with the counting times set to be sufficiently long so as to ensure a relative counting error $\sigma_c \leq 1$ % for all the elements. In the case of BTC the corrected *k*-ratios for Ce-Lα_1 were firstly calculated using true net Ce-intensities obtained from Eqs. (1) and (2) and then processed within the matrix correction. The quantification was performed using the $\Phi(\rho z)$-PROZA (Bastin) matrix correction with the oxygen content being calculated using the stoichiometry to nominal cation valencies [6]. To improve the accuracy and to minimise the matrix effects, stable stoichiometric oxide compounds were chosen as the standards [7]. A summary of the experimental conditions for the WDS analyses is given in table 1. The corresponding detection limits (C_{DL}) were calculated using the Ziebold equation [8].

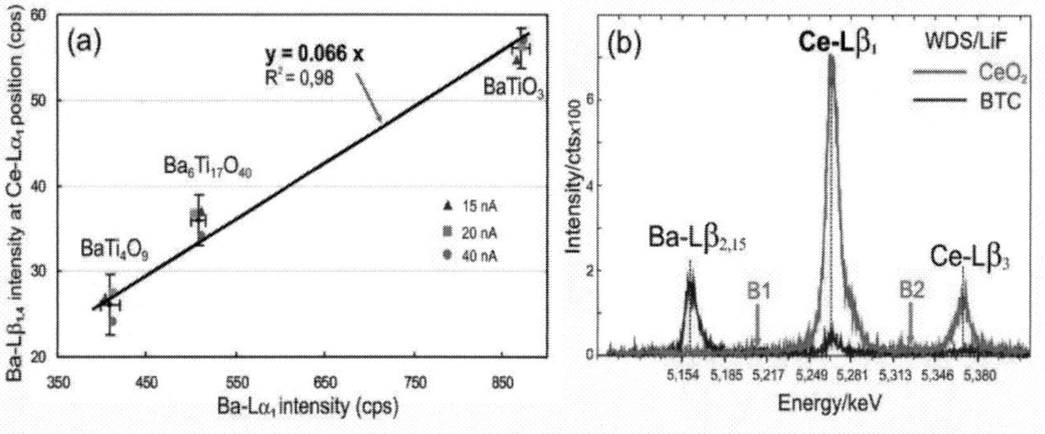

Figure 5. (a) Calibration curve used for the estimation of the Ba-L$\beta_{1,4}$ contribution to the Ce-Lα_1 intensity (data for 15 nA and 20 nA were corrected to nominal current of 40 nA); (b) LiF wavelength scans for the Ce-Lβ_1 line on the CeO_2 standard and on the BTC sample.

Table 1. Experimental set-up for the WDS analyses.

Element	Line	Crystal	P/B ratio	C_{DL} (m.f.%)	Standard
Ba	Lα_1	LiF	143	0.058	$BaTiO_3$
Ti	Kα_1	LiF	162	0.012	$BaTiO_3$
Ce	Lα_1	LiF	120	0.058	CeO_2
Ce	Lβ_1	LiF	57	0.101	CeO_2
Pb	Mα	PET	150	0.044	PbO
Ti	Kα	PET	107	0.005	$SrTiO_3$
Bi	Mα	PET	150	0.035	Bi_2O_3
Nb	Lα	PET	360	0.023	Nb_2O_5
Mg	Kα	TAP	800	0.003	MgO

IOP Conf. Series: Materials Science and Engineering **109** (2016) 012015 doi:10.1088/1757-899X/109/1/012015

3. Results and discussion

3.1. Nb-doped $BaBi_4Ti_4O_{15}$ (BBTN)

The results of the quantitative analysis for the BBTN samples that were prepared with the nominal addition of 5, 8 and 10 mol% of Nb are given in table 2, with the analysis of the undoped sample (BBT) added for comparison.

Table 2. Average atomic fractions (%) from N = 10 measurements on the BBTN samples

Sample	Ba	Bi	Ti	Nb	Ti+Nb	Ba+Bi
BBT	4.26 ± 0.12	16.71 ± 0.16	16.57 ± 0.08	-	16.57	20.97
BBTN5	4.93 ± 0.23	15.94 ± 0.17	15.72 ± 0.15	0.90 ± 0.14	16.62	20.87
BBTN8	5.21 ± 0.24	15.71 ± 0.21	15.20 ± 0.15	1.34 ± 0.16	16.54	20.92
BBTN10	5.56 ± 0.13	15.33 ± 0.18	14.85 ± 0.07	1.72 ± 0.12	16.57	20.89

The results reveal that the concentrations of the Ti and Nb in the doped samples agree well with the nominal composition described by the formula $BaBi_4Ti_{4-x}Nb_xO_{15}$; however, the Ba- and Bi-concentrations differ significantly from the nominal values. The change of the Ba- and Bi-concentrations in the BBTN samples versus the dopant concentration is shown in figure 6a. It is clear that the Ba-concentration increases and the Bi decreases in proportion to the Nb-concentration. Consequently, a systematic change in the Ba/Bi ratio with an increase in the dopant concentration is present, as shown in figure 6b, whereas the total [Ti]+[Nb] remains practically constant and equal to the Ti-concentration in the undoped sample (see table 2), directly verifying that the Nb^{5+} incorporates/substitutes at the Ti^{4+} sites. The excess charge introduced by the substitution of Ti^{4+} with Nb^{5+} is thus compensated by the change in the Ba^{2+}/Bi^{3+} ratio [9]. According to the obtained elemental concentrations, the composition of the BBTN solid solution is reliably described by the general formula $Ba_{1+x}Bi_{4-x}Ti_{4-x}Nb_xO_{15}$.

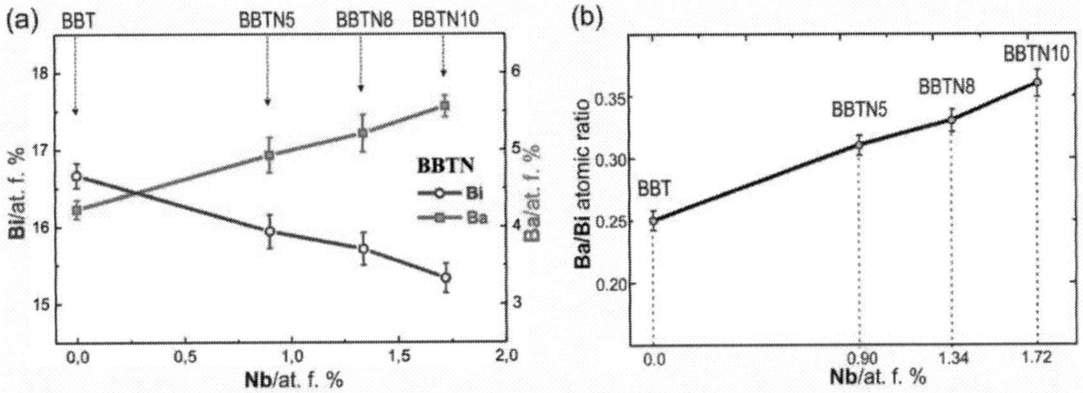

Figure 6. (a) Change of the Bi and Ba concentrations in the BBTN samples; (b) variation of the Ba/Bi ratio versus the Nb-dopant concentration.

3.2. Solid-solution $Pb(Mg_{1/3}Nb_{2/3})O_3$–$PbTiO_3$ *(PMN-PT)*

The PMN-PT single crystal was produced from the starting 70PMN-30PT polycrystalline precursor. After the synthesis, the final composition of the PMN-PT single crystal, which was several millimetres in size, was analysed in 10 microscopic areas that were positioned randomly over the crystal. In each of these areas 10 analytical points were selected within a circle of about 20 µm in diameter. This sampling made it possible to study the composition of the crystal on both the macroscopic mm-scale (between the areas) and the microscopic µm-scale (within the areas). The total experimental uncertainty was then calculated from the corresponding components of variance as

$$\sigma_{exp} = [(\sigma_{wp})^2 + (\sigma_{bp})^2 + (\sigma_{ba})^2]^{1/2}$$

where $(\sigma_{wp})^2$ is the variance within the points associated with the counting statistics, $(\sigma_{bp})^2$ is the variance between the points within an individual area, and $(\sigma_{ba})^2$ is the variance between the areas [10]. The quantitative results for one representative microscopic area on the PMN-PT sample are presented in table 3.

Table 3. Quantitative results for one typical area with 10 analytical points.

	Mass fractions %					
Point	**Pb**	**Mg**	**Nb**	**Ti**	**O**[a]	Total[b]
1	65.25	1.69	13.20	4.97	15.12	100.23
2	65.29	1.70	13.15	4.93	15.09	100.16
3	65.39	1.70	13.20	4.95	15.10	100.34
4	65.65	1.71	13.15	4.96	15.07	100.54
5	65.46	1.70	13.12	4.95	15.08	100.31
6	64.99	1.69	13.13	4.95	15.11	99.87
7	65.24	1.69	13.20	4.97	15.12	100.22
8	65.56	1.69	13.13	4.97	15.07	100.42
9	65.43	1.70	13.14	4.96	15.09	100.32
10	65.30	1.71	13.14	4.95	15.10	100.20
Area average	**65.36**	**1.70**	**13.16**	**4.96**	15.09	100.26
σ_{exp}	± 0.23	± 0.01	± 0.04	± 0.02	± 0.02	± 0.18
± σ_{exp}^{rel} (%)	**0.35**	**0.75**	**0.34**	**0.34**	0.11	0.18
σ_{wp}^2	0.01760	0.00010	0.00101	0.00013		
σ_{bp}^2	0.03463	0.00006	0.00100	0.00016		
Average atomic f. %	**20.00**	**4.43**	**8.98**	**6.57**	**60.02**	
σ_{exp}	± 0.04	± 0.02	± 0.02	± 0.01	± 0.02	
Mg/Nb ratio	0.493 ± 0.003					

[a] By stoichiometry.

[b] Non-normalized analytical total.

Within the microscopic areas all the elemental concentrations are uniform, with minimal scattering between the points and with the relative analytical uncertainties σ_{exp}^{rel} below 1 %, indicating that the composition of the PMN-PT material within the areas is homogeneous. The achieved, non-normalized analytical totals close to 100 % verify the correctness of the analyses. In addition to this, another parameter for the consistency of the analytical procedure was introduced on the basis of the crystal chemistry of the PMN-PT perovskite. Namely, in this case the B-sites in the perovskite structure are occupied by multiple ions Ti^{4+}, Mg^{2+} and Nb^{5+}, and the Ti^{4+} ions are substituted for the complex $(Mg_{1/3}Nb_{2/3})^{4+}$. Thus, it is reasonable to expect that the atomic ratio Mg/Nb = 0.5 should remain unchanged. The Mg/Nb ratio was then calculated and monitored as a significant parameter for the quantitative data [11]. For the results given in table 3, this ratio is 0.493 ± 0.003, which is, taking into account the experimental uncertainty, in very good agreement with expected ratio of 0.5. The elemental atomic fractions fit perfectly to the ABO_3-type perovskite formula for the PMN-PT solid solution, given by $Pb_{1.000\pm0.002}(Mg_{0.221\pm0.001}Nb_{0.449\pm0.001})Ti_{0.328\pm0.001}O_3$.

The quantitative results for 10 analysed areas are given in table 4. Increased scattering σ_{exp}^{rel} between the area averages is evidently present for the Ti (± 2.3 %), for the Mg (± 1.4 %) and for the Nb (± 1.3 %); however, it remains as only ± 0.5 % for the Pb, indicating that the Pb-concentration is uniform over the entire single crystal. The between-the-area variances σ_{ba}^2 for the Mg, Nb and Ti were considerably higher, revealing that a certain degree of heterogeneity over the crystal is present for the B-site cations. The statistical evaluation of the σ_{bp}^2 and σ_{ba}^2 variances has also shown that the differences between the area averages for the Mg, Nb and Ti were statistically significant, while the differences for the Pb were found to be insignificant. The PMN-PT solid-solution formula calculated from the overall average values is given by $Pb_{0.994\pm0.004}(Mg_{0.220\pm0.003}Nb_{0.447\pm0.005})Ti_{0.334\pm0.007}O_3$ with a consistent Mg/Nb ratio of 0.494 ± 0.005.

Table 4. Average quantitative results for 10 areas over the PMNT crystal.

Area	Mass fractions %				
	Pb	**Mg**	**Nb**	**Ti**	**O**
1	65.86	1.72	13.08	5.28	15.18
2	65.36	1.70	13.16	4.96	15.09
3	65.71	1.70	13.31	5.08	15.15
4	65.47	1.70	13.27	5.22	15.22
5	65.76	1.73	13.37	5.05	15.15
6	65.89	1.71	13.29	5.09	15.14
7	65.76	1.68	12.95	5.25	15.13
8	65.40	1.74	13.44	4.97	15.17
9	66.01	1.68	13.11	5.10	15.08
10	65.60	1.73	13.37	5.05	15.16
Grand average	**65.68**	**1.71**	**13.23**	**5.10**	15.15
σ_{exp}	± 0.33	± 0.02	± 0.17	± 0.12	± 0.04
$\pm\ \sigma_{exp}^{rel}$ /%	**0.50**	**1.43**	**1.28**	**2.27**	0.26
σ_{ba}^2	0.04811	0.00044	0.02394	0.01292	
Average atomic f. %	**19.90**	**4.42**	**8.94**	**6.69**	**60.05**
σ_{exp}	± 0.08	± 0.05	± 0.10	± 0.14	± 0.03
Mg/Nb ratio	0.494 ± 0.005				

Taking into account all the quantitative results, the composition of the PMN-PT single crystal is described by the formula $Pb(Mg_{1/3}Nb_{2/3})_{0.67}Ti_{0.33}O_3$, i.e., 67PMN-33PT, which is significantly different from the starting formula. The variations that were measured for the B-site cation concentrations between the analysed areas over the crystal indicate that an opposite trend in the occupancy between the Ti^{4+} ions, on the one hand, and the Mg^{2+} and Nb^{5+} ions, on the other, is present, as shown in figure 7a. The graph of the relative B-site occupancy, calculated by means of the Ti/(Mg+Nb) atomic ratio (figure 7b), confirms that a slightly unequal occupation of the B-sites is present over the entire grown PMN-PT crystal, with an average ratio of 0.50 and a variation of ± 5 % relative. The unequal B-site occupation is a known phenomenon that takes place in Pb-based relaxor ferroelectrics. In this work the fluctuations observed for Ti (and also for Mg and Nb) were quantitatively evaluated by

EPMA, so directly verifying the suggestion of an inhomogeneous distribution of the Ti^{4+} concentration in the PMN-PT, as given in [12].

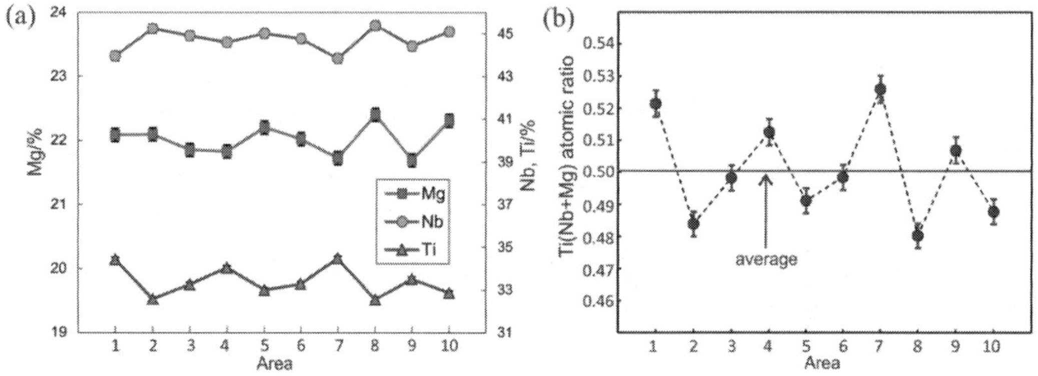

Figure 7. (a) The occupancy of the Ti^{4+}, Mg^{2+} and Nb^{5+} B-site ions between 10 areas; (b) the relative occupancy between the areas, given by the Ti/(Mg+Nb) ratio.

3.3. Ce-doped $BaTiO_3$ (BTC)

A summary of the quantitative results for 10 randomly selected BTC grains is given in table 5, where the correction of the Ba-Ce peak overlap was performed using equation (1). The good counting statistics achieved (σ_c^{rel} < 1 %) and the high analytical sensitivities ensured that the elemental concentrations were measured with great precision and accuracy. The between-the-grains scattering (σ_{bg}) for the Ba and Ce is evidently larger than the uncertainty within the individual grains (σ_{wg}), whereas the values of σ_{bg} and σ_{wg} for Ti remain similar. This suggests that certain compositional variations for Ba and Ce between the grains are present, whereas the Ti-concentration remains uniform, as shown in detail in figure 8.

Table 5. Quantitative analysis of 10 BTC grains.

n = 10 grains	Mass fractions %		
	Ba	Ce	Ti
average C ± σ_{wg}	54.38 ± 0.17	4.63 ± 0.08	20.13 ± 0.10
between grains σ_{bg}	± 0.59	± 0.67	± 0.12
analytical sensitivity	0.14	0.03	0.05
relative σ_{wg}	0.3 %	1.5 %	0.5 %
rel. count. error σ_c^{rel}	0.3 %	0.8 %	0.3 %

Since the sum [Ba]+[Ce] remains constant (within the range ± σ_{wg}) the observed opposite trend between the Ba and Ce variations indicates that in this sample Ce ions incorporate at the Ba-sites in the $BaTiO_3$. The results of the WDS quantitative analyses obtained using the three, above-described analytical approaches (A), (B) and (C) as well as using the analysis of the Ce-Lα_1 line without an overlap correction (D) are given in table 6.

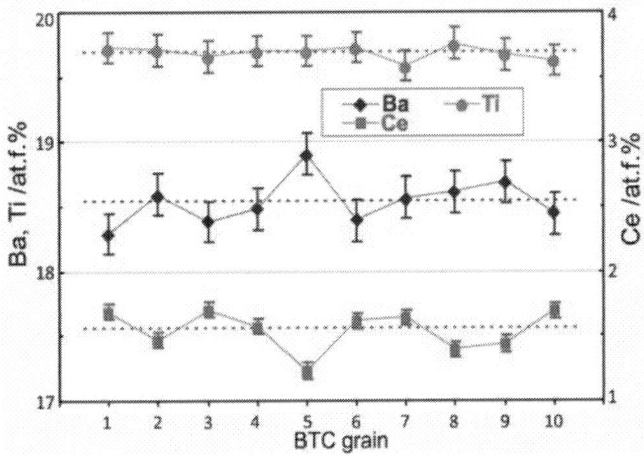

Figure 8. Elemental concentrations between the BTC grains showing an opposite change between the Ba- and Ce-values, and uniform Ti-values.

Table 6. Comparison of the quantitative results of the BTC sample obtained with different methods of analysis.

Method	Mass fractions (%)					Atomic fractions (%)			
	Ba	**Ti**	**Ce**	**O**	Total[a]	**Ba**	**Ti**	**Ce**	**O**
(A)	54.38	20.13	4.63	20.64	99.78	18.54	19.69	1.55	60.23
(B)	54.37	20.13	4.73	20.64	99.87	18.52	19.67	1.58	60.23
(C)	54.69	20.16	4.39	20.63	99.88	18.62	19.69	1.47	60.22
(D)	54.16	20.13	7.22	20.49	102.00	18.10	19.30	2.36	60.24

[a] Non-normalized analytical total (m. f. %).

The presented results reveal that both methods that were introduced for the correction of the Ba-L$\beta_{1,4}$–Ce-Lα_1 peak overlap, i.e., (A) with eq. (1) and (B) with eq. (2), provide similar quantitative results. The elemental concentrations fit perfectly to the perovskite solid-solution formula $Ba_{0.923\pm0.004}Ce_{0.077\pm0.001}Ti_{0.981\pm0.002}(V_{Ti})_{0.019\pm0.001}O_3$ or in general form by $Ba_{1-x}Ce_xTi_{1-x/4}(V_{Ti})_{x/4}O_3$, where $x = 0.08$ and V_{Ti} are the vacancies at the Ti-sites. This result directly confirms that in the selected sample Ce incorporates into the $BaTiO_3$ as donor Ce^{3+} at the Ba^{2+} sites. In this case, the excess donor charge is compensated with the formation of ionized vacancies at the Ti-sites, V_{Ti} [13]. Direct evidence for this mechanism of charge compensation is that the measured deficit (from the nominal 20.00 at.f. %) for the Ti-concentration, which is attributed to $[V_{Ti}]$, is equal to $[Ce^{3+}]/4$.

The non-normalized analytical totals close to 100 % verify the correctness of the WDS analytical procedures with the proposed methods (A), (B) and (C). However, it should be noted that the quantification of the Ce-Lβ_1 (method C) is less correct and gives a slightly underestimated (by 6 % relative) Ce-concentration, which is believed to be due to an inaccurate absorption correction caused by the uncertainty in the MAC values for the Ce-Lβ_1 line. The analysis performed by measuring the Ce-Lα_1 line without applying the peak-overlap correction (method D) is obviously erroneous, with a significantly overvalued Ce concentration and, consequently, an excessive analytical total.

4. Conclusion

Quantitative EPMA-WDS microanalyses were applied to determine the chemical compositions of selected complex perovskite ferroelectric materials that were modified by dopants. It was shown that each material requires specific and detailed optimisation of the analytical procedure in order to achieve the best results in terms of the precision and accuracy. The results made it possible to obtain accurate chemical formulae for the compounds/solid-solutions, to define the solubility of the dopants in the host materials, and, indirectly, to evaluate the mode of dopant incorporation and the charge-compensation mechanisms as well. Reliable quantitative compositional data are very important for studies of the structure-composition-property relations of these materials and to provide a correct interpretation of their ferroelectric properties.

References

[1] Bhalla A S, Guo R and Roy R 2000 *Mater. Res. Innovat.* **4** 2-26
[2] Jonker G H and Havinga E E 1982 *Mater. Res. Bull.* **17** 345-350
[3] Roy R 1954 *J. Am. Ceram. Soc.* **37** 581-588
[4] Xu Y 1991 *Ferroelectric materials and their applications.* (Amsterdam: North Holland)
[5] Samardžija Z, Čeh M, Makovec D and Kolar D 1996 *Mikrochim. Acta* **Suppl. 13** 517-523
[6] Bastin G F, Dijkstra J M and Heijligers H J M 1998 *X-ray Spectrom.* **27** 3-10
[7] Lifshin E and Gauvin R 2001 *Microsc. Microanal.* **7** 168-177
[8] Ziebold T O 1967 *Anal. Chem.* **39** 858-861
[9] Samardžija Z, Makovec D and Čeh M 2002 *Mikrochim. Acta* **139** 159-163
[10] Marinenko R and Leigh S 2004 *Microsc. Microanal.* **10** 491-506
[11] Samardžija Z, Jeon J-H and Čeh M 2007 *Mater. Charact.* **58** 534-543
[12] Tu C-S, Chen L-F, Schmidt V H and Tsai C-L 2001 *Jpn. J. Appl. Phys.* **40** 4118-4125
[13] Makovec D, Samardžija Z and Kolar D 1996 *J. Solid State Chem.* **123** 30-38

IOP Conf. Series: Materials Science and Engineering **109** (2016) 012016 doi:10.1088/1757-899X/109/1/012016

Improved spectrum simulation for validating SEM-EDS analysis

P Statham[1], C Penman[1] and P Duncumb[2]

[1] Oxford Instruments NanoAnalysis, High Wycombe HP12 3SE, Great Britain
[2] Willow House, Great Shelford, Cambridge CB22 5LZ, Great Britain

E-mail: peter.statham@oxinst.com

Abstract. X-ray microanalysis by SEM-EDS requires corrections for the many physical processes that affect emitted intensity for elements present in the material. These corrections will only be accurate provided a number of conditions are satisfied and it is essential that the correct elements are identified. As analysis is pushed to achieve results on smaller features and more challenging samples it becomes increasingly difficult to determine if all conditions are upheld and whether the analysis results are valid. If a theoretical simulated spectrum based on the measured analysis result is compared with the measured spectrum, any marked differences will indicate problems with the analysis and can prevent serious mistakes in interpretation. To achieve the necessary accuracy a previous theoretical model has been enhanced to incorporate new line intensity measurements, differential absorption and excitation of emission lines, including the effect of Coster-Kronig transitions and an improved treatment of bremsstrahlung for compounds. The efficiency characteristic has been measured for a large area SDD detector and data acquired from an extensive set of standard materials at both 5 kV and 20 kV. The parameterized model has been adjusted to fit measured characteristic intensities and both background shape and intensity at the same beam current. Examples are given to demonstrate how an overlay of an accurate theoretical simulation can expose some non-obvious mistakes and provide some expert guidance towards a valid analysis result. A new formula for calculating the effective mean atomic number for compounds has also been derived that is appropriate and should help improve accuracy in techniques that calculate the bremsstrahlung or use a bremsstrahlung measurement for calibration.

1. Introduction

X-ray microanalysis by SEM-EDS requires corrections for the many physical processes that affect emitted X-rays for elements present in the material. These corrections are only applicable under the following conditions: specimen composition must be uniform within the region excited by scattered electrons, the surface must be flat over the region where any X-rays exit towards the detector, the geometry for incident beam, surface and X-ray detector must be known, the incident focussed electron beam must not be affected by specimen charging, the correct elements must be identified and suitable line emissions must be excited and measurable. As analysis is pushed to achieve results on smaller features and more challenging samples it becomes increasingly difficult to determine if all these conditions are upheld and whether the analysis results can be trusted.

If results are calibrated by a standard or beam current measurement, an analytical total close to 100 % gives one level of reassurance. When results are normalized for speed and convenience, an alternative "Check Total" can be used [1] but neither method is infallible and even when the total is bad, neither measure provides any information to diagnose the cause. Therefore, we have investigated an idea suggested by Duncumb [2] that it would be useful for the operator to be presented with a theoretical simulated spectrum based on the analysis result. Provided the simulation was accurate, it could indicate problems with the analysis and thus prevent serious mistakes in interpretation. For example, when more than one line emission is generated (e.g., K, L or M), the simulation would show if the observed L/K or M/L ratio was consistent with the theory. If the wrong elements were identified or an element was missing then line ratios or peak/background levels would be different from the observed spectrum. Likewise, specimen charging, inhomogeneity and geometry issues may also be exposed.

2. Current accuracy of theoretical spectrum simulation

To be useful for validating results, the theoretical simulation has to be sufficiently accurate. DTSA-II is a well-known public domain programme that provides SEM-EDS spectrum simulation [11] (Indeed, Fiori's original DTSA was the first available programme to offer this capability). The 2014 version (Halley 2014-08-01) provides an "instantaneous" simulation based on a series of formulae ("Analytical Model") or a longer Monte Carlo (MC) simulation that sums the emissions due to many randomized electron trajectories based on various theories for fundamental electron interactions. Figure 1 shows an example of DTSA-II results for a pure Ni specimen compared with experiment where the spectra have been normalized to have the same intensity across a region spanning the Ni K peaks. In figure 1a, whereas the background for the MC simulation is fairly close to that measured, the analytical model shows a large deficit for most of the energy range. In figure 1b, it is apparent that both simulations overestimate the Ni-Lα intensity and the Ni-Lα/Lℓ ratio is higher than that measured. In figure 2, the results for simulations for pure Au (this time normalized using a region spanning the Au L peaks) again show that the MC simulated background is close to that measured but the analytical model is poor and both types of simulation give a poor match to the Au M intensity in figure 2b.

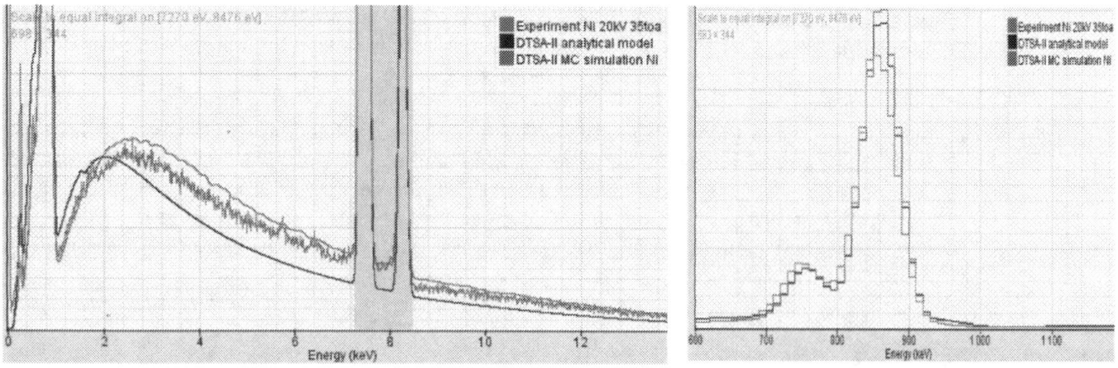

Figure 1. a) DTSA-II (2014) Ni simulations 20 kV compared with experiment (X-max80 SDD, red curve). b) Same, low energy region.

For bremsstrahlung background modelling DTSA-II MC looks promising but for characteristic line intensities even the most elaborate MC programmes are limited in accuracy by uncertainty in fluorescence yields and transition probabilities. Furthermore, for a real-time interactive tool, an analytical model is considerably faster to compute so we chose to refine an analytical model to see what could be achieved in terms of accuracy.

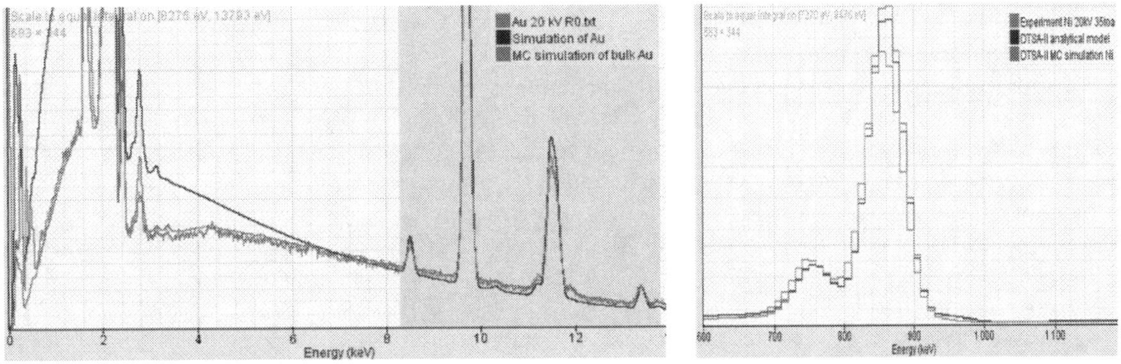

Figure 2. a) DTSA-II (2014) Au simulations 20 kV compared with experiment (X-max80 SDD, red curve). b) Same, low energy region.

3. Improved theoretical model

We started with an earlier parameterized analytical model [3] and extended it to incorporate differential absorption of emission lines and differential excitation of shells, including the effect of Coster-Kronig transitions. Thus, for an element A of mass fraction C_A, the average characteristic intensity emitted per electron per unit solid angle for a given line is

$$I_A = (\omega_A \, p_A) \bullet R_A N_0 C_A / A_A \int Q_A/S.dE \bullet f(\chi)_A \bullet (1 + F) \bullet (1 + G) \bullet CK_{ss} \qquad (1)$$

where ω_A is the fluorescence yield for element A for the appropriate shell, and p_A is the proportion of emission for the line. A_A is the atomic weight of element A, and N_0 is Avogadro's number. $(1 - R)$ is the intensity fraction lost by backscatter. $S = -dE/d(\rho x)$ is the stopping power or rate of electron energy loss with mass thickness ρx where ρ is the density. Absorption factor $f(\chi)_A$ is for radiation travelling towards the detector and fluorescence enhancement factors $(1 + F)$ and $(1 + G)$ are for characteristic and bremsstrahlung continuum radiation respectively. The cross-section used follows the form in [3],

$$Q = q_s \bullet m \bullet (\ln U) / (E_c^2 \bullet U^m) \qquad (2)$$

where E_c is the critical excitation potential for the analysed peak and U is the overvoltage ratio E/E_c. m is a constant adjusted for each shell and q_s is a multiplier set by experiment to match the observed intensities from each shell to one another and to those results for which the intensities are known in absolute terms. CK_{ss} is the enhancement factor for primary vacancies due to Coster-Kronig non-radiative transitions. For the L1, L2, L3 sub-shells, $CK_{L1} = 1$, $CK_{L2} = 1 + f_{12} \bullet N_1 / N_2$ and $CK_{L3} = 1 + (f_{13} + f_{23} \bullet f_{12}) \bullet N_1 / N_3 + f_{23} \bullet N_2 / N_3$, where f_{ij} is the probability for shifting a vacancy from subshell i to j and N_i is the relative number of primary vacancies in subshell i [4]. As in [3], the bremsstrahlung continuum intensity is represented by the following parameterized form

$$I_v = q_c \bullet Z \bullet F \bullet (U - 1)^P \qquad (3)$$

where F and P are parameterized functions of Z and E_0 and q_c is an overall multiplier used to match predicted intensity to a specific absolute measurement. Additional corrections for absorption and backscatter are applied as in [3]. In [3] the parameterisation was optimized to achieve a best-fit to a large set of measurements of the ratio of peak intensity to total background intensity thus avoiding the need to correct for beam current. In the present work we have made frequent reference measurements on a pure element standard so that we could independently optimize the models for characteristic emissions and bremsstrahlung background.

A source of inaccuracy in previous published attempts to generate a parameterized spectral form is the assumed detector efficiency as a function of energy. Although a parameterisation may work well for one particular detector, if the efficiency is not correct the analytical formulae may not work so well on a different detector. In the current study, rather than rely on an estimate based on an assumed structure for the detector and window, an efficiency characteristic was measured using the method described in [5] for a reference X-Max 80 SDD detector. Using this efficiency-calibrated detector, 124 spectra were acquired from a large set of reference material standards at both 5 kV and 20 kV covering elements Z = 4 - 92.

Published tables (e.g., Johnson and White [6]) only provide "nominal" relative line intensities because these depend on specimen and excitation method. For previous work to improve accuracy for peak shapes at low kV [7], 603 line intensities were optimized using experimental data and these results were incorporated in the current work to improve simulation accuracy. To deal with differential absorption of individual lines, 56 empirical mass attenuation coefficients were also determined by fitting to spectra obtained at different kV. For 233 element emission series (K, L or M), a computed background was fitted, interpolated and subtracted to obtain total intensity cps for the series. The fluorescence yield for Co-K was fixed at the value deduced from previous measurements at known beam current using a grid-free Be window detector and a machined collimator at measured distance from the specimen. The parameterisation for ionisation cross-section and remaining fluorescence yields were optimized to give the best fit to the ratio of intensity to that for a pure element measured at the same beam current (Si for 5 kV data, Co for 20 kV data). Fluorescence yields were interpolated for elements where no measurement was available. For lines with sufficient intensity, (K for Z >3, L for Z > 20, M for Z > 46), the mean absolute relative error in the new simulation was 3.4 %.

For absolute intensities, the new simulation was tested against a few experimental data points derived from plots obtained by Procop [8] and table 1 shows that the agreement is within a few percent.

Table 1. Predicted counts/electron/steradian compared with measurements corrected for detector efficiency and absorption.

X-ray peak	kV	New simulation	Procop expt [8]
Cu-Kα	18	6.8 E-05	7.1 E-05
Cu-Kα	20	9.5 E-05	9.8 E-05
Ge-Kα	20	5.0 E-05	5.3 E-05
Si-Kα	20	7.3 E-04	7.4 E-04

For pure elements, the bremsstrahlung background intensity depends on atomic number, Z. For compounds a common approach is to use the mass fraction average Z instead, but this does not take into account the physics of bremsstrahlung generation. Calculations of the bremsstrahlung per atom using a screened Coulomb potential [9] show that the emission is not quite proportional to Z^2. The data from [9] for Al and Au for incident electron energies of 10 keV and 50 keV suggests that the intensity for 5 keV bremsstrahlung varies approximately as $Z^{1.75}$. When an electron passes through a thin layer of material, the bremsstrahlung emission is the sum of contributions weighted according to the number of atoms of each element and the total emission along the whole electron path is found by integrating the product of this sum with ($1/S$). The stopping power S_{cmpd} for a compound is the mass fraction average of the stopping power for individual elements j and using Joy and Luo's form for S [10], the stopping power is given by

$$S_{cmpd} \propto \Sigma_j \left[C_j \cdot Z_j / A_j \ln\left(1.166\left[E/J + k\right]\right)\right] \qquad (4)$$

If, following [10], the values $k = 0.8$, $J = 9.76Z + 58.8Z^{-0.19}$ for $Z >= 7$ and $J = 11.5Z$ for $Z < 7$ are used, the term $\ln(1.166[E/J + k]$ can be approximated by $0.66 \cdot E^{0.273} / Z^{0.25}$ $(E \text{ in keV})$ so that the terms involving Z_j can be separated from those involving E when forming the path integral. Since the number of atoms of element j per unit volume is given by $\rho.C_j .N_0/ A_j$, the Z-dependence of the bremsstrahlung can be expressed by using an effective Z for a compound given by

$$Z_{eff} = (\Sigma_j [C_j \cdot Zj^{1.75}/ A_j])/ \Sigma_j [C_j \cdot Z_j^{0.75} / A_j] \qquad (5)$$

This expression was tested by measuring the integrated background intensity from 7.4 - 8.2 keV in spectra from pure elements and compounds (BN, Al_2O_3, CaF_2, NaCl, MgO, GaP, FeS_2, KBr, InAs, SrF_2, HgTe, BaF_2, LaB_6, CeO_2, PbTe) obtained at 20 kV and normalising to the Co-K intensity from the reference standard. At these energies, there is little matrix absorption and the results should fall on a smooth curve that is a function of Z_{eff}. It is clear from figure 3a that this is not the case when Z_{eff} is represented by the concentration-weighted average Z. However, using the new expression (5) for Z_{eff} brings the results for compounds into line with those for pure elements. The improvement is most noticeable for compounds involving elements with large difference in Z (LaB_6, BaF_2, CeO_2).

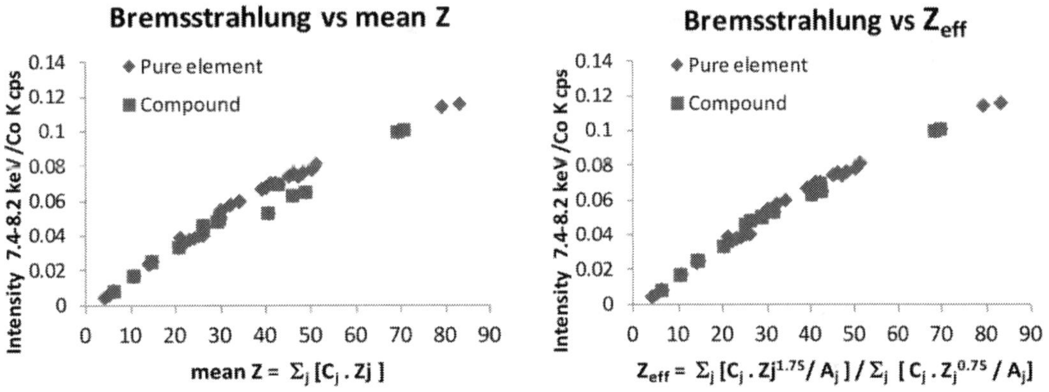

Figure 3. a) Use of conventional mass average Z. b) Eq. (5) used for Z_{eff}.

All peak-free regions were identified in spectra and the parameterisation optimized, making use of the new expression for Z_{eff}, to minimize sum of mean absolute relative errors over all spectra and for all peak-free energy regions. Thus, a mean absolute relative error of 6.3 % was achieved in the predicted background intensity for energies from 90 eV up to half the beam energy E_0.

When the optimized new model is used to simulate the same Ni and Au spectra as used for figures 1 and 2, the background shape and line intensities are now close to the experimental spectra as seen in figures 4 and 5.

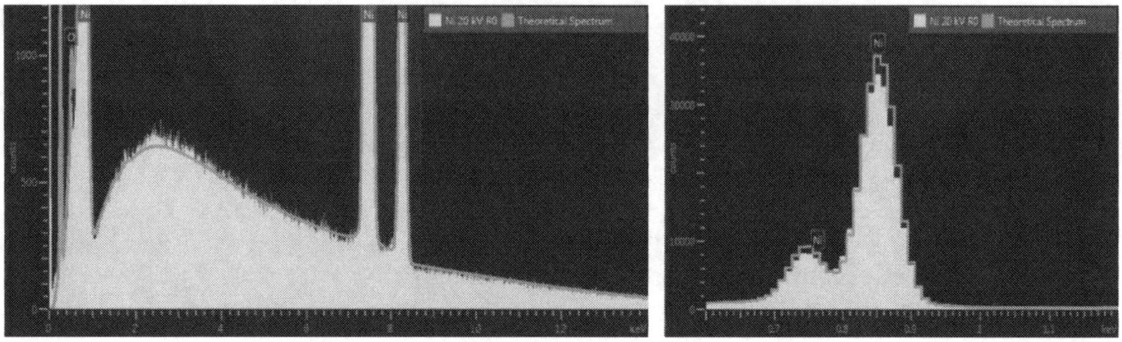

Figure 4. a) New theoretical spectrum for Ni 20 kV compared with experiment (X-max80 SDD, yellow bars). b) Same, low energy region.

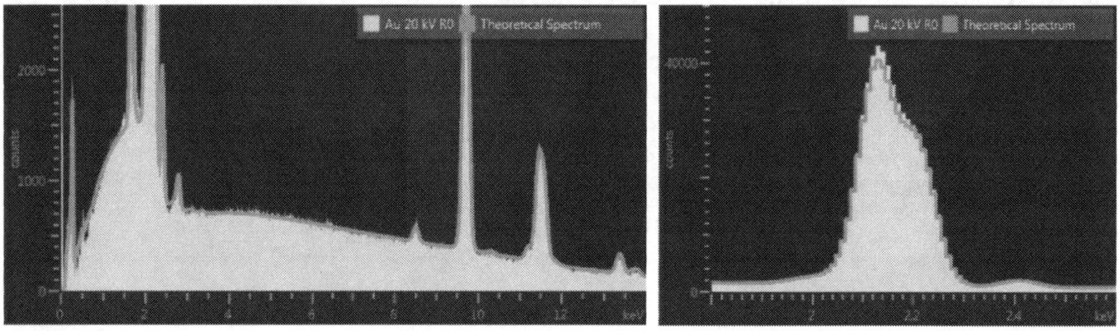

Figure 5. a) New theoretical spectrum for Au 20 kV compared with experiment (X-max80 SDD, yellow bars). b) Same, low energy region.

4. Examples of application

When analysing an unknown, the measured composition given by the analytical results can now be used to simulate a "Theoretical Spectrum" overlay (cyan line) that is accurate enough to show up inconsistencies exceeding 10 %. This overlay differs from the conventional "Fitted Spectrum" (magenta line, as in figure 9) that shows the result of least squares fitting of peaks and background to the spectrum. The following examples show how a theoretical spectrum can be useful in avoiding mistakes in routine microanalysis.

Figure 6 shows a colour image derived from spectrum image data and figure 7 shows a single spectrum that has been reconstructed from pixels from a rectangular region that seems to be within what appears to be a homogeneous band of material. However, the theoretical spectrum based on the analysis result is inconsistent with the Ga and In peak heights. This warns that the analysis is unreliable and warrants further investigation. In this case 20 keV scattered electrons can excite X-rays in layers adjacent to the layer in question and some emerging X-rays also travel through different material on the way to the detector (which is on top in this field of view). In the Ga-K X-ray map in figure 8, the blurring due to electron scattering is more apparent. However, when many maps are combined with the sharp secondary electron image, the layers look distinct and imply a small information volume. The theoretical spectrum mismatch is therefore a useful prompt to check out the other data to see if X-rays are confined to the layer.

IOP Conf. Series: Materials Science and Engineering **109** (2016) 012016 doi:10.1088/1757-899X/109/1/012016

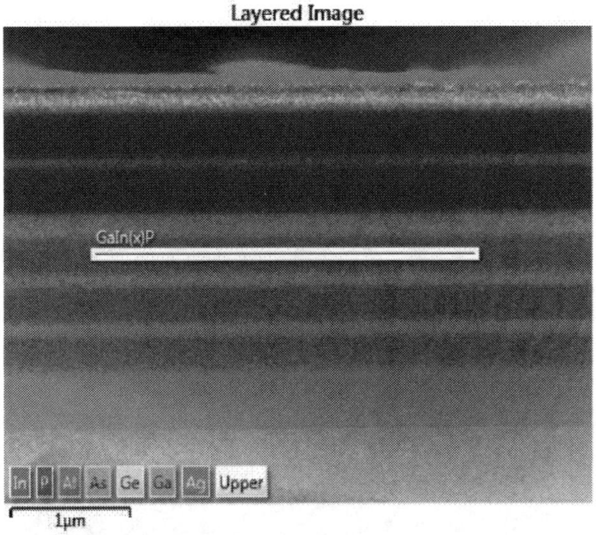

Figure 6. Series of colour X-ray maps overlaid on a secondary electron image of a sectioned semiconductor sample recorded at 20 kV. The white rectangle shows a region of interest and the spectrum in figure 7 is the result of summing all pixel spectra within this region.

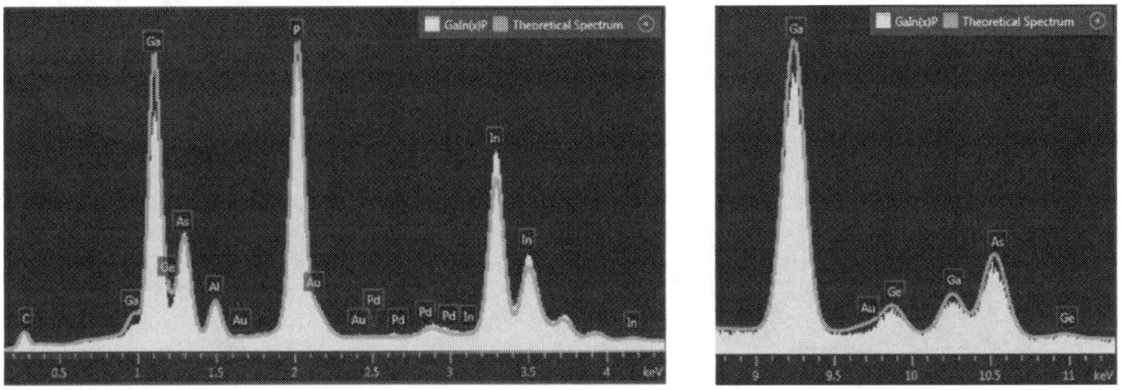

Figure 7. a) Sum spectrum for boxed region in figure 6. b) Same, region near Ga-K peak.

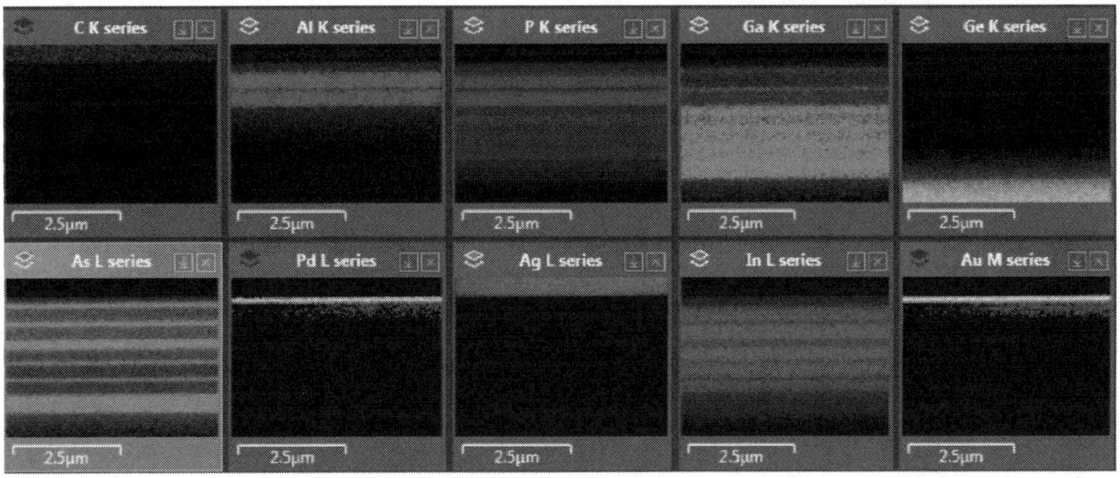

Figure 8. Individual X-ray maps for same field of view as for figure 6.

IOP Conf. Series: Materials Science and Engineering **109** (2016) 012016 doi:10.1088/1757-899X/109/1/012016

Figure 9 shows a spectrum of Ni-based superalloy where the conventional fitted spectrum overlay (magenta) is a good match. Including Sr-L improves the fit at 1.8 keV and Zn-K has been chosen in preference to Re-L at 8.6 keV. The analysis total in table 2 is only slightly high, but the theoretical spectrum in figure 10 shows that if Zn were present at 0.92 %, there should be a Zn-L peak at 1 keV. When Zn and Sr are removed and replaced with Re, figure 11 shows much better agreement with the theoretical spectrum and the analytical total in table 3 is also improved.

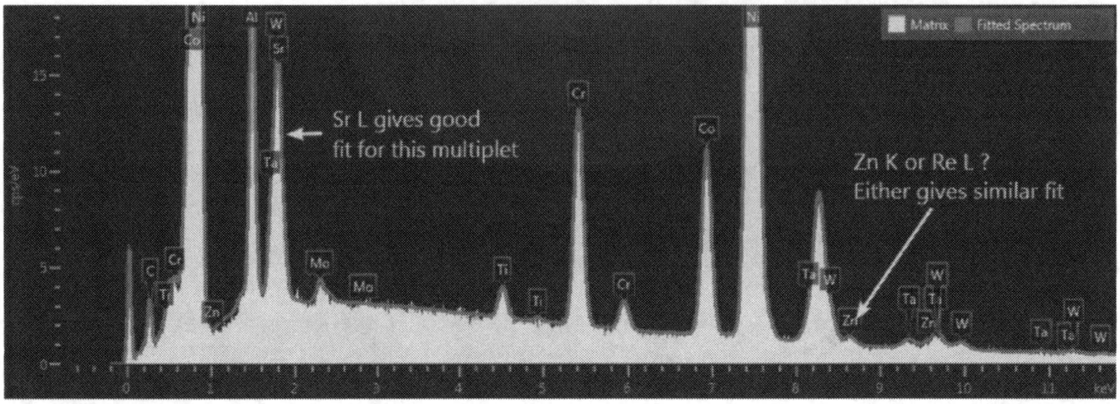

Figure 9. Spectrum from Ni-based superalloy (yellow) overlaid with result of spectrum fitting (magenta).

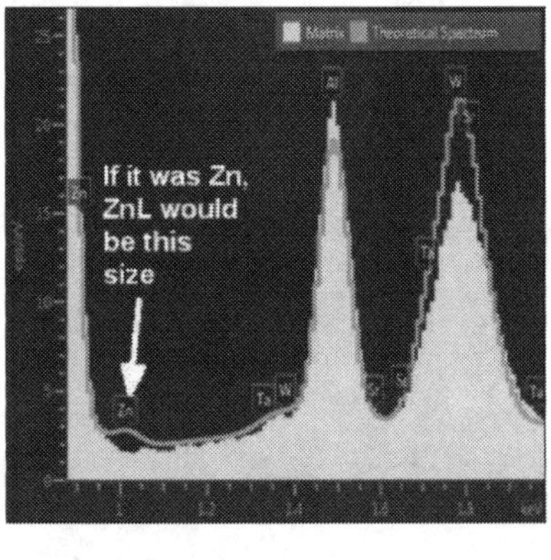

Figure 10. Theoretical spectrum for table 1 composition overlaid on original spectrum. Note differences near Zn-L, Ta-M.

Elmt	Wt%
Al	5.22
Ti	0.72
Cr	6.31
Co	9.63
Ni	62.36
Zn	0.92
Sr	2.93
Mo	0.71
Ta	4.72
W	9.25
Total	102.77

Table 2. Results for figure 10.

At low kV, some lines may not even be excited so that analysis results normalized to 100 % can conceal gross errors. The conventional fitted spectrum (magenta) is a good match in both figures 12 and 13. In figure 12, a 4 kV spectrum of Quartz (SiO_2), the theoretical spectrum is also a good match. However, for figure 13 when the analysis results, Si: 44 % O: 56 %, are used to calculate the theoretical spectrum, the inconsistent peak heights warn of a serious problem so the operator should not assume this is also SiO_x. The analysis is wrong here because at 4 kV the Ca-K peak from Wollastonite ($CaSiO_3$) is not excited so Ca is not even detected.

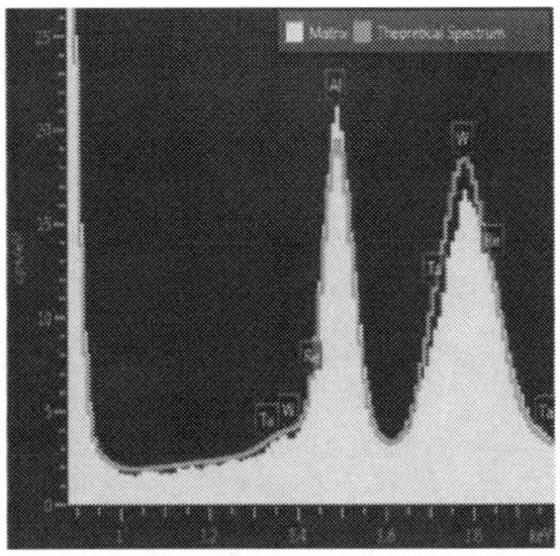

Figure 11. Theoretical spectrum with Re instead of Zn and Sr.

Elmt	Wt%
Al	5.26
Ti	0.72
Cr	6.30
Co	9.62
Ni	62.30
Mo	0.67
Ta	4.63
W	8.76
Re	2.37
Total	100.63

Table 3. Results for figure 11.

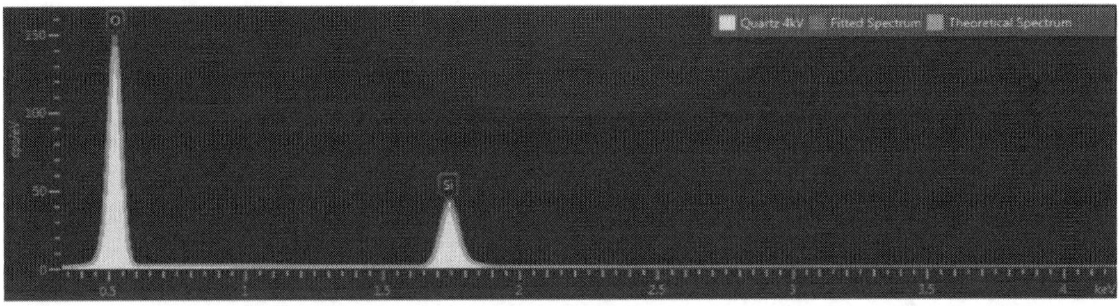

Figure 12. SiO_2 4 kV spectrum overlaid with fitted (magenta) and theoretical spectrum (cyan).

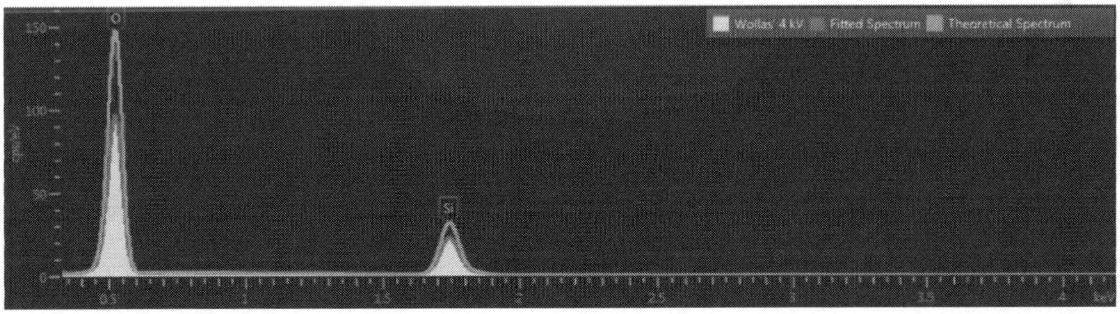

Figure 13. $CaSiO_3$ 4 kV spectrum overlaid with fitted (magenta) and theoretical spectrum (cyan).

5. Conclusion

To give some confidence in a compositional analysis result, the operator is usually presented with the result of fitting peaks and background to the spectrum but a good fit can often be achieved even when the elements or composition results are wrong. However, with a sufficiently accurate theoretical model, a simulated spectrum corresponding to the measured elemental composition can be overlaid on the measured spectrum and, unlike the curve showing results of fitting, this theoretical spectrum can expose some non-obvious mistakes and help avoid bad conclusions. In this work, we have developed a parameterized theoretical model that is sufficiently accurate to provide useful diagnostics and some expert guidance towards a valid analysis result. A new formula for calculating the effective mean

IOP Conf. Series: Materials Science and Engineering **109** (2016) 012016 doi:10.1088/1757-899X/109/1/012016

atomic number for compounds has also been derived that is appropriate and should improve accuracy for techniques that calculate the bremsstrahlung or use a bremsstrahlung measurement for calibration.

References

[1] Statham P J 2004 *Microchim. Acta* **145** 229-235
[2] Duncumb P 1994 *Mikrochim. Acta* **114/115** 3-20
[3] Duncumb P, Barkshire I R and Statham P J 2001 *Microsc. Microanal* **7** 341-355
[4] Bambynek W, Craseman B, Fink R W, Freund H U, Mark H, Swift C D, Price R E and Rao P V 1972 *Rev. Mod. Phys.* **44** 716-813
[5] Statham P J 2010 *Microsc. Microanal.* **16** (Suppl. 2) 1304-1305
[6] Johnson G G Jr. and White E W 1970 *X-ray emission wavelengths and keV tables for nondiffractive analysis.* ASTM Data Series DS 46 (Philadelphia, PA: American Society for Testing and Materials)
[7] Statham P J 2014 *Microsc. Microanal.* **20** (Suppl. 3) 690-691
[8] Procop M 2004 *Microsc. Microanal.* **10** 481-490
[9] Reimer L 1998 *Scanning electron microscopy, 2nd edition* Springer Series in Optical Sciences, vol. 45. (Heidelberg: Springer) 384
[10] Joy D C and Luo S 1989 *Scanning* **11** 176-180
[11] Ritchie N W M 2009 *Microsc. Microanal.* **15** 454-468

IOP Conf. Series: Materials Science and Engineering **109** (2016) 012017 doi:10.1088/1757-899X/109/1/012017

Development of soft X-ray emission spectrometer for EPMA/SEM and its application

H Takahashi[1], T Murano[1], M Takakura[1], S Asahina[1], M Terauchi[2], M Koike[3], T Imazono[3], M Koeda[4] and T Nagano[4]

[1] JEOL Ltd, 1-2 Musashino, 3 chome, Akishima, Tokyo 196-8558, Japan
[2] Tohoku Universities, Sendai 980-8577, Japan
[3] Japan Atomic Energy Agency 8-1 Umemidai, Kizu 619-0215, Japan
[4] SHIMADZU Corp., 1, Nishinokyo-Kuwabaracho, Nakagyo-ku, Kyoto 604-8511, Japan

E-mail: takahash@jeol.co.jp

Abstract. A newly developed wavelength-dispersive soft X-ray emission spectrometer (WD-SXES) with two kinds of gratings, JS50XL and JS200N, were installed on electron probe microanalysers (EPMA) and scanning electron microscopes (SEM). The new detector covers the energy range from 50 to 210 eV with an energy resolution of better than 0.2 eV at Al-L emission on Al metal. With this low energy range and high energy resolution, various kinds of X-ray lines of K, L, M, N emission spectra from lithium to uranium could be observed and chemical state analysis carried out. This WD-SXES has also a high potential for analysing trace light elements under 100 ppm. The design, having no mechanically scanning components, allows parallel spectral acquisition over the entire energy range of each grating (50 to 170 eV and 70 to 210 eV).

1. Introduction

Due to the ongoing evolution of the hardware and, later, the software [1] of EPMA/SEM since their development over half century ago, these instruments have continued to be applied to not only scientific but also industrial fields. According to the development and association of various detectors and analysers, instrument capabilities have been expanded to observe and analyze into the microscale and even the nanoscale. The drive for further improvements are continuing for many kinds of fields. On the other hand, the history of transmission electron microscope (TEM) is longer and it provides invaluable capabilities for the characterisation and understanding new materials by techniques such as high resolution observation up to atomic resolution, combined with elemental analysis by SDD (silicon drift detector), crystalline information by electron diffraction and chemical state analysis by electron energy loss spectrometer (EELS). However, sample preparation for TEM is often time intensive compared to that of EPMA/SEM.

Up until now, a conventional method of chemical state analysis with EPMA/SEM has largely been lacking. Our newly developed wavelength-dispersive soft X-ray emission spectrometer (WD-SXES, JEOL SS-94000 SXES) for EPMA/SEM has a high energy resolution of 0.2 eV for Al-L emission spectrum of Al metal [2-4]. This resolution is competitive to that of EELS and XPS. This capability of chemical state analysis realizes the full characterisation of bulk samples with EPMA/SEM using three techniques: elemental analysis by EDS/WDS, crystalline analysis by EBSD and chemical state analysis by WD-SXES. In this report, the principle and the development of the new WD-SXES spectrometer is presented. Application examples for practical samples and their several functions are also demonstrated. The applicable EPMA and SEM are as follows: EPMA; JXA-8500F, -8100, -8200, 8530F, -8230, SEM; JSM-7100F, -7800F, -7800F*Prime*, -7200F.

2. Principle

The diffraction grating has been well used as a device of soft X-ray spectroscopy for many years. However, due to the poor sensitivity and narrow X-ray wavelength (energy) region of spectrometers, this method was not widely employed. Due to the recent innovations of very sensitive back-illuminated CCD cameras and a varied line spacing grating, in which the pattern of line spacing is designed to optimize X-ray detection, a new spectrometer has been developed. In this new spectrometer, the project designed the specification which covered the energy range from 50 to 170 eV. It is possible to detect Li-K emission, and the energy resolution is at least 0.3 eV at Al-L emission on Al metal. Moreover, this analysis system provides parallel detection of X-rays over its whole energy range. For the purpose of detection from 50 to 170 eV with high energy resolution, the new type of grating is named JS50XL. For the requirement for more trace element analysis for B, C, and N, a 2nd grating, JS200N, was developed, which covers the energy range from 70 to 210 eV. The spectrometer can have both kinds of gratings installed and can be alternated by a simple switch. When the gratings are switched, the position reproducibility is very high, resulting in no requirement for recalibration.

2.1. Spectroscopic principle due to varied line spacing grating

In order to take parallel detection for several X-rays at the same time, this system does not apply the Rowland circle geometry but uses the fixed grating with varied line spacing (5). The groove is 1,200 lines per mm at the centre of grating. Intermediate between the grating and sample, the focal mirror was applied. It is flexible to adjust the optimum focal condition, comparable to poly-capillary optics. The reproducibility of spectra is expected to be under 1 or 2 pixels difference. It has extremely good performance which is capable of providing chemical state spectra. The spectroscopic principle of the laminar type varied line spacing grating is described as shown in eq. 1 and figure 1 in which the incident angle is constant and the exit angle is dependent on the wavelength of the X-rays. In this laminar type grating, the high order X-ray intensity is smaller than that of brazed type of grating. Normally it is preferred to avoid interference with higher order X-rays. However, these higher order reflections are useful to allow the detection of higher energy X-rays within the energy ranges of the gratings: Using 1st order X-rays, the detection energy range is from 50 to 210 eV but, using higher order X-rays, this is extended in practice up to approximately 700 eV.

$$d(\sin\alpha + \sin\beta) = m\lambda \qquad (1)$$

where d = width, α = incident angle, β = exit angle, λ = wavelength, and m = order of diffraction.

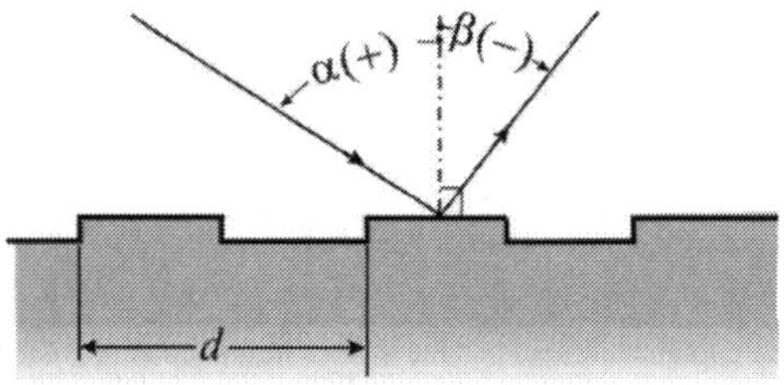

Figure 1. Schematic illustration of X-ray diffraction from a grating surface.

In conventional WDS, the X-rays are dispersed by Bragg's law using a Rowland circle geometry of X-ray source, analysing crystals and detector. In this new design, the focal plane is flat field at the CCD device using the varied line spacing grating. This grating pattern, which is produced by a holographic exposure method [7], was optimized for high energy resolution and small scatter. The grating is slightly concave with correction for aberration and greater groove spacing closer to the X-ray generation point. As shown in figure 2, X-rays with the same energy but emitted in different directions are diffracted by the grating to the same height on the CCD detector, and exhibited as a spectrum in the PC system. This is very similar to the EELS system.

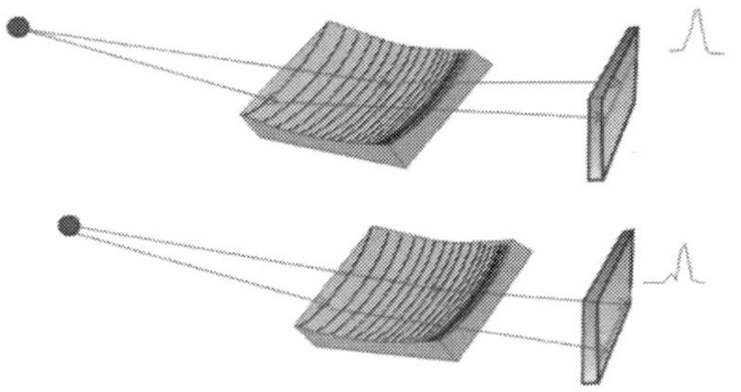

Figure 2. Illustration of parallel detection method of different energy X-rays.

The geometry of this spectrometer is shown in figure 3. In case of JS50XL, the incident angle, α, is 86° and emergent angle, β, is 87°. The size of grating is 50 x 30 mm. From these parameters, the solid angles are approximately estimated to be 0.017 mstr (JS200N) and 0.024 mstr (JS50XL), respectively.

2.2. Back-illuminated CCD
In order to detect low energy X-rays (soft X-rays), a very sensitive non-coated back-illuminated CCD camera has been utilized. The camera is a full-frame back-illuminated CCD camera, 2048 x 2048 pixels, pixel size 13.5 x 13.5 µm, dynamic range of electric charge; 80,000 electrons. The soft X-rays are detected at the Si base plate at the front side of the CCD device without X-ray absorption. This type of CCD is superior for low noise and good sensitivity in detecting low energy X-rays from 10 eV to approximately 1,000 eV than that of a conventional CCD.

3. Feature
According to the specification of the spectrometer, soft X-rays from 50 eV to 210 eV can be detected with a high energy resolution. From the X-ray diffraction eq. 1, higher order X-rays can also be detected. From practical experimental results, X-rays up to approximately 700 eV can be detected.

3.1. What kinds of X-rays
In this low energy range, the possibility of detection for soft X-rays using conventional WDS or EDS is very poor. However, the developed SXES can provide a range of possibilities for these kinds of soft X-ray lines. As these X-ray lines are unfamiliar for microanalysis users, the Bearden's wavelength table is useful to give an overview and to consider the different types of soft X-ray lines that could be detected [8]. From this table, the relationship between atomic number and various X-ray energies were plotted and shown in figure 4. The region from 50 eV to 210 eV is coloured light grey and that from 210 to 700 eV is coloured dark grey. From this figure, it is understood that SXES that it is

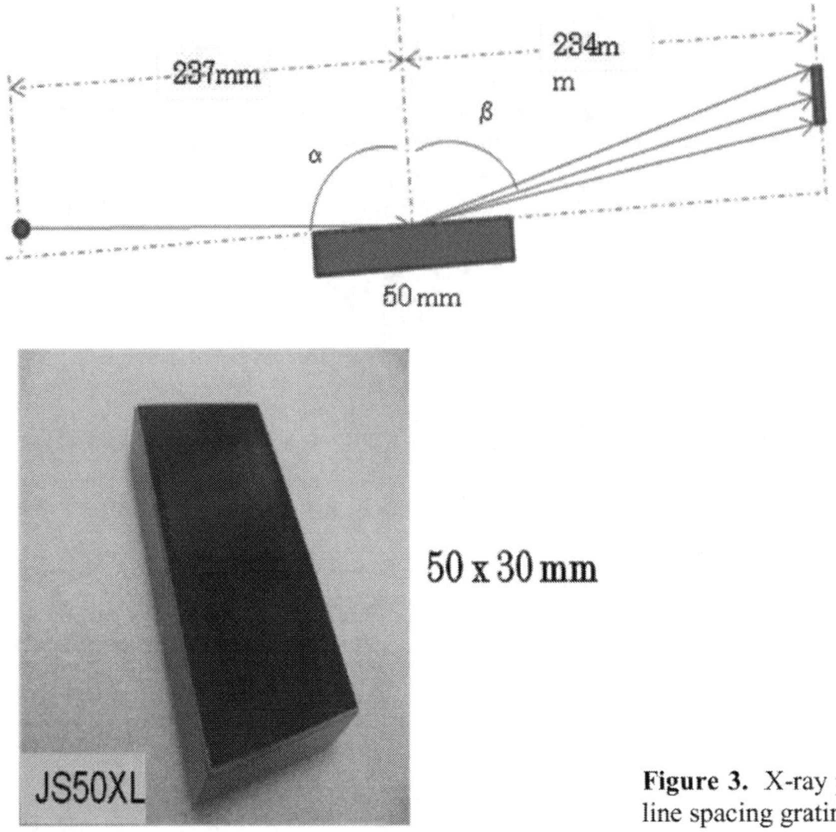

Figure 3. X-ray path of SXES and varied line spacing grating JS50XL.

possible to analyse almost every element from lithium to uranium, with the exception of only a few elements.

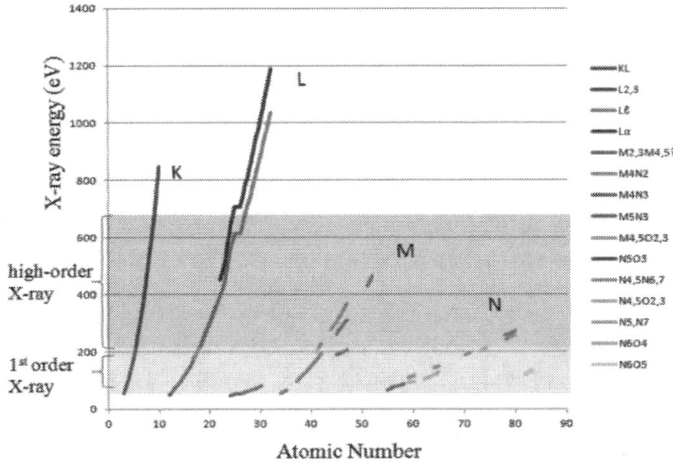

Figure 4. Plotted graph from table 5 in Ref. [5].

3.1.1. K emission spectra. Observable 1st order K emission spectra of SXES are Li, Be and B. Using the higher order K emission spectra, C, N, O and F can be observed. With the high energy resolution of these spectra, better than that of WDS, chemical bonding state analysis is achievable. In figure 5, the difference between Li-K emission spectra obtained from metallic lithium and lithium fluoride can be seen. In the case of lithium metal, without corrosion by exposure to air, Li-K emission is easily observed [9]. In the case of LiF sample, using a soft beam condition, it is difficult to detect Li-K emission. However, if the electron beam is increased, F-K emission can be recorded.

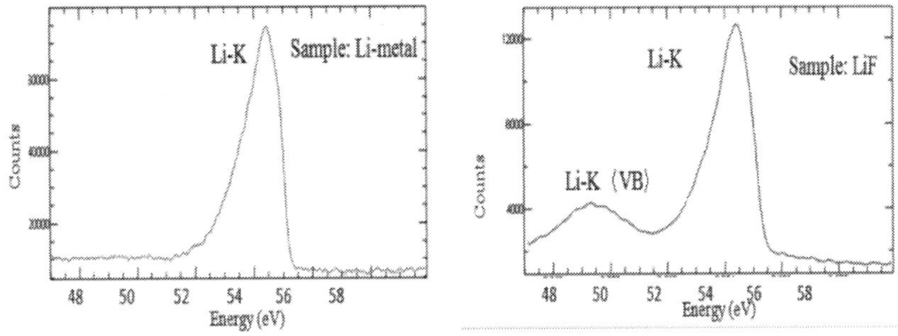

Figure 5. Li-K spectra of metallic lithium (left) and lithium fluoride (right).

3.1.2. L emission spectra (1): $L_{2,3}$-M_1 (Mg, Al, Si, P, Cl). Using a standard specification, Al-L emission spectrum can be detected. By adjusting the lower limit to less than 50 eV, Mg-L emission spectrum can be also detected. However, if the spectrometer is adjusted in an attempt to achieve less than 45 eV, the spectrometer's condition would be out of the ideal diffraction condition and the energy resolution would be dramatically reduced. For metallic Mg and Al, the peak profile corresponding to the Fermi-edge can be observed very sharply like a steep wall, as shown in figure 6. Using this sharpness, the energy resolution of this spectrometer can be evaluated. Since the resolution of the spectrometer is better than the natural width of the X-ray line, a full width half maximum (FWHM) specification, like the EDS resolution method, is unsuitable. The energy resolution of SXES is specified as the energy width between 75 % and 25 % of the distance from top to background of the Fermi-edge. This definition takes into account the broadening of the Fermi-edge due to finite temperature. From this determination method, a resolution of 0.3 eV is guaranteed whilst in practice 0.2 eV is attainable.

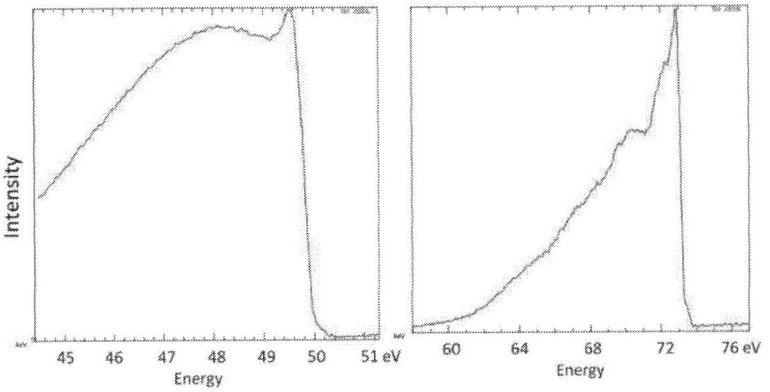

Figure 6. Mg-L and Al-L spectra for pure element.

This sharply defined Al-L Fermi-edge is always observed, on both Al-compounds and Al metals. The lower profile form of the Fermi-edge corresponds to the energy density of state. In other words, as described in Fabian's textbook, this profile corresponds to the 1st to 4th Brillouin zone [10]. A similar examination can be performed on Si-compounds and P-compounds under the same conditions. In the case of a Si-wafer, a sharp Fermi-edge is not observed since it is modified by sp3 combination band structures.

3.1.3 L emission spectra (2): Lℓ; L_3-M_1 (Ti, V, Cr, Mn). In the case of the transition elements such as Ti, V, Cr, and Mn, the Lℓ X-ray line has a higher intensity than that of the Lα and Lβ. L_ℓ is produced by an L_3-M_1 transition whilst Lα and Lβ are from L_3-M_5 and L_2-M_4 transitions. The Lℓ is a higher energy than Lα and Lβ, and also has a lower absorption problem than that of the absorption edge between Lα and Lβ. For these reasons, Lℓ is expected to provide chemical state information and allow for quantification by low voltage analysis. Lℓ-spectra of Ti-compounds are shown in figure 7. In the case of Ti-oxides, peak shift and sub shoulder peak shapes are observed in correspondence with valence bonds. For Ti^{4+}, such as $SrTiO_3$, the peak shift and peak shoulder at the higher energy side is smaller than for Ti-metal (ground state). These practical spectra will be considered in comparison with chemical calculations and discussed in a future study. From the understanding of the Ti-spectra, the low voltage analysis of the V- and Cr-spectra can then be applied in a similar manner.

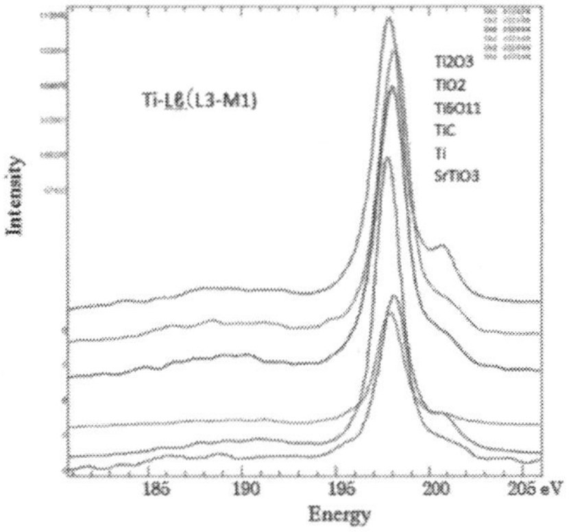

Figure 7. Comparison of Ti-Lℓ spectra for Ti-compounds (Courtesy of CSIRO, Colin MacRae and Nick Wilson).

3.1.4. M emission spectra. For the 5th period elements of Zr, Nb, Mo, Ag, Cd, Sn, and Sb, the Mz-line of the spectra gives the strongest and sharpest peak. The peak profiles of Zr-Mz and Nb-Mz are shown in figure 8. For Zr and Mo, experiment condition was 5 kV, 100 nA, 5 min JS200N, and for Nb, was 5 kV, 50 nA, 5 min JS200N. The peak height for higher order X-rays for elements with a higher atomic number reduces to a greater degree than Lℓ of the lighter transition metals. As with the Lℓ peak, a similar low voltage analysis capability of the Mz is expected.

3.1.5. N-band emission spectra. Each spectrum of the three elements Pt, Au and Bi in the sixth row of periodic table of the elements shows a distinct peak consisting of a few sub-peaks in the energy range between 120 eV and 150 eV, as shown in figure 9. As far as the present authors are aware, these peaks have not been reported in any literature to date and do not appear in any known data tables. They are tentatively assigned to the second order emission from N-bands. The shapes of the Pt and Au peaks look rather similar. In fact, if the sub-peak at 140 eV is ignored, the peaks consist of two main sub-peaks, the position of which shifts to the higher energy by about 50 eV with the increasing of the atomic number from Pt to Au. The similar systematic energy shift of the two main sub-peaks

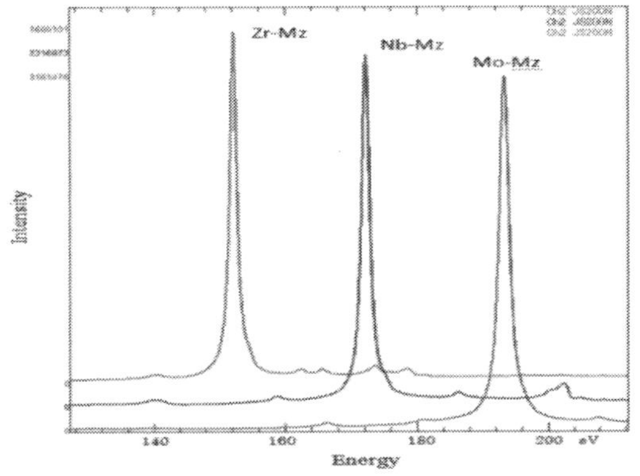

Figure 8. Comparison of Zr-Mz, Nb-Mz and Mo-Mz lines.

has been observed in spectra of a series of the sixth row elements; starting from Hf, continuing to Ta, W, Re, Os, Ir, Pt until Au. Although the shapes of the peaks of Pt and Au are similar, slight difference can be noticed. The Bi-peak shape was observed to be different to that from the other sixth row elements. Many features in these series of spectra, including the peaks described above, will also be studied in greater detail in a further study.

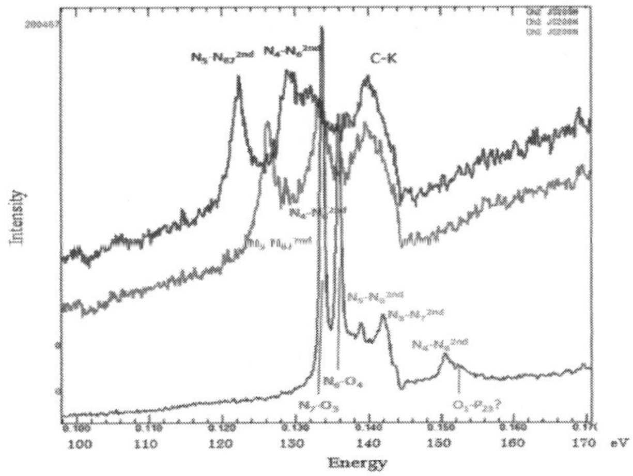

Figure 9. Various kinds of N-band emission spectra of Au, Pt and Bi.

4. Application

4.1. Chemical state spectral mapping of a lithium ion battery

Recently the spectral mapping of chemical state by WD-SXES has been developed and applied to various kinds of materials. For example, three samples A, B and C of a lithium-ion battery anode with different charging conditions: 0 % (A), 30 % (B), and 100 % (C), were supplied separately sealed in three glove bags filled with Ar gas. They were transferred to a sample holder in a glove-bag type of wrapper filled with Ar gas, which covered both the sample holder and the insertion chamber of the EPMA as shown in figure 10a. Though cumbersome, this precaution was necessary to avoid chemical reactions of the sample surface in atmospheric air. The three samples, each of which had a length of 50 mm and a width of 30 mm, were mounted on the sample holder side by side as shown in figure 10b. A large area, 30 x 80 mm, was selected for spectral mapping of the chemical state of the lithium. The mapping area consisted of: A and B, with a part length of 30 mm and the full width of 30 mm, and C, with a part length of 30 mm and a part width of 20 mm. This area was divided into 30 x 50 pixels so that the inter-pixel distance was 1.0 mm in the length direction and 1.6 mm in the width direction. Using a diameter of 1 μm and a beam current of 0.8 μA at an accelerating voltage of 2 kV the intensities of X-rays in the energy range between 48 and 165 eV were acquired for 60 s/pixel. Three maps were constructed by using the X-rays in three energy ranges corresponding to sub peak of Li-K (49 - 51 eV), Li-K (53 - 55.5 eV) and C-K of the 3rd order (nominally 93 - 96 eV) and are shown in figure 11. The map due to the sub peak of Li-K on the left hand side showed an increase in relative intensity with increasing charge in the anode material. The middle map in figure 11 indicates that the metallic lithium formed locally and distributed heterogeneously under the fully charged condition. The right hand side map due to C-K of the 3rd order in figure 11 indicates the intensity distribution due to graphite; the other important constituent of the anode. Under the fully discharged condition, the intensity was strongest. The intensity reduced substantially when the amount of charging increased by 30 %. However it stayed at this level, when the amount of charging increased to 100 %.

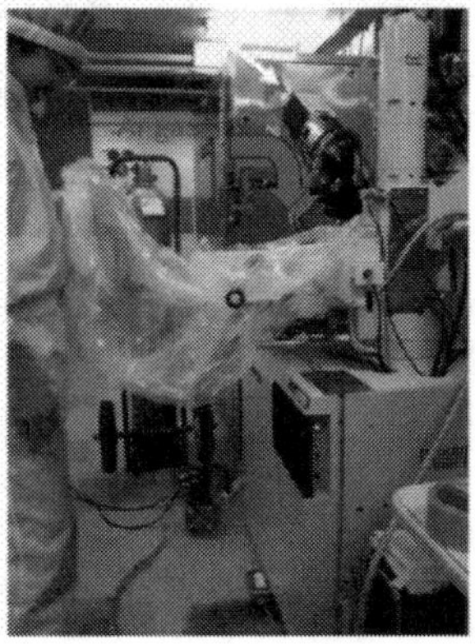

a)

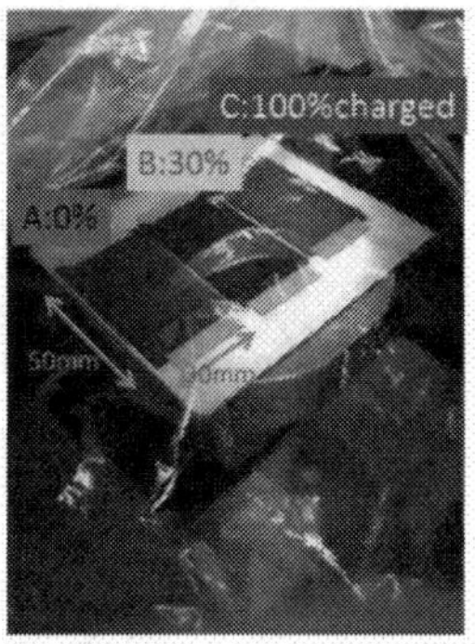

b)

Figure 10. Two important steps are illustrated in a procedure for transferring the samples of a lithium-ion battery anode and mounting them on a sample holder of the EPMA. a) Both the sample holder and the insertion chamber were covered with a wrapper to avoid the direct contact of the sample surface to atmospheric air. b) The three samples with different amounts of charging were mounted on the sample holder side by side.

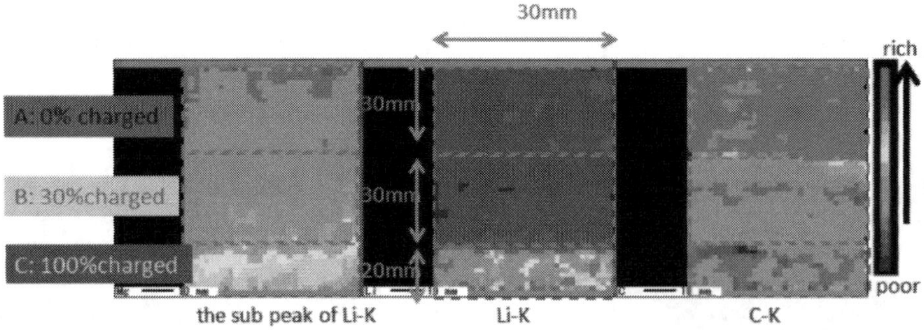

Figure 11. Three maps constructed by using the X-rays in the three energy ranges corresponding to sub peak of Li-K (left), Li-K (middle) and C-K of 3rd order (right). Each map covered an area as large as 30 x 80 mm^2 consisted of three samples with different amount of charging: 0 % (A), 30 % (B), and 100 % (C).

Three typical spectra corresponding to the three different conditions are shown in figure 12. Figures 12a and 12b show the spectra in the energy range between 48.5 and 59.5 eV and between 92 and 97 eV, respectively. The former energy range covers the sub peak Li-K (50 eV) and Li-K emission (54 eV) with the partially overlapped C-K of the 5th order (56 eV), whereas the latter covers

C-K of the 3rd order. In spite of the close energy between the Li-K emission and C-K of the 5th order, these are distinguished by high energy resolution of this spectrometer. The spectra were extracted from the typical pixels observed in the three maps shown in figure 10. As shown in figure 12a, the sub-peak of Li-K and the Li-K were not observed in the fully discharged anode, whereas the sub-peak of Li-K was observed in the charged anodes at around 50 eV and the Li-K was clearly observed in the fully charged anode. The change in the intensity of the C-K the 3rd order shown in figure 12b. Additional bonding information can be observed in the shoulder around 96 eV due to the π bond in the profile of the fully charged anode. The sub-peak intensity corresponds to the charging state of the battery. It is estimated that the intercalation phenomena of Li-ion to the anode graphite layers is the meaning of Li-ion and graphite layers are interaction and bonding with ionic combination. So this distribution corresponds to the electron charging. On the other hand, metallic Li exists in the fully charged area and has a dendrite structure. C-K is decreased as the graphite layers distance is gradually expanded by the intercalation with Li-ions. Compared with TEM-SXES, this large area distribution of chemical state spectral map can be provided by EPMA/SEM-SXES.

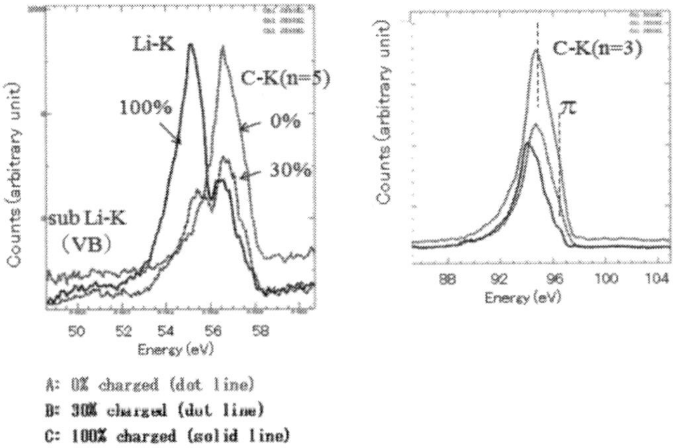

Figure 12. Three typical spectra of a lithium ion battery anode corresponding to the three different conditions: 0 % (A), 30 % (B), and 100 % (C). a) The energy range between 48.5 and 59.5 eV covered the emission peaks of the sub-peak of Li-K, Li-K and C-K of the 5th order; and b) between 84 and 104 eV covered the one of the C-K of the 3rd order

4.2. Trace element analysis

The WD-SXES has a very high potential for analysing trace elements in steel. With precise calibration, it is possible to extend the range of trace element analysis to very low concentrations. In the ordinary EPMA, the calibration method is usually used in the lower concentration ranges and requires a series of reliable standard specimens. The standard deviations of calibration curves determine the concentration limit as well as the accuracy of measured concentrations. With the WD-SXES, a better peak to background ratio can be obtained. Furthermore, the better energy resolution of the WD-SXES produces a more genuine target spectrum, which is well separated from overlapped spectra that often occur, and well defined background intensities resulting in the target spectra of the element of interest for the standards being very similar to that taken from the sample.

With preliminary experiments, calibration curves of boron, carbon and nitrogen in steels were obtained within the concentration range between 10 and 100 ppm by mass. The signal of each of the characteristic X-rays from the trace elements; B-K, C-K of the 2nd order and N-K of the 2nd order, could be unambiguously observed. The calibration curves for the three trace elements were linear. These straight lines indicated that the concentration range of 10 ppm by mass could be detected using this method. The results of the present experiments strongly suggest that a similar method can also be used for the analysis of trace elements in semiconductors. As shown in figure 13, the detection limit of lithium was estimated in a 5 mass% Li-Al alloy to be 80 ppm by mass using Eq. (2):

IOP Conf. Series: Materials Science and Engineering **109** (2016) 012017 doi:10.1088/1757-899X/109/1/012017

$$D.L. = \frac{2\sqrt{2}}{\sqrt{P \times P/B} \cdot \sqrt{t}} \cdot C\% \tag{2}$$

Compared with TEM-SXES this SXES-EPMA/SEM is possible to quantify these kinds of trace elements for metals.

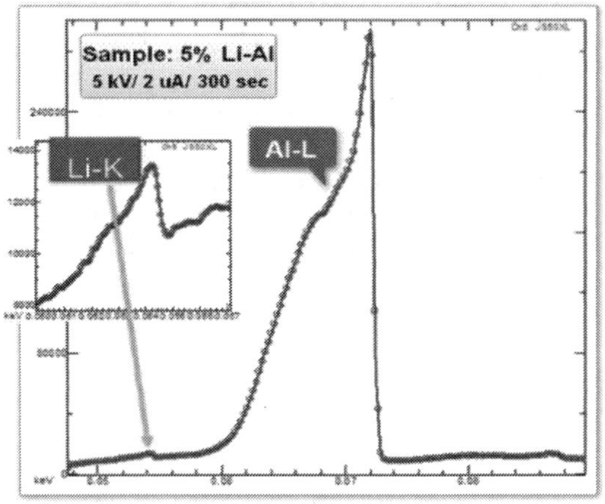

Figure 13. Li-K and Al-L emission of 5 % Li-Al.

4.3. Ti-N overlapping

Using EDS and conventional WDS, N-K and Ti-L overlapping spectra are difficult to separate. To overcome this problem, deconvolution of the spectra is often applied. For WD-SXES, with its higher energy resolution, there is no requirement for deconvolution. For example, figure 14 shows a small TiCN grain in steel that was analysed using WD-SXES. For quantitative analysis, deconvolution is better to get more correct results. But as a primary result without deconvolution, we can distinguish the N-K peak. It is important to observe with native information, because sometimes, deconvolution include the artefact by calculation. The N-K and Ti-L peaks can be seen as clearly separated. Line step analysis was carried out with a step size of 100 nm. The N-K can be seen to be stronger on the outside edge (blue) than that of the inside (red) of the TiCN grain. The C-K intensity is the inverse than that of nitrogen.

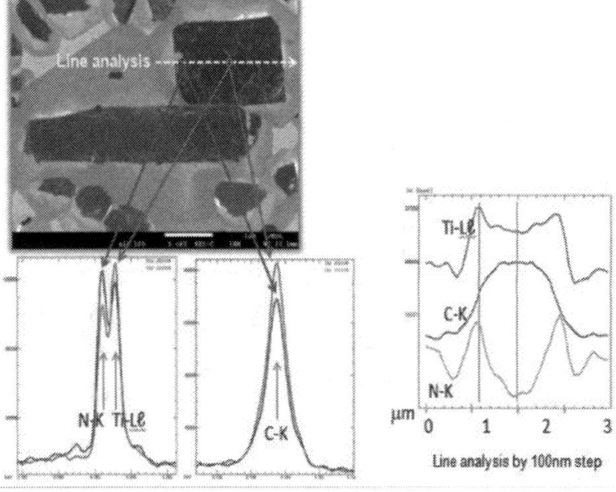

Figure 14. Backscattered electron image of TiCN, N-Kα, Ti-Lℓ and C-Kα spectra at the rim and core of grain. Three line analysis of N-K, Ti-Lℓ and C-K onto TiCN grain in steel.

5. Conclusion

A newly developed wavelength dispersive soft X-ray emission spectrometer (WD-SXES) with two kinds of gratings; JS50XL and JS200N have been installed on both EPMA and SEM. The spectrometer covers the energy range between 50 and 210 eV. X-ray s of up to 700 eV can be detected by using higher order reflections. Experimentally the lower end of the energy range could be extended down to 46 eV. The energy resolution of this SXES was 0.2 eV which was evaluated from the Fermi-edge of the Al-L spectrum.

The valence band Li-K and ground state Li-K emission spectra due to lithium metal and lithium fluoride were observed for the first time in EPMA. The sub-peak of Li-K and Li-K emission spectra from samples of a lithium ion battery anode under different charge states were measured and mapped in an area as large as 30 x 80 mm^2. A clear distinction in spectra due to chemical states of lithium shows a high potential for the characterisation of lithium ion battery anodes.

For the trace element analyses of boron and nitrogen in steel, their calibration curves could successfully be obtained in the composition range between 10 and 100 ppm by mass. This method could detect their concentrations as small as 10 ppm by mass. The trace element analysis of light elements in materials other than boron and nitrogen in steel could possibly be performed at these levels of concentrations.

Acknowledgements

This work is supported as a programme of Collaborative Development of Innovative Seeds (Practicability Verification Stage) by Japan Science and Technology Agency. We appreciate useful discussions and advice from professor Makoto Kikuchi of Tokyo Institute of Technology and Dr Calum Dickinson of JEOL UK.

References

[1] Scott V D and Love G 1983 *Quantitative Electron-Probe Microanalysis.* (New York, NY: John Wiley & Sons)
[2] Terauchi M *et al.* 2012 *J. Electron Microsc.* **61** 1
[3] Imazono T *et al.* 2012 *Appl. Opt.* **51** 2351
[4] Takahashi H *et al.* 2013 *Microsc. Microanal.* **19** (Suppl. 2) 1258
[5] Harada T and Kita T 1980 *Appl. Opt.* **19** 3987
[6] Namioka T and Koike M 1995 *Appl. Opt.* **34** 2180
[7] Bearden J A 1967 *Rev. Mod. Phys.* **39** 78
[8] Takahashi H *et al.* 2014 *JEOL News* **49** 73
[9] Takahashi H *et al.* 2015 *Proc. Microsc. Microanal.* 1014
[10] Fabian D J 1968 *Soft X-ray band spectra.* (London: Academic Press) 62

Physics-based simulation models for EBSD: advances and challenges

A Winkelmann[1], G Nolze[2], M Vos[3], F Salvat-Pujol[4] and W S M Werner[4]

[1] Bruker Nano GmbH, Am Studio 2D, 12489 Berlin, Germany
[2] BAM, Federal Institute for Materials Research and Testing, Unter den Eichen 87, 12205 Berlin, Germany
[3] The Australian National University, Research School of Physics and Engineering, Atomic and Molecular Physics Laboratories, 0200 Canberra, Australia
[4] Vienna University of Technology, Wiedner Hauptstrasse 8-10, 1040 Vienna, Austria

E-mail: aimo.winkelmann@bruker.com

Abstract. EBSD has evolved into an effective tool for microstructure investigations in the scanning electron microscope. The purpose of this contribution is to give an overview of various simulation approaches for EBSD Kikuchi patterns and to discuss some of the underlying physical mechanisms.

1. Introduction

Access to suitable microscopic characterisation methods of crystalline phases is a necessary prerequisite in order to develop new applications of advanced materials. In this context, diffraction methods are essential tools. The method of electron backscatter diffraction (EBSD) is based on the measurement of Kikuchi patterns [1], which appear in the angular distribution of backscattered electrons (BSE) in the scanning electron microscope (SEM). In the context of the EBSD method, these patterns are also called electron backscatter diffraction patterns (EBSP), or backscattered Kikuchi patterns (BKP) with the extension to transmission Kikuchi diffraction (TKD) in the SEM. Over the last two decades, EBSD has evolved into an effective tool for microstructure investigations [2-7] with a wide field of applications in orientation mapping [8], grain- and phase-boundary characterisation, local phase identification, and lattice strain determination [9-11]. Because a range of physical and chemical properties of modern materials is directly affected by the crystal structure, there is an evident need for characterisation techniques like EBSD which are locally sensitive to crystal symmetries on the relevant length-scales of the material's microstructure, possibly down to the nanoscale [12-15].

Experimentally, EBSD is conceptually simple: in principle only a phosphor screen imaged by a sensitive CCD camera is needed, see figure 1a. The geometry of the Kikuchi patterns is governed by the gnomonic projection of the Bragg reflection conditions of a point source inside a crystal [16, 17]. However, this does not give any information on the details of the observed intensity distribution near a Kikuchi band, and the explanation of the exact physical Kikuchi pattern formation process is considerably more complicated. Kikuchi pattern formation is based on the concept of atomic emitters of electron waves which are incoherent with respect to the incident beam [18] and the multiple scattering of these waves inside a crystal gives rise to the characteristic Kikuchi pattern intensity distribution. Because of the strong effects of multiple scattering and absorption, the quantitative calculation of the backscattered diffraction pattern has to be based on the dynamical theory of electron diffraction.

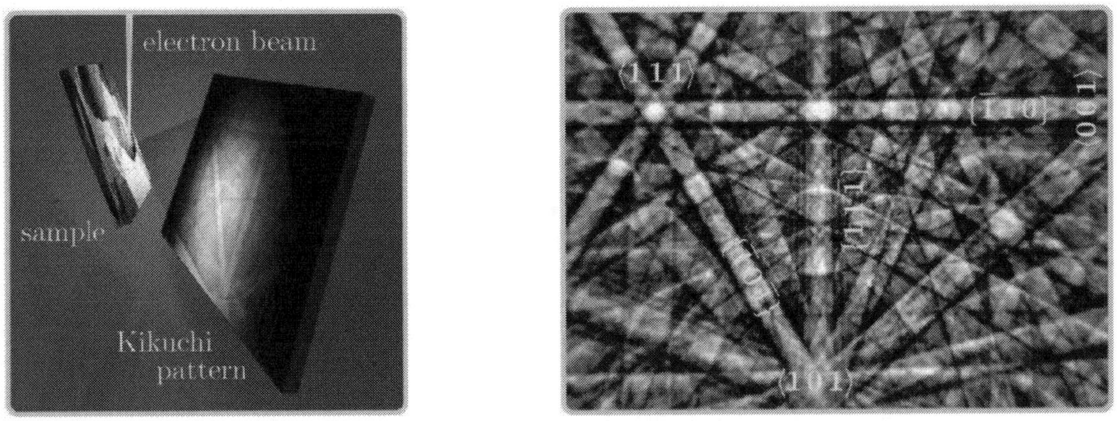

a) Schematic of an EBSD setup.

b) Kikuchi pattern of GaP with the main zone axes and lattice planes indicated.

Figure 1. Principle of Kikuchi pattern measurement in the SEM. The primary electron beam is focussed on a tilted sample, and the Kikuchi pattern is measured in gnomonic projection on a planar detector. The combined local resolution of the electron beam and the interaction volume for Kikuchi pattern formation determines the resolution that is available for spatially resolved orientation mapping.

Due to the increased information content which is provided by dynamical electron diffraction in EBSD Kikuchi patterns, the detailed physics-based simulation and analysis of the respective effects is becoming more and more important [19, 20]. For example, Kikuchi patterns are intrinsically sensitive to the point-group symmetry of a crystal [21, 22]. With the huge increase of the commonly available computing power, it has become possible to routinely carry out quantitative simulations of high-resolution Kikuchi patterns with an efficiency that now makes it possible to directly compare complete experimental Kikuchi patterns with theoretical simulations. Typical applications of dynamical diffraction simulations include the theoretical analysis of EBSD-based lattice strain determination [11, 23], EBSD image segmentation and orientation determination [24{26], and the analysis of orientation relationships in phase transformations [27]. With respect to non-centrosymmetric phases, point-group sensitive orientation mapping has also been demonstrated [28]. The range of possible applications of dynamical EBSD pattern simulations includes even space-group sensitivity, which has been demonstrated by discrimination of left-handed and right-handed quartz using EBSD Kikuchi patterns [29].

SEM based diffraction can also be observed via electron channelling patterns (ECP) [30], which measure Kikuchi patterns of backscattered electrons resulting from the diffraction of the incident beam. Due to its close relationship to EBSD, the results presented here also largely apply to ECP, which is a particularly useful method for direct imaging of dislocations in the SEM [4, 30]. This type of application which has recently received renewed interest [31, 32] with decisive advances in the necessary quantitative image simulations [33].

The purpose of this contribution is to give an overview of simulation approaches for Kikuchi patterns and some of the underlying physical mechanisms. In this way, the present paper aims to show that EBSD (and also ECP) in combination with Kikuchi pattern simulations provides an important tool for advanced analysis of crystalline phases in the SEM.

2. Models for Kikuchi pattern simulation

In this section we review the existing models for EBSD-related Kikuchi pattern simulation in order to critically point out their specific advantages and limitations.

The calculation of all the simulated Kikuchi patterns was carried out using the software ESPRIT DYNAMICS (Bruker Nano), which implements the Bloch wave approach for EBSD pattern simulation [19, 20]. In detailed Kikuchi pattern simulations, a typical number of about 1,000 or more reciprocal lattice vectors *hkl* is taken into account. These are selected according to the reciprocal lattice vector length $d^*_{hkl} = 1/d_{hkl}$ (typically $d^*_{hkl} < 1/0{:}05 \ldots 1/0{:}035$ nm^{-1}) and the relative strength with respect to the largest structure factor amplitude $|F|_{max}$ (typically 15 % ... 10 % or less). The simulations presented here neglect experimental excess-deficiency effects, which are related to the incident beam direction [34]. In order to compare the diffraction patterns obtained with different simulation models, we use use the normalized cross-correlation coefficient r to quantify the agreement between two Kikuchi patterns [35, 36]:

$$r = \frac{\sum\limits_{i,j} \left[f(i,j) - \bar{f}\right] \cdot \left[w(i,j) - \bar{w}\right]}{\sqrt{\sum\limits_{i,j} \left[f(i,j) - \bar{f}\right]^2} \cdot \sqrt{\sum\limits_{i,j} \left[w(i,j) - \bar{w}\right]^2}} \tag{1}$$

where $f(i, j)$ and $w(i, j)$ are the pixel intensity values of the corresponding regions of interest (ROI) in the two patterns to be compared, while $\bar{f}$ and $\bar{w}$ are the mean values in these ROIs. The absolute value of r is in the range between 0 and 1 and does not depend on scale changes in the intensity of both patterns. From our experience, values of $r > 0.6 \ldots 0.7$ indicate visually convincing fits between experimental and simulated values. For pattern matching or model discrimination via r, the statistically necessary levels of r could be lower or higher than these values, depending on the details of the application.

In the following sections we proceed to evaluate the capability of different theoretical models to describe the experimentally observed Kikuchi patterns.

2.1. *The geometrical model*

In the simplest case, the Kikuchi patterns are interpreted with the help of the gnomonic projection of Kossel cones [16, 37] which are formed by all possible directions of Bragg reflection from a lattice plane with normals $(h\ k\ l)$ and $(\bar{h}\ \bar{k}\ \bar{l})$. This does not include any information about band intensities. In this model, the width of a Kikuchi band is proportional to twice the Bragg angle θ_B:

$$\sin\theta_B = \frac{n \cdot \lambda}{2 \cdot d_{(h\,k\,l)}} \tag{2}$$

In the case of EBSD the resulting Bragg angle is in the order of only a few degrees because of the relative values of the electron wavelength λ and the unit cell dimensions. Diffraction orders with n > 1 in (2) describe the higher-order interferences that form additional parallel lines outside the main band. A geometrical simulation for GaP displaying the Kossel cone intersections of all possible Bragg reflections with $d_{hkl} > 0.05$ nm and $|F|_{hkl} > 0.15|F|_{max}$ can be seen in figure 2 (the same reflection selection criteria will be applied in all the following simulations shown here). While the actual agreement with the experiment is very limited, the entangled network of Kossel cone projections gives a peculiarly indirect view of the structure of reciprocal space and its effects on EBSD Kikuchi patterns. The most notable features are the characteristic circular areas surrounding some important zone axes like {1 1 1}, which are related to the "Higher Order Laue Zone" (HOLZ) rings seen also in the experimental GaP pattern [38]. It is useful to note that all the reflections shown in figure 2 will also enter into the dynamical simulation discussed later, where the influence of each possible reflection is considered quantitatively, in contrast to the purely geometrical drawing of Kossel-cone hyperbolas. The simple Bragg-reflection interpretation also neglects any non-centrosymmetric effects since the Bragg angle is the same for $d_{(h\,k\,l)}$ and $d_{(\bar{h}\,\bar{k}\,\bar{l})}$ and the respective Kossel cones are mirror-symmetric with respect to the relevant lattice plane.

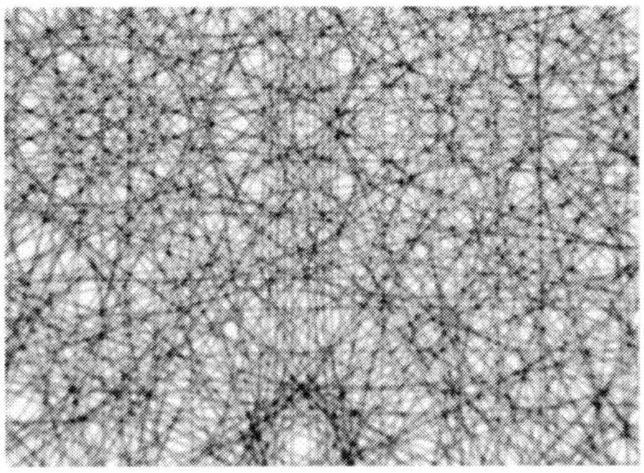

Figure 2. Gnomonic projection of the possible Bragg conditions in GaP for $d_{hkl} > 0.5$ Å and $|F|_{hkl} > 0.15|F|_{\max}$ for the same experimental conditions as in figure 1b. As no information about the intensity of a reflection is available in this model, all reflections are drawn with the same grey level.

2.2. The kinematical model

As can be seen by comparison of figures 1b and 2, not all of the geometrically possible bands are equally visible in the experimentally measured Kikuchi pattern. In order to account for this, the kinematical intensity I_{hkl} of a Bragg reflection *hkl* is often used empirically to describe the different relative intensities of Kikuchi bands. As is well known from X-ray diffraction, I_{hkl} is given by the square of the absolute value of the structure factor $|F|^2_{hkl}$ in the absence of *hkl*-dependent absorption effects:

$$I_{hkl} \propto F^*_{hkl} F_{hkl} \tag{3}$$

A corresponding scaling of the Bragg reflection lines is shown in figure 3, which allows to roughly estimate the relative intensities of the experimentally observed Kikuchi bands in figure 1b.

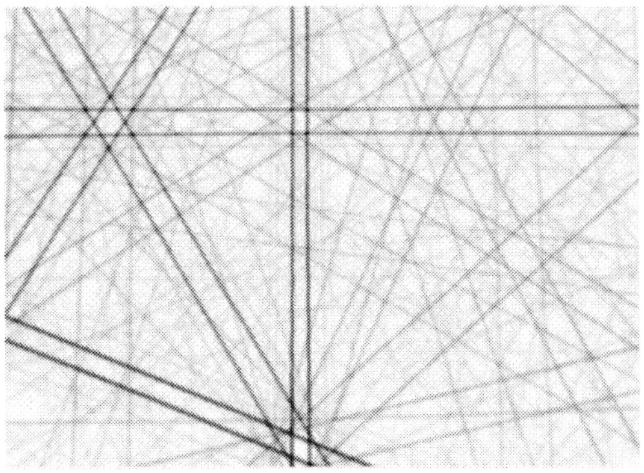

a) Intensity at Bragg reflection scaled by $|F|^2_{hkl}$.

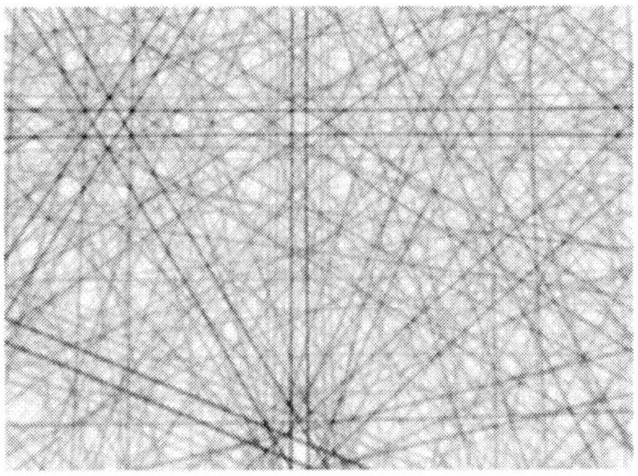

b) Intensity at Bragg reflection scaled by $|F|_{hkl}$.

Figure 3. Gnomonic projection of the Bragg condition in GaP. The darkness level of the lines at the Bragg conditions is scaled differently as a function of the structure factor amplitude.

Assignment of a kinematic intensity I_{hkl} and $I_{(\bar{h}\,\bar{k}\,\bar{l})}$ to the two related Bragg reflections of a Kikuchi band is likewise still neglecting non-centrosymmetric effects. In the kinematic model, there is no possibility to differentiate between I_{hkl} and $I_{(\bar{h}\,\bar{k}\,\bar{l})}$ due to Friedel's rule, which states that both these intensities are equal in the absence of anomalous absorption. Thus, the kinematical EBSD model is sensitive to Laue-symmetry only.

The angle-dependent Kikuchi band profile intensity can be described by a function $B(\vartheta)$, where ϑ is measured from the relevant lattice plane. The general form of $B(\vartheta)$ is determined by the distance of ϑ to the Bragg angle. For large angles, we observe the average background intensity, since we are away from diffraction effects, which are strongest near the Bragg angle. In the region near the Bragg angle, the Kikuchi band intensity is modulated and can go below or above the background for different angles. To approximate the complicated experimental Kikuchi band profile, a constant intensity I can be empirically assigned to the full geometrical band width, resulting in bands with step profiles $B_K(\vartheta)$:

$$B_K(\vartheta) = \begin{cases} I, & \text{if } |\vartheta| \leq \vartheta_B \\ 0, & \text{otherwise} \end{cases} \tag{4}$$

This formula introduces an additional experimental observation, namely that the Kikuchi bands in EBSD usually (but not always) show a higher intensity than the background. From within the kinematical model itself, however, the actual intensity distribution B(ϑ) of a Kikuchi band cannot be predicted at all. This also includes the constant I for the band intensity. The obvious choice seems to be the kinematical intensity I_{hkl}. In order to check the visual agreement of this assumption with the experiment, in figure 4 we have displayed the Kikuchi bands B_K with grey values in one case proportional to the kinematical intensity $I_{hkl} \propto |F|^2$, and in the other case proportional to the structure factor amplitude $|F|$. We found correlation coefficients of $r(|F|^2) = 0.51$ and $r(|F|) = 0.58$ for both these types of simulations (compared to the experimental pattern) showing that the linear dependence is the better approach and that assignment of I_{hkl} is not optimal (The correlation coefficients have been calculated for a ROI covering only the right side of the patterns, because the high intensity zone axes on the left side would otherwise be weighted too strongly).

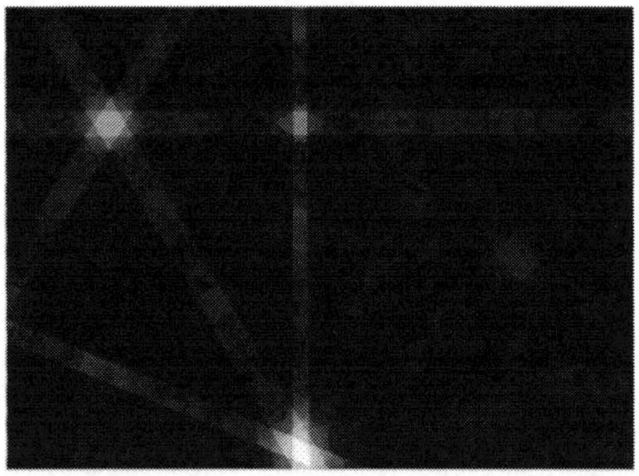

a) Band intensity scaled by $|F_{hkl}|^2$, $r_{ROI} = 0.507$.

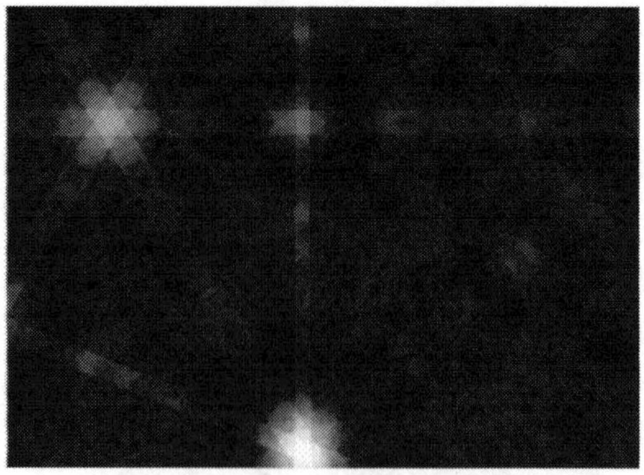

b) Band intensity scaled by $|F_{hkl}|$, $r_{ROI} = 0.580$.

Figure 4. Gnomonic projection of GaP Kikuchi bands with step profiles B_K according to eq. 4. The band intensity is assumed to be a) $\propto |F_{hkl}|^2$, or b) $\propto |F_{hkl}|$. Correlation coefficients r are taken with respect to the experimental pattern in figure 1b.

We would like to stress that the kinematical EBSD models are limited in several important aspects:

(i) The kinematical intensity I_{hkl} actually describes the intensity of an incident plane wave beam which comes from infinity, is coherently Bragg reflected by single scattering processes at the crystal atoms and goes again to infinity to be detected in a typical spot-like diffraction pattern. This is a very different experimental situation than the case of an internal point source inside the crystal which emits an electron wave that is multiply scattered and then detected at infinity outside the crystal. Extensions of results from one situation to the other are not necessarily valid.

(ii) The kinematical model cannot provide an actual formula for the Kikuchi band profile. Instead, the kinematical intensity I_{hkl} is assigned purely empirically. While this simple model might usefully approximate a typical standard situation in EBSD and gives a rough impression about the intensity distribution within a Kikuchi pattern, it totally fails to describe effects like Kikuchi band contrast inversion where one actually observes dark Kikuchi bands [1, 39].

(iii) The pairs of Bragg reflections hkl and $\bar{h}\bar{k}\bar{l}$, which are relevant for a Kikuchi band, independently contribute to the final pattern in a simple sum of all "step band" intensities B_K. This results in unrealistic intensity maxima especially along low-indexed zone axes where many bands are crossing each other.

(iv) The kinematical intensity scales with the square of the structure amplitude $I_{hkl} \propto |F|^2$. When compared to experimental Kikuchi patterns, however, a linear dependence on the structure factor amplitude actually seems to provide a better fit, as we have seen above. The fact that the kinematical intensity I_{hkl} does not give the best agreement of the simulations using B_K-profiles exposes the severe limitation of the kinematical theory in the description of the Kikuchi band profile. Depending on the choice of I, the modified intensity decay from the stronger to the weaker reflections causes a different relative importance of specific simulated pattern features. This could influence the results when the B_K-model is applied in correlation-based strain determination, where various dependencies on F_{hkl} can be selectively chosen [40-43]. In view of the general deficiencies of the kinematical model, we have not looked for other functional dependencies that might provide an even better empirical fit to the experiment.

2.3. The two-beam dynamical profile

While the kinematical EBSD model neglects most of the physics which is involved in the detailed Kikuchi band formation, the dynamical theory of electron diffraction can provide a detailed description of experimentally measured Kikuchi patterns. A simplified dynamical theory of the Kikuchi pattern profile is given in [18], where an analytical band shape is derived based on the two-beam approximation. Assuming only single-element backscatterers placed on Bravais lattice positions, this band shape expresses the variation of the backscattered electron intensity near a Bragg reflection with reciprocal lattice vector **g** qualitatively as:

$$B_A(\vartheta) \propto -\frac{w_{\mathbf{g}} + a_{\mathbf{g}}}{1 + w_{\mathbf{g}}^2 - a_{\mathbf{g}}^2} + \dots \tag{5}$$

where $w_{\mathbf{g}} = w_{\mathbf{g}}(\vartheta; F_{hkl}, \lambda)$ is a normalized deviation parameter with $w_{\mathbf{g}} = 0$ at the Bragg condition, and $a_{\mathbf{g}}$ contains absorption parameters for the reflection **g**. Bands with higher order reflections are approximated by the sum of two-beam profiles for the corresponding set of reflections **g**, 2**g**, 3**g**, ... which need to be included in the list of reflectors. In the dynamical model, absorption parameters turn out to be important for the actual shape of the Kikuchi band profile and they can even contribute to an inverted band shape. This illustrates another deficiency of the kinematical model, which does not consider these absorption effects at all. For more details on the simplified two-beam profile, we refer the reader to Chapter 9 in [18].

In figure 5b, the Kikuchi pattern of GaP is simulated using a sum of analytical two-beam profiles B_A according to eq. 5. While the computational effort of this approach is still rather low, the resulting Kikuchi pattern already shows a much better agreement with the experiment than any of the

IOP Conf. Series: Materials Science and Engineering **109** (2016) 012018 doi:10.1088/1757-899X/109/1/012018

kinematical approximations discussed above. A main deficiency of this approach is again seen in the zone axes, where the fine structure is not reproduced faithfully due to the simple sum of band intensities with no interference. Also, in the theoretical derivation of this model, the actual distribution of the different chemical elements in the unit cell is not taken into account exactly. In view of the limits of the kinematical models discussed above, the analytical two-beam profile is a better standard choice for simplified pattern simulations in the absence of the full many-beam dynamical calculations which are discussed next.

a) Experiment.

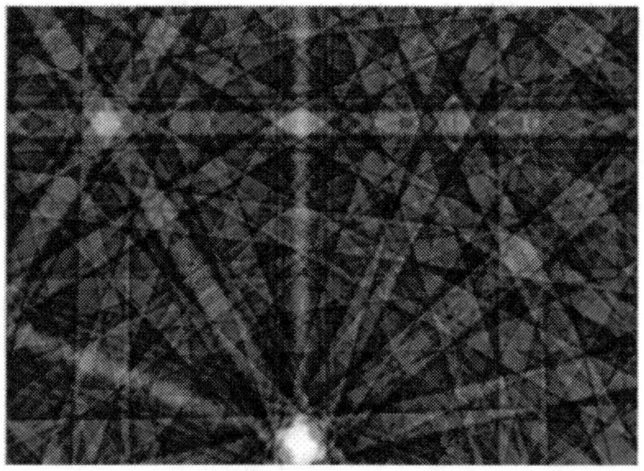

b) Sum of two-beam dynamical profiles, $r = 0.71$.

c) Many-beam dynamical simulation, best fit for point group $\bar{4}3m$, $r = 0.81$.

d) Many-beam dynamical simulation, $r = 0.78$, Laue-group $m\bar{3}m$ equivalent orientation (90° rotation around $\bar{4}$-axis).

Figure 5. Comparison of dynamical simulations to an experimental GaP pattern.

2.4. Many-beam dynamical pattern simulation

Realistic intensity simulations of EBSD patterns have been introduced using the Bloch wave approach of the dynamical theory of electron diffraction [19, 20, 23, 25]. This approach calculates the probability that an electron will leave the crystal in a specific direction after having been incoherently scattered from the incident beam at an atomic position inside the crystal and then having undergone multiple coherent scattering in the crystal in the presence of absorption effects. In order to compare the typical quality of dynamical EBSD simulations using the many-beam approach, in figure 5c the resulting best fit intensity distribution is displayed in comparison to the experimental pattern in figure 5a. As can be seen, the numerous specific features around zone axes and within bands are well reproduced by the many-beam simulation.

In the middle of the pattern, {1 1 1} becomes visible as a vertical band. The intensity of this band is asymmetrically shifted to the right, as can be even seen by eye on close inspection. This specific asymmetry is an effect of the crystal polarity in the non-centrosymmetric zinc-blende type structure of GaP, which can also be seen quantitatively in the cross-correlation coefficients: the best-fit simulation gives a normalized $r = 0.81$. The Laue-group equivalent orientation in figure 5d has an additional 90° rotation around the c-axis and gives a worse value of $r = 0.78$, with the difference of $\Delta r = 0.03$ being statistically significant.

The different values of r with respect to the two Laue-group equivalent orientations clearly demonstrates the sensitivity of the dynamical simulations to the actual point group of the crystal in which these two orientations are not equivalent. Such a point-group sensitive orientation analysis is beyond the limits of the conventional procedure that is applied up to now in EBSD systems. As a unique new application of dynamical diffraction simulations, in figure 6 we show results of absolute orientation mapping for polycrystalline GaP [28]. The conventional Laue-group-specific analysis of EBSD Kikuchi patterns has to leave the absolute orientation of the GaP crystal undetermined by 90° around {0 0 1}. This indeterminacy is resolved by the point-group sensitive pattern matching approach.

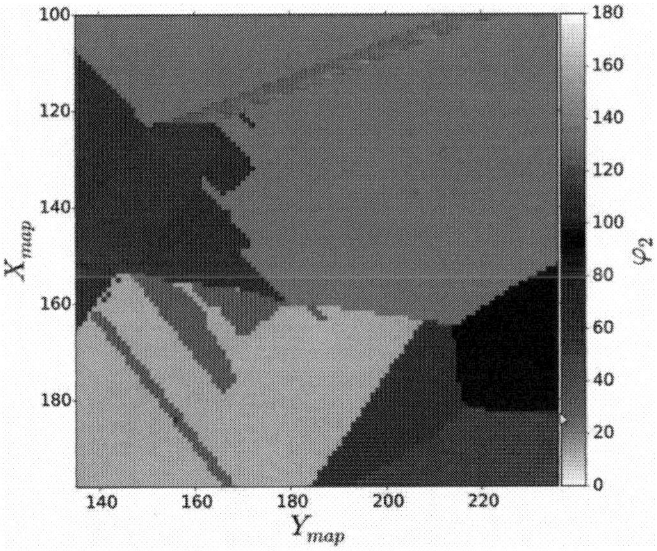

Figure 6. Point-group sensitive orientation map for polycrystalline GaP [28], showing the value of the Euler angle φ_2. In the GaP point group $\bar{4}3m$, φ_2 can assume values from 0 to 180°, whereas the corresponding Laue group m$\bar{3}$m limits φ_2 to the range 0 ... 90°.

Summarizing this section, the kinematic EBSD interpretation is limited to a description of the Kossel-cone geometry of Bragg angles and to a rough estimation of the ordering of reflections according to their kinematical intensity. Still, this is useful for the selection of the strongest reflectors *hkl* for indexing of EBSD patterns. The kinematical models for EBSD are completely insensitive to the absence of an inversion centre in the crystal structure. To realistically describe EBSD Kikuchi patterns, one needs to use models which apply the dynamical theory of electron diffraction. The lack of an inversion centre in the crystal structure is manifested in non-symmetric Kikuchi band profiles for polar lattice planes.

3. Physical mechanisms

The dynamical diffraction simulations presented above are based on certain assumptions about the physical pattern formation mechanisms. In this section, some of these fundamental effects are presented.

3.1. Transfer of momentum in crystal scattering

The electrons which are observed on the EBSD screen have undergone a substantial change in their direction of movement. This implies a corresponding transfer of momentum q. Coherent elastic scattering from the crystal describes the situation when the electron does not excite any internal degrees of freedom of the crystal and a macroscopic mass M_C of the crystal takes up the momentum. This results in an infinitesimal value of the recoil energy $E^C{}_R = |q|^2 / 2M_C$. However, there is also the possibility that a single atomic mass M_A takes up the momentum, which then results in a measurable recoil energy even for electron scattering at typical SEM energies [45]. The probability of coherent elastic scattering is strongly suppressed if the respective single-atom recoil energy $E^A{}_R = |q|^2 / 2M_A$ exceeds typical phonon energies [46]. This is the case for a typical EBSD geometry, where coherent backscattering from the crystal can thus be usually completely neglected and spot diffraction patterns of the incident beam are absent. Exceptions can be observed by RHEED (reflection high energy electron diffraction) patterns in EBSD at glancing angles or in transmission Kikuchi diffraction (TKD) near the incident beam direction, when spot patterns can be observed that result from coherent elastic scattering. At low electron energies, in the order of 0 to 1,000 eV, the recoil energy can be low enough for coherent backscattering to be observable in the method of low energy electron diffraction (LEED) for surface analysis [47]. Due to the dependence of the recoil energy on the atomic mass, electron spectroscopy can be applied to differentiate between different scattering atoms. This is shown in figure 7, where Au atoms on a sapphire (Al_2O_3) surface have been measured for two exit angles [44]. The three peaks corresponding to Au, Al, and O can be clearly distinguished at their expected positions. The two different curves in figure 7 show the influence of diffraction effects due to different measurement geometries. In order to prove that Kikuchi patterns result from these incoherent recoil scattering events, we measured element-resolved Kikuchi band profiles and compared them with dynamical simulations [44, 48]. In principle, the recoil loss in the electron-atom collision makes it possible to measure separate Kikuchi patterns for different chemical elements in a compound phase.

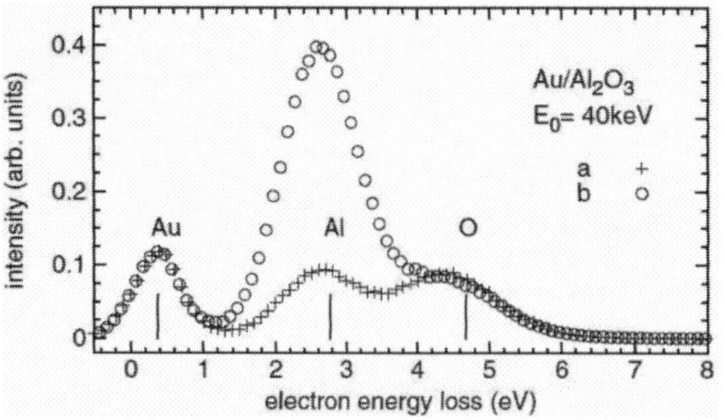

Figure 7. Measured energy spectra of backscattered electrons from Au/Al_2O_3 for two different exit directions "a" and "b". [44]. The vertical lines indicate the energy transfer that would be expected from a classical elastic "billiard ball" collision between a single electron of 40 keV and the respective atom.

3.2. Energy spectrum of backscattered electrons

While the incoherent atomic recoil is the smallest possible energy loss for backscattered electrons, inelastic scattering effects contribute to the formation of Kikuchi patterns in a range of energy losses of up to several hundred eV [51]. This is why a correct description of the energy spectrum of the backscattered electrons is a key ingredient in a more complete model of Kikuchi pattern formation.

Monte-Carlo (MC) simulations provide a useful tool for the description of electron scattering in materials [52], neglecting, however, the influence of diffraction effects. Inelastic scattering is often taken into account by assigning the loss energy based on the path length travelled in the crystal. This "continuous slowing down approximation" (CSDA) neglects the quantized character of inelastic loss processes, as a continuous loss energy is always assigned to every path length (no matter how small) and thus the possibility of having no energy loss at all is excluded also (i.e., the coherent elastic backscattering discussed above).

The quantized character of inelastic loss processes can be included by introducing the differential inverse inelastic mean free path (DIIMFP) into MC simulations, which takes into account the specific distribution of possible energy losses for an electron in a material [53].

In figure 8, we calculated the spectrum of electrons backscattered from a Si sample with a primary energy of 10 keV at an incidence angle of 70° [54]. The integrated spectral intensity is normalized to 1.0 for the range of 50 eV to 10 keV. In the figure, we show the result of the CSDA approximation as implemented in the CASINO [49] code compared to a simulation using the DIIMFP [50]. We can see that the DIIMFP approach reproduces the correct experimental spectrum with an elastic peak and with quantized plasmon losses (17 eV for Si) [55]. These spectral details are important to exactly model the energy-dependent diffraction effects, which are concentrated in the spectral region rather near to the elastic peak. This means that the CSDA has strong limitations with respect to the combined problem of electron transport and diffraction, as it neglects the elastic peak, the plasmon features and other energy loss signatures in the backscattered electron energy spectrum and their specific localisation in the crystal structure.

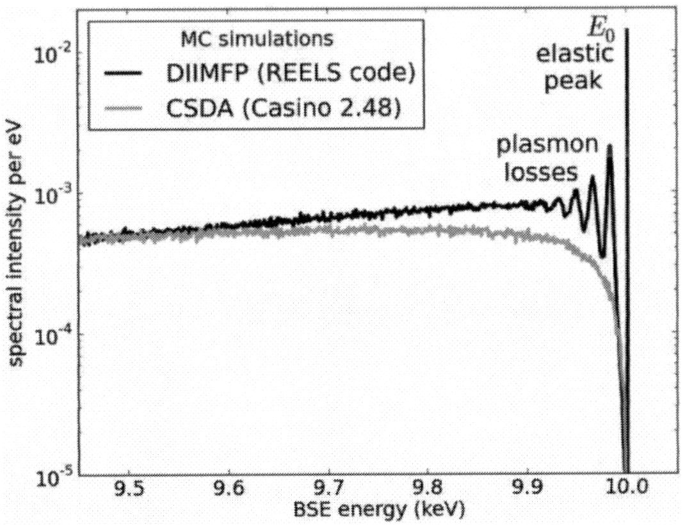

Figure 8. Simulated energy spectra of backscattered electrons from silicon in the continuous slowing down approximation (CSDA) using CASINO [49] and, alternatively, using the differential inverse inelastic mean free path (DIIMFP) approach [50]. E_0 = 10 keV, 70° incidence angle.

The DIIMFP approach can also be used for the simulation of the angular dependence of the backscattered electrons and the intensity distribution on the phosphor screen, as is shown in figure 9. For the conversion of electrons into photons, the energy-dependent response function of the phosphor has to be taken into account [56]. A wealth of information is contained in such angle- and energy-resolved MC simulations, for instance concerning the correlation between scattering angle and the energy-distribution of the backscattered electrons [25], which influences the energy-dependent diffraction pattern structure. This type of MC simulation is also relevant to the use of the EBSD screen as an angle-resolved detector of backscattered electrons independently of the actual diffraction effects [57-59].

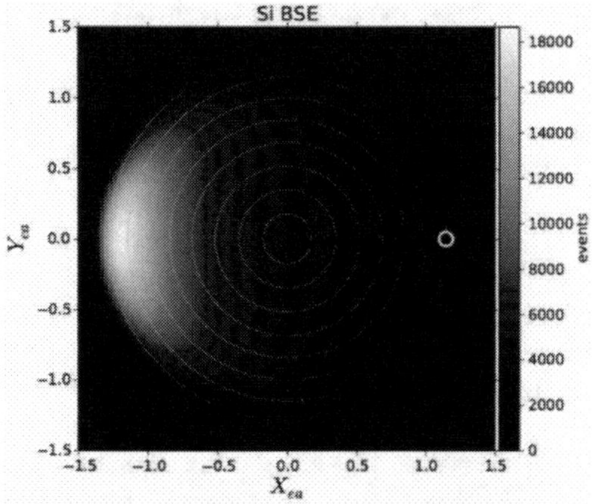

a) Angular distribution of BSEs in the hemisphere above the sample, equal-area projection with circles at 10° divisions. The incidence direction at 70° is shown as the small circle. Most intensity is scattered into the forward direction near low take-off angles.

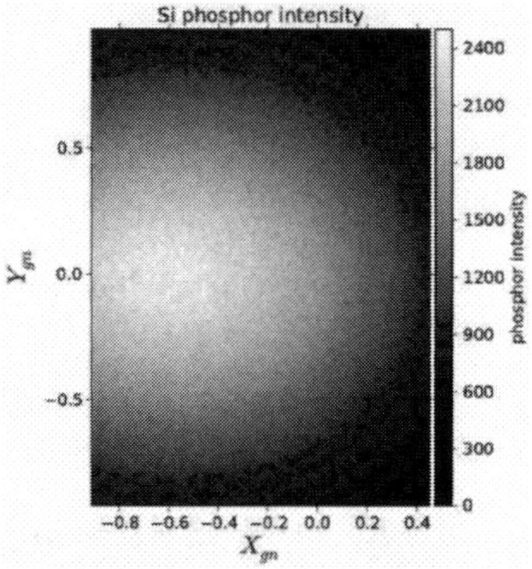

b) Light intensity from a phosphor screen in a typical EBSD setup near the maximum of the angular BSE distribution (gnomonic projection, screen orientation consistent with a)).

Figure 9. Monte Carlo simulation of electron backscattering from Si. The primary beam energy is 20 keV for 50×10^6 incident electrons. Backscattered electrons with energy losses up to 10 kV are considered.

The Monte-Carlo simulations are based on the particle picture of electron scattering and thus do not contain diffraction information in a direct way. One of the remaining challenges is to consistently combine the MC model results with the diffraction simulations [25, 54, 60]. This concerns, for example, the depth sensitivity of EBSD. Based on observations of the width of measured diffraction lines, the energy spread and correspondingly the related depth sensitivity of electrons contributing to an EBSD pattern has been estimated to be in the range between 10 nm and 40 nm at 20 kV, with the lower values reached for dense materials [6]. Experimental observations of the disappearance of Kikuchi pattern diffraction contrast when depositing amorphous layers on crystalline samples are consistent with this estimation [61, 62]. A successful model also needs to explain the available experimental data on the energy dependence of Kikuchi band contrast in different scattering geometries [51, 55]. An example for the complex energy dependence of Kikuchi band contrast from a Si sample is shown in figure 10, where the data has been measured using a high-resolution electron energy spectrometer [55]. This data shows that depending on the incidence angle, Kikuchi patterns can be formed at energy losses up to several hundred eV. Due to the concentration of the backscattered intensity near the elastic peak, effectively only a rather narrow energy spectrum contributes to experimental EBSD patterns. This justifies the approximation of simulating the dynamical diffraction effects at a fixed "effective" energy near the primary beam energy. For the quantitative reproduction of energy-sensitive effects in the fine structure of experimental EBSD patterns, however, an extended energy range has to be taken into account [20].

Summarizing, it is clearly an important question how the spatial distribution and the energy spectrum of the backscattered electrons in the sample are influencing the EBSD patterns because the inclusion of these effects in dynamical simulations could possibly allow to extract additional information from experimental EBSD measurements. In the general case, this will require a fully quantum-mechanical treatment of electron backscattering from crystals [63].

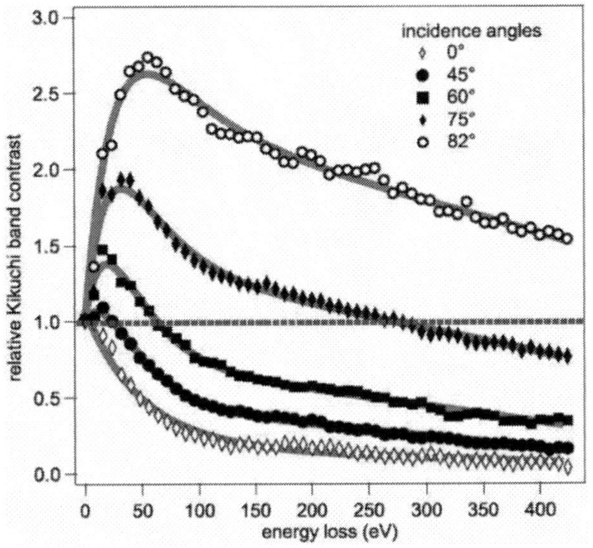

Figure 10. Diffraction contrast of backscattered electrons from Si in dependence on the energy loss and the incident beam direction [55]. The contrast is measured relative to the quasi-elastically backscattered electrons near zero energy loss (horizontal line, blue). The diffraction contrast is calculated as a normalized difference between the highest and lowest intensity in the energy-resolved Kikuchi band profile, see [55].

References

[1] Alam M N, Blackman M and Pashley D W 1954 *Proc. R. Soc. Lond. A* **221** 224-242
[2] Venables J A and Harland C J 1973 *Philos. Mag.* **27** 1193-1200
[3] Schwartz A J, Kumar M, Adams B L and Field D P (Eds.) 2009 *Electron Backscatter Diffraction in Materials Science 2nd edition.* (Berlin: Springer)
[4] Wilkinson A J and Hirsch P B 1997 *Micron* **28** 279-308
[5] Schwarzer R A 1997 *Micron* **28** 249-265
[6] Dingley D 2004 *J. Microsc.* **213** 214-224
[7] Randle V 2008 *Adv. Imaging Electron Phys.* **151** 363-416
[8] Adams B L, Wright S I and Kunze K 1993 *Metall. Trans. A* **24** 819-831
[9] Wilkinson A J, Meaden G and Dingley D J 2006 *Ultramicroscopy* **106** 307-313
[10] Maurice C and Fortunier R 2008 *J. Microsc.* **230** 520-529
[11] Wilkinson A J and Britton T B 2012 *Mater. Today* **15** 366-376
[12] Small J A, Michael J R and Bright D S 2002 *J. Microsc.* **206** 170-178
[13] Keller R R and Geiss R H 2012 *J. Microsc.* **245** 245-251
[14] Trimby P W 2012 *Ultramicroscopy* **120** 16-24
[15] Brodusch N, Demers H and Gauvin R 2013 *J. Microsc.* **250** 1-14
[16] Kossel W 1936 *Ann. Phys.* **417** 512-526
[17] Gajdardziska-Josifovska M and Cowley J M 1991 *Acta Cryst. A* **47** 74-82
[18] Reimer L 1998 *Scanning Electron Microscopy - Physics of Image Formation and Microanalysis.* (Berlin-Heidelberg-New York: Springer Verlag)
[19] Winkelmann A, Trager-Cowan C, Sweeney F, Day A P and Parbrook P 2007 *Ultramicroscopy* **107** 414-421
[20] Winkelmann A 2009 in: *Electron Backscatter Diffraction in Materials Science.* (Schwartz A J, Kumar M, Adams B L and Field D P; Eds.) (Berlin: Springer Verlag) chap 2
[21] Marthinsen K and Høier R 1988 *Acta Cryst. A* **44** 700-707
[22] Baba-Kishi K Z 1991 *J. Appl. Crystallogr.* **24** 38-47
[23] Villert S, Maurice C, Wyon C and Fortunier R 2009 *J. Microsc.* **233** 290-301
[24] Park S U, Wei D, De Graef M, Shah M, Simmons J and Hero A O 2013 in: *2013 IEEE Int. Conf. Image Process.* pp 3780-3784
[25] Callahan P G and De Graef M 2013 *Microsc. Microanal.* **19** 1255-1265

[26] Chen Y H, Park S U, Wei D, Newstadt G, Jackson M A, Simmons J P, De Graef M and Hero A O 2015 *Microsc. Microanal.* **21** 739-752
[27] Nolze G and Winkelmann A 2014 *Cryst. Res. Technol.* **49** 490-501
[28] Winkelmann A and Nolze G 2015 *Appl. Phys. Lett.* **106** 072101
[29] Winkelmann A and Nolze G 2015 *Ultramicroscopy* **149** 58-63
[30] Joy D C, Newbury D E and Davidson D L 1982 *J. Appl. Phys.* **53** R81
[31] Naresh-Kumar G, Hourahine B, Edwards P, Day A, Winkelmann A, Wilkinson A, Parbrook P, England G and Trager-Cowan C 2012 *Phys. Rev. Lett.* **108** 135503
[32] Zaefferer S and Elhami N N 2014 *Acta Mater.* **75** 20-50
[33] Picard Y N, Liu M, Lammatao J, Kamaladasa R and Graef M D 2014 *Ultramicroscopy* **146** 71-78
[34] Winkelmann A 2008 *Ultramicroscopy* **108** 1546-1550
[35] Gonzalez R C and Woods R E 2002 *Digital Image Processing.* (Upper Saddle River, NJ: Prentice-Hall)
[36] Tao X and Eades A 2005 *Microsc. Microanal.* **11** 341-353
[37] Young C T and Lytton J L 1972 *J. Appl. Phys.* **43** 1408-1417
[38] Michael J and Eades J 2000 *Ultramicroscopy* **81** 67-81
[39] Winkelmann A and Nolze G 2010 *Ultramicroscopy* **110** 190-194
[40] Abou-Ras D, Gibmeier J, Nolze G, Gholinia A and Konijnenberg P 2008 *Cryst. Res. Technol.* **43** 234-239
[41] Kacher J, Landon C, Adams B L and Fullwood D 2009 *Ultramicroscopy* **109** 1148-1156
[42] Britton T B, Maurice C, Fortunier R, Driver J H, Day A P, Meaden G, Dingley D J, Mingard K and Wilkinson A J 2010 *Ultramicroscopy* **110** 1443-1453
[43] Alkorta J 2013 *Ultramicroscopy* **131** 33-38
[44] Winkelmann A and Vos M 2011 *Phys. Rev. Lett.* **106** 085503
[45] Went M R and Vos M 2008 *Nucl. Instr. Meth. Phys. Res. B* **266** 998-1011
[46] Lipkin H J 1973 *Quantum Mechanics: New Approaches to Selected Topics.* (Amsterdam: North-Holland Publ. Co.)
[47] Van Hove M A, Weinberg W H and Chan C M 1986 *Low Energy Electron Diffraction.* (Berlin-Heidelberg-New York: Springer Verlag)
[48] Vos M, Winkelmann A and Nolze G 2015 *Ultramicroscopy* **156** 50-58
[49] Drouin D, Couture A R, Joly D, Tastet X, Aimez V and Gauvin R 2007 *Scanning* **29** 92-101
[50] Salvat-Pujol F and Werner W S M 2013 *Surf. Interface Anal.* **45** 873-894
[51] Deal A, Hooghan T and Eades A 2008 *Ultramicroscopy* **108** 116-125
[52] Joy D C 1995 *Monte Carlo Modeling for Electron Microscopy and Microanalysis.* (Oxford: Oxford University Press)
[53] Werner W S M 2001 *Surf. Interface Anal.* **31** 141-176
[54] Winkelmann A, Salvat-Pujol F and Werner W S M 2013 *Microsc. Microanal.* **19** 738-739
[55] Winkelmann A, Aizel K and Vos M 2010 *New J. Phys.* **12** 053001
[56] Ozawa L and Itoh M 2003 *Chem. Rev.* **103** 3835-3856
[57] Payton E J and Nolze G 2013 *Microsc. Microanal.* **19** 929-941
[58] Wright S I, Nowell M M, de Kloe R, Camus P and Rampton T 2015 *Ultramicroscopy* **148** 132-145
[59] Schwarzer R A and Hjelen J 2015 *Microscopy Today* **23** 12-17
[60] Winkelmann A 2010 *J. Microsc.* **239** 32-45
[61] Yamamoto T 1977 *Phys. status solidi* **44** 467-476
[62] Zaefferer S 2007 *Ultramicroscopy* **107** 254-266
[63] Dudarev S L, Rez P and Whelan M J 1995 *Phys. Rev. B* **51** 3397-3412

IOP Conf. Series: Materials Science and Engineering **109** (2016) 012019 doi:10.1088/1757-899X/109/1/012019

Low voltage imaging and X-ray microanalysis in the SEM: challenges and opportunities

R Wuhrer[1] and K Moran[2]

[1] Western Sydney University, Advanced Materials Characterisation Facility (AMCF), P.O. Bag 1797, AU-2751 Penrith, New South Wales, Australia
[2] Moran Scientific Pty Ltd., 4850 Oallen Ford Road, 2580 Bungonia, New South Wales, Australia

e-mail: richard.wuhrer@uws.edu.au

Abstract. Low voltage imaging, X-ray microanalysis and X-ray mapping has become very important for the investigation of nanomaterials and their surfaces. This is especially true for low voltage imaging of non-conductive samples and beam sensitive samples. Operating the SEM at lower accelerating voltage allows for greater surface sensitivity, the ability to minimize charging effects, nanometre scale lateral X-ray spatial resolution and nanoscale X-ray depth sensitivity. Determining the correct accelerating voltage for imaging in a SEM is dependent on the instrument's operating performance at low voltage, the material being viewed, and other factors that limit effectiveness of low voltage microanalysis, which will be discussed in this paper.

1. Introduction

There have been significant advances in the field of scanning electron microscope (SEM), X-ray microanalysis, X-ray mapping (XRM), gunshot residue (GSR) analysis and electron back-scattered diffraction (EBSD) over the last five to ten years. The development of a stable field emission gun (Schottky emitter) and a new generation of secondary electron (SE) and backscattered electron (BSE) detectors has made high resolution SEM imaging on the nanometre level possible (figure 1) [1-8]. With the development of silicon drift detectors (SDD) this has allowed for greater X-ray collection efficiencies with greater sensitivity for X-rays in the low energy part of the spectrum with the SEM operating at lower accelerating voltages.

For decades, high energy accelerating voltages, ranging from 15kV to 30kV, have been used by microscopists for imaging and analysing samples because of the sub-nanometre image resolution and the ability to excite the well-known K-lines in the atomic spectra for the identification of the elements and quantification of their concentrations.

Conventional SEMs that use high accelerating voltages and conventional "lower" or "below-lens" detectors require a thin conductive coating on the sample to avoid charging artefacts if they are non-conductive. When the electron beam impinges on an uncoated non-conductive sample, a charge can build up inside the specimen, which can make imaging unstable, difficult and even in extreme cases, impossible [1]. The development of the variable pressure and/or low-vacuum SEM (VPSEM or LVSEM) and environmental SEM (ESEM) [9-11] allows samples to be imaged in their natural state

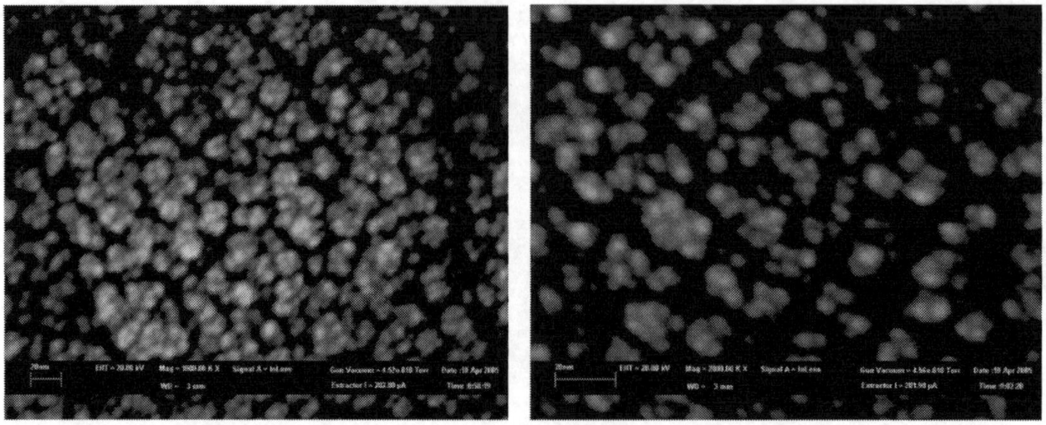

a) b)

Figure 1. High resolution "in-lens" SE images of an "opal eye", (sample prepared by B Griffin). Image taken at 20 kV, 3 mm working distance using a Zeiss Supra operated in high vacuum mode. Images shown at a) 500,000 and b) 2,000,000 times magnification.

without any sample modification or sample coating. These systems utilize gas ionisation (charge compensation) in the SEM chamber, eliminating charging artefacts usually seen with imaging non-conductive samples under high vacuum conditions [9-11].

More recently, low voltage imaging has become popular and easier to perform due to the development of a new generation of field emission SEM's (FEGSEM or FESEM). These have better electron optic systems and better performing (new generation) SE and BSE detectors. Operating the SEM at lower accelerating voltages allows for increased surface information (greater surface sensitivity), sometimes reduced beam damage (depending on the sample), and also gives the ability to minimize charging effects (figure 2) [1-4, 6-8]. By operating at low accelerating voltages, the problem of charging can be minimized and even eliminated [1, 12-18] as can be seen in figure 3, which is an image of a non-conductive polymer that is beam sensitive. The polymer was deposited on a silicon wafer, which can also assist in carrying away any charge.

Low voltage operation is very important in the observation of non-conductive charging samples, imaging of beam sensitive samples and obtaining greater information of the sample surface whether it is conductive or non-conductive [1]. Low voltage imaging generally refers to the application of SEM techniques at beam energies below 5 kV. On the other hand, very low voltage imaging or sometimes referred to as ultra-low voltage imaging refers to beam energies below 1 kV [1, 3, 12, 13].

This paper will initially cover low voltage imaging in the SEM and how to determine the ideal accelerating voltage (beam energy) for particular specimen types. Furthermore, the paper will also cover the challenges and opportunities for low voltage microanalysis with particular emphasis on low voltage X-ray mapping and post-processing X-ray maps.

2. Low voltage imaging: electron-specimen interactions

In the SEM instrument, lowering the beam energy of the electrons has a profound effect on the interactions that occur within the sample. The elastic and inelastic scattering processes critically depend on the energy of the incident electrons. The mean-free-paths of these elastic and inelastic processes rapidly decrease with decreasing beam energy [18]. At low voltages, both the electron range and the interaction volume are significantly reduced [12, 18]. Figure 4 shows the Monte Carlo simulations of electron scattering in carbon, iron and gold at 20 kV and 5 kV beam energies. At low

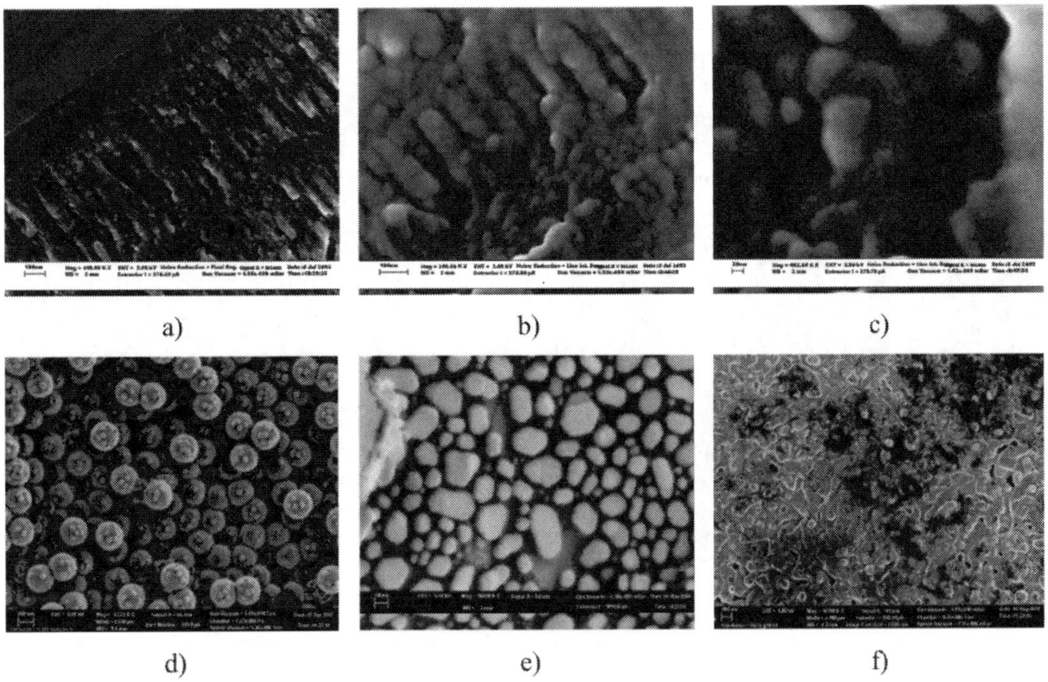

Figure 2. a-c) Multilayer carbon coating. Imaged at 2 kV, using an "in-lens" SE detector. Sample was freshly broken and imaged in cross-section immediately. Image c) reveals fine structure detail that would easily be missed operating at higher accelerating voltages; d) Image of "Opal Eye" sample taken at 50 eV on a Zeiss FEGSEM revealing fine surface detail; e) Gold on carbon image taken at 100 eV; and f) Image of material surface taken at 1 kV revealing a thin layer of contamination. Image micrometre bar lengths represent a) 100 nm, b) 100 nm, c) 20 nm, d) 200 nm, e) 100 nm, and f) 200 nm, respectively.

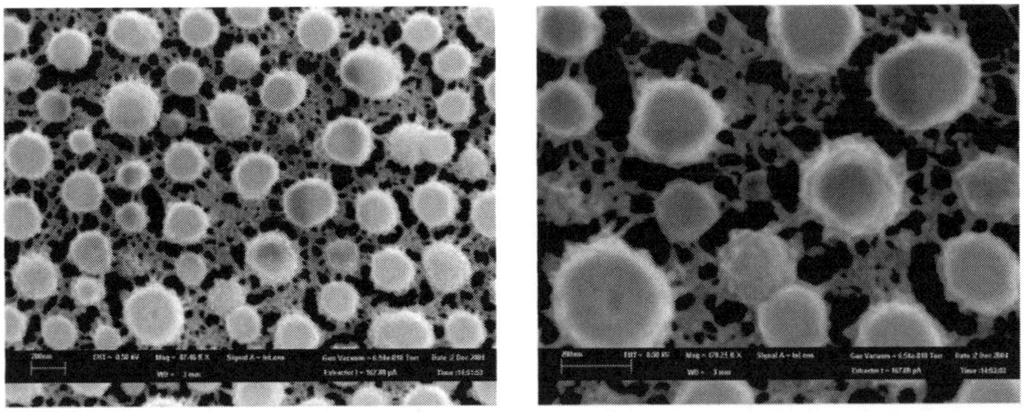

Figure 3. Polymeric material deposited on silicon wafer. Sample imaged at 0.5 kV, using an "in-lens" SE detector.

energies the electron range (Re) falls from the micrometre level at 15 kV to the nanometre range at 500 V beam energies. For example, the typical electron range of 1 kV beam with carbon is 7 nm (figure 4b). This rapid decrease in electron range causes a large characteristic change in the secondary electron yield (δ) for all specimens, which rises from 0.1 at 30 keV to 1 at 1 - 4 keV, and even greater than 1 for some materials [12-14].

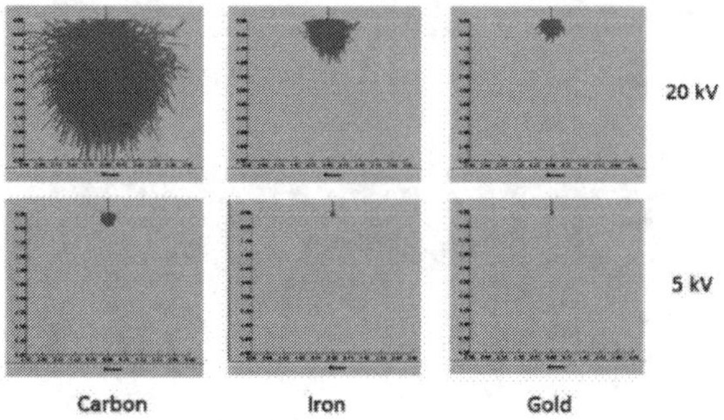

a)

Typical electron ranges (1 kV)	
Carbon (C)	33 nm
Aluminium (Al)	25 nm
Copper (Cu)	7 nm
Silver (Ag)	6 nm

b)

Figure 4. a) Monte Carlo simulations [19]. Electron beam diffusion varies with accelerating voltage and specimen density. All are scaled to the same depth and width. b) Table of typical electron ranges for different materials at 1 kV accelerating voltage [19].

Below ~ 1 kV, the secondary electron yield falls as there is less energy available to produce secondary electrons [12-14]. Figure 5a shows the effect on the electron yield as a function of the accelerating voltage [20]. As can be seen in figure 5a, there exists an electron beam energy (E_1 and E_2) where the secondary yield is unity. These E_1 and E_2 points are called the crossover energies where the E_2 point is a stable operating condition and typically exists between 0.5 kV and 4 kV [14, 19, 21]. The E_2 point is dependent on the instrument and material [12, 18, 19].

There are many sources of electron signals in the SEM, not all of which come from the sample, as can be seen in figure 6. Some of these include secondary electrons generated by [19, 21]:

1. The primary beam (SE_1).
2. By backscattered electrons (SE_2).
3. By backscattered electrons interaction with chamber walls as well as other parts of the SEM (SE_3).

So, with the SEM operating at low voltage:

1. The spatial resolution improves for all emission signals SE_1, SE_2, BSE, X-ray, cathodoluminescence (CL) as the interaction volume shrinks as $1/E^{1.6}$;
2. The signal to noise ratio (S/N) improves as secondary yield rapidly increases;
3. The information carried by all emissive signals is related to the near-surface of the specimen; and
4. Less background secondary electrons (SE) as SE_2 signal are generated from the same volume as SE_1.

A clean sample is absolutely critical for low voltage imaging [1, 19]. At low voltage conditions the specimen generally looks more solid than translucent as emitted signals only contain information about the near surface of the specimen. In addition, the reduced electron beam penetration enhances edge-effect emission but reduces the topographic contrast, as most SE signal is generated in the escape depth irrespective of the surface tilt. Therefore, low voltage SEM images appear flatter and less 3D compared to high-voltage SEM images.

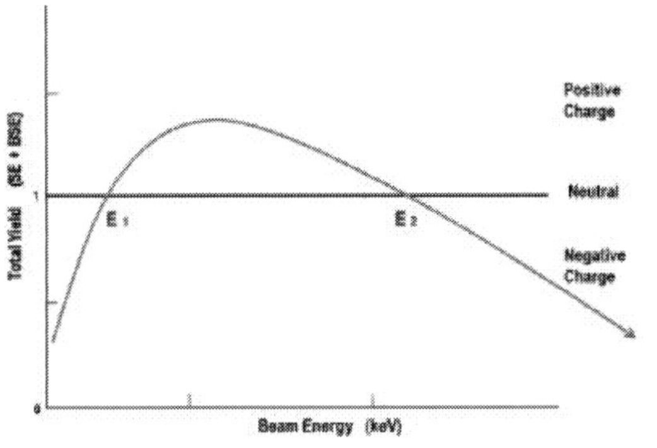

a)

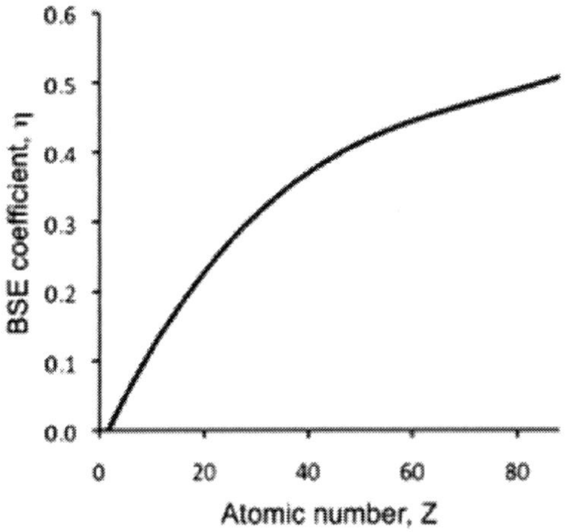

b)

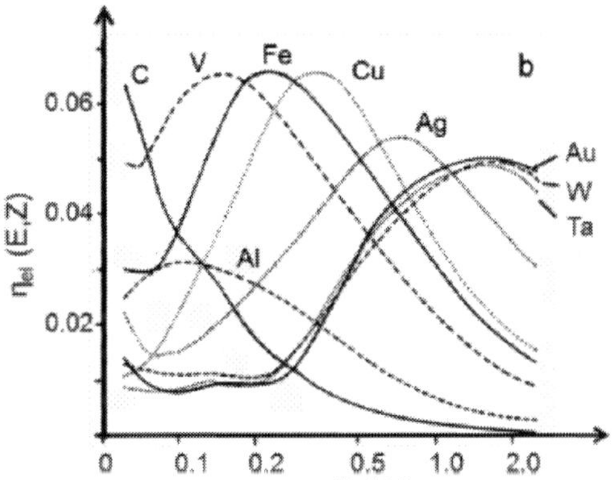

c)

Figure 5. a) Electron yield as a function of beam energy, reproduced from [19, 21]. b) Backscatter coefficient as a function of atomic number [1, 21]. c) Backscatter coefficient versus accelerating voltage [1, 21].

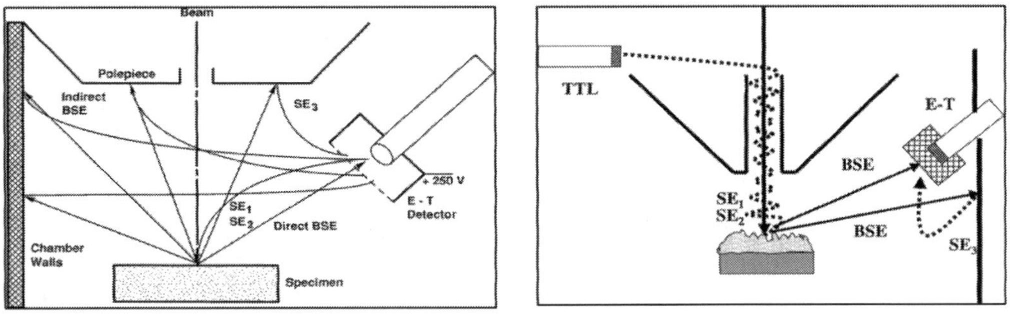

Figure 6. Schematic images showing the sources of electron signals in the SEM. Schematic modified from [19, 21]

With regard to the back-scatter signal, the BSE yield (η) is considered independent of the accelerating voltage above ~ 2 keV and is usually shown as in figure 5b [6-8]. However, below 2 keV, the situation is more complex as η rises for low-Z materials but falls for high-Z materials (figure 5c) [1, 6-8, 22-24]. At very low voltages, a small amount of charging can significantly change the image contrast, especially if you are looking at backscattered electrons. This is also a point with using a stage bias/beam deceleration, a sharp or pointy area can emit a lot of secondary electrons and with stage bias these can be accelerated up the column and appear as backscattered electrons [6, 7]. When using stage bias and backscattered (in-lens) together, careful interpretation of the images is required.

3. What is the right accelerating voltage for imaging in a SEM

There is no one correct low voltage that can be used to image a sample surface and balance the charge [12, 13]. The correct low voltage beam energy is dependent on the instrument operating conditions and the material, and should be selected on a case-by-case basis [12, 13, 20]. Imaging of rough samples, multiphase samples, composites and insulators is not simple, as the charge dissipation may be different for the different phases present [1, 12]. As can be seen in table 1, every material has a unique E_2 value, which is the point where there is a charge balance (current in equal's current out).

Table 1. Approximate E_2 values reported for different materials [12, 13, 19].

Material	E_2 (keV)	Material	E_2 (keV)
Nylon	1.18	PMMA	1.6
5%PB7/nylon	1.40	NaCl	2.0
Acetal	1.65	LiF	1.9
Polyvinyl chloride	1.65	Pyrex glass	1.9
Teflon	1.82	GaAs	2.6
Quartz	3.0	Alumina	2.9 - 4.2

4. Low voltage imaging of non-conductors: determining E2 charge balance point

With low voltage imaging in a SEM, the goal is to eliminate charging or at least to control and minimise it. The paper written by Joy and Joy [12] has an excellent section discussing and explaining the why and how of low voltage SEM imaging. According to Kirchoff's law on conservation of charge, currents at a point must sum to zero [7, 8, 12, 13].

$$I_B = (\eta + \delta)I_B + I_{SC} + (dQ/dt) = I_{BSE} + I_{SE} + I_{SC} + (dQ/dt)$$

where δ and η are the SE and BSE yields, Q is the charge deposited in the specimen after some time (t), and I_{SC}, I_{BSE}, and I_{SE} are the currents for the specimen, backscattered and the secondary electrons.

If the sample is a conductor, then it cannot charge and the dQ/dt equals zero. However, with an insulator, I_{SC} is zero and we require the sample not to charge, so dQ/dt equals zero. If this condition is achieved, then a dynamic charge balance exists where:

$$I_B = I_B(\eta + \delta) \qquad \text{i.e., } \eta + \delta = 1$$

If $\eta + \delta = 1$, then a dynamic charge balance exists;
If $\eta + \delta < 1$, then negative charging will occur;
If $\eta + \delta > 1$, then positive charging will occur.

The variation of the ($\eta + \delta$) yield curve is much the same for all materials with the E_2 being stable and the E1 not stable [12, 13].

The method that is employed to determine the E_2 point of a material with low voltage imaging has been well published [1, 12-14, 18]. A schematic of the procedure is shown in figure 7 and basically involves the following [12]:

1. Set up the microscope at a particular accelerating voltage say ~ 2 kV, focus surface and allow image to stabilize.
2. Set the SEM magnification to ~ 300 x at a fast rate (say TV rate) (figure 7a).
3. Increase the magnification to ~ 1,000 x as quickly as possible and wait for around 5 seconds (figure 7b).
4. Then drop the magnification back to the original magnification (~ 300 x) and look at the scan square that is visible in the centre of the screen (figure 7c).

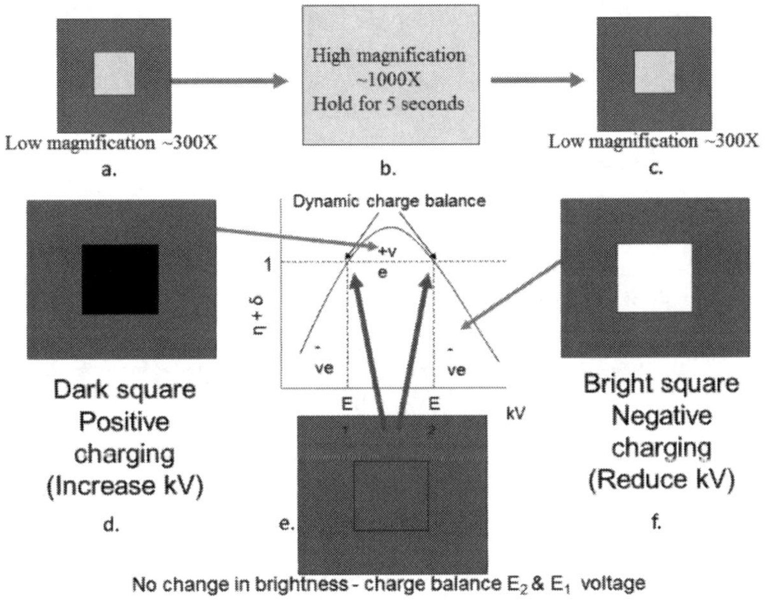

Figure 7. Schematic representation of procedure for determining the E_2 point for materials. Reproduced from [12].

If the area scanned (scan square) in the lower magnification image appears brighter than the background (figure 7f), then the sample is charging negative and the beam energy is greater than the E_2 point. On the other hand, if the area scanned (scan square) in the lower magnification image appears darker than the background (figure 7d), then the sample is charging positive and either the beam energy is less than the E_2 point (figure 7e) or there could be contamination on the surface. Be aware that sample preparation and cleanliness is critical with any low voltage imaging [1, 12, 19].

When performing the above procedure, charge can build up in the sample more easily than it can discharge, which will cause fluctuating electric fields and disturb the electron beam as well as the secondary electrons [12-15]. Implanting charge into the sample must be avoided, and if it does occur, then 1) move to another area of the sample, 2) tilt the sample [12], or better still 3) vent the chamber and then re-pump the chamber as this aids in removal of charge from the sample. To avoid this, prepare two samples (same samples) in the microscope at the same time and also place conductive paint all the way around the sample edges. Use the first sample to determine the low voltage conditions and then go to the second sample to perform imaging. This minimizes the chance of charge storage.

On a final note, low voltage imaging and reducing charge not only involves adjusting the accelerating voltage to a lower value but also involves lowering the probe current and changing the scan speed to acquire the image [1, 12].

5. Low voltage SEM instrumentation developments

Besides the development and improvement of field emission SEM's, there has also been a continual improvement and development of a new generation of detectors. These detectors have many generic names including "in-lens" electron detector, "upper" electron detector, and "through-the-lens" (TTL) detector [1, 3, 6 7, 15, 19]. These detectors, depending on their location and design principles, collect different signals as shown in figure 6.

Another development has been the use of beam deceleration techniques that can provide excellent image resolution at very low landing energies [1, 19]. This is done by maintaining a higher accelerating voltage in the gun and column (and therefore reducing chromatic aberrations and therefore increasing resolution at low voltages) and placing a negative bias on the sample [1, 6]. This firstly slows the beam down (decelerates), therefore, increasing pure surface topographic detail due to reduced beam penetration, and secondly pushes more secondary electrons off the sample improving the signal-to-noise ratio (especially at very low accelerating voltages) as well as reducing charging and contamination [1, 6].

6. Low voltage microanalysis

Traditional microanalysis using energy dispersive X-ray spectroscopy (EDS) involves using higher accelerating voltages. The accelerating voltage of the SEM determines the energy of the electron beam which consequently determines which X-rays are excited during the interaction with the material. The SEM is usually operated at accelerating voltages in the range of 15 kV to 25 kV in order to excite the well-known K-lines in the atomic spectra, however, the use of higher accelerating voltages usually results in larger interaction volumes, which consequently degrades the spatial resolution of the X-ray image and increases the analytical volume (figure 4a).

It is generally not possible to obtain high spatial resolution chemical microanalysis from bulk materials when using high accelerating voltages [18]. However, qualitative X-ray maps can highlight small features on the nanometre level, especially if averaging and filtering techniques are employed to resolve the nano-features.

With the development and advancement of SDD's, the sensitivity for X-rays in the low energy part of the spectrum has substantially improved. Manufacturers are now producing SDD's with such large detector areas, that greater X-ray collection efficiencies are possible. This, and high count rate throughput, is now allowing the possibility of operating the SEM at much lower accelerating voltages and subsequently reducing the interaction volume of the electron beam with the material as well as achieving higher spatial resolution information (figure 4). Furthermore, SDD's can now be purchased with varying windows [30] and also windowless systems [25-29] allowing greater sensitivity of the lower energy X-rays. Figure 8 shows the EDS spectrum from an Amptek detector with two different window materials (C1 and C2). These windows are made from 90 nm and 40 nm silicon nitride (Si_3N_4) with a very thin aluminium coating to extend the low energy response [26]. The C2 window has excellent transmission for low Z elements with ten times more carbon X-rays (43.9 %) and 29 % transmission of Li [26].

Just becoming commercially available are new soft X-ray spectroscopy (SXES) systems, with a measurable energy range of 50 to 210 eV, an order of magnitude better energy resolution (with 0.3 eV resolution reported), and detection limits for light elements of less than 100 ppm [30].

By lowering the accelerating voltage a major improvement in the spatial resolution and image quality occurs. This forces the selection of X-ray lines with low excitation energy such as L- and M-family lines between 0 - 4 kV, rather than using K- and L-lines between 0 - 10 kV [1, 18, 25]. The

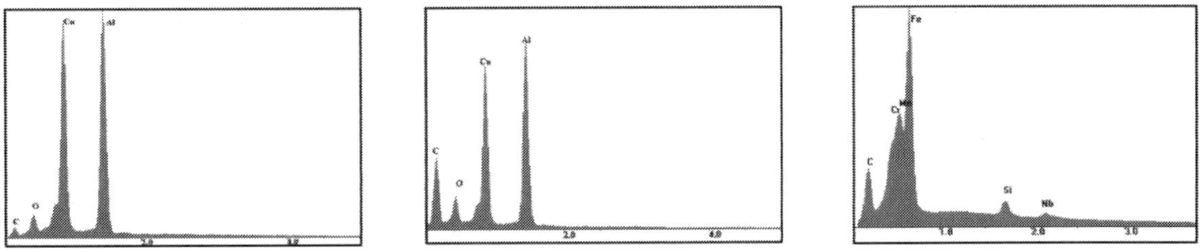

a) b) c)

Figure 8. Intensity versus energy spectra from an Amptek Fast123 detector, collected at E_0 = 8kV. a) C1 window of a contaminated Cu-Al grid, b) C2 window of a contaminated Cu-Al grid (same area), and c) Spectra (from C2 window detector) of a white cast iron hard-facing material that has NbC particles present. X-axis represents energy (kV).

measurement of low energy L-line X-rays is complicated by low fluorescence yield, an increase in X-ray absorption, numerous X-ray interferences from other elements within the sample, and less accurately determined mass absorption coefficients [25-29]. The overlaps of K-line X-rays from light elements with L- and M-line X-rays from heavier elements limit the low voltage analysis and mapping capabilities of conventional microanalysis systems [35].

The use of transition element L-lines generally results in poor accuracy as well as poor sensitivity [21, 31, 32]. Furthermore, operating at low voltages has limitations due to the large quantification errors existing on the lower energy X-ray lines [34]. Extensive work is still required on fundamental parameters [31], algorithms and mass absorption coefficients before any accurate quantitative analysis can be performed [32-34], especially with the many overlaps that exist in the low voltage regime of multi-element materials. With multielement materials that have overlaps, at best low voltage microanalysis is only good for qualitative analysis.

For quantitative analysis, standards based analysis is the better way to go [31]. As can be seen in table 2, standardless analysis has many more parameters to be determined for accurate analysis compared to standards based analysis. All low voltage analysis will give better results if using standards based analysis [31], however there are still errors. Wavelength-dispersive X-ray spectroscopy (WDS) also has problems in the light element region due to high order overlaps.

Table 2. Factors affecting standards and standardless analysis.

Standards	**Standardless**	
Common to both:		
Gain and Zero, kV, Take Off Angle, Overlaps, Beam Current and Count Rate		
	Resolution	**Fano Factor**
Dead Time	**Dead Layer**	**Detector Thickness**
Live Time	**Gold Layer**	**Window Thickness**
Intensity Factors	**Sum Peak Calculations**	**Escape Peaks**
Standards	**Carbon Coating Thickness**	
	Incomplete Charge Collection	
	Other absorbing layers	

7. Advantages of low voltage X-ray mapping

Characterisation is an essential aspect of materials research and of quality control in materials production. Characterisation of materials frequently involves the determination of point-to-point variation in composition, structure and microstructure, so a variety of imaging and analysis techniques come into play [36, 37]. Understanding the distribution of elements and phases in structures is critical to optimising the performance of all materials. Subtle changes in composition have been studied for many years to determine the properties of many materials and aid in the production of these materials [35-37]. This is quite often done by performing microanalysis and also XRM [35, 40-45]. XRM is an extremely useful problem solving tool which involves the collection of characteristic X-rays as a function of the position of the scanning electron beam on the specimen [36-39]. XRM provides an image related to the distribution and relative abundance of elements within a given specimen and makes XRM particularly useful for; i) identifying the location of individual elements, and ii) mapping the spatial distribution of specific elements and chemical phases within a sample (figure 9) [40]. This is especially true with low voltage X-ray mapping, due to the higher spatial resolution [1], allowing nanolevel features being detected [46, 47]. However, the problems with multi-element materials are that overlaps can exist, especially with the L-lines. As can be seen in figure 8c, the EDS spectra shows the hard facing material of NbC particles in white cast iron, has numerous L-lines overlapping each other. The subsequent generation of X-ray maps collected at 8 kV using K-lines (figure 9b-e) and L-lines (figure 9f, h-j) shows major errors for Mn L-line. Furthermore, the quantification results using the L-lines are totally incorrect. This can minimized by more precise calibration of the detector parameters (due to non-linearities in the detector electronics) to enable better correction of the overlaps in the low energy range.

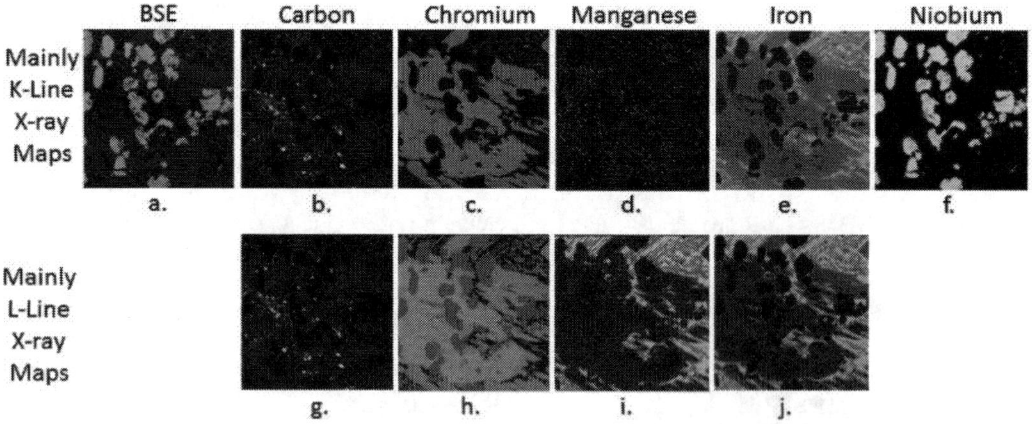

Figure 9. X-ray maps taken at 8 kV of hard-facing material with NbC particles in white cast iron. a) BSE image, b) Carbon K-line, c) Chromium K-line, d) Manganese K-line, e) Iron K-line, f) Niobium L-line, g) Carbon K-line, h) Chromium L-line, i) Manganese L-line, and j) Iron L-line. Notice the Mn X-ray map for the K-line and L-line X-ray maps are totally different, due to the linearity and overlap problems occurring at the L-lines for Cr, Mn, and Fe.

If utmost care is taken with collecting and processing X-rays maps, especially at low voltages, then further post processing (data manipulation) techniques may be performed [39], even if it is only used as a first approximation to determine a qualitative result. These post-processing techniques have been extensively covered by the authors in previous publications and include [36-45]:

1. Generating scatter diagrams and rotational scatter diagrams [39, 40] for locating phases.
2. Pseudo colouring (figure 10) for locating phases. The colours of the elements can be rotated, which can often reveal further information (and other phases) in the material. Indeed, the rotating of colours between elements at low voltage -ray mapping has aided revealing fine precipitates and small boundary interfaces that would be otherwise missed.
3. Filtering to make phase selection easier, to produce cleaner images and aid in trace element location as well as aiding in finding small precipitates.
4. Elemental ratio maps. This is a powerful technique that helps to reveal subtle chemical variations in a sample. Ratio maps can also be created from different combinations of elements and from the different background regions near elements of the spectra [38, 39].
5. Totals mapping. The totals at each point of the map can be calculated and mapped (when standards based analysis is performed). These total maps give an idea of the quality of the analysis and also make it easy to determine problems such as missing elements and interaction volume effects [39, 43, 45].

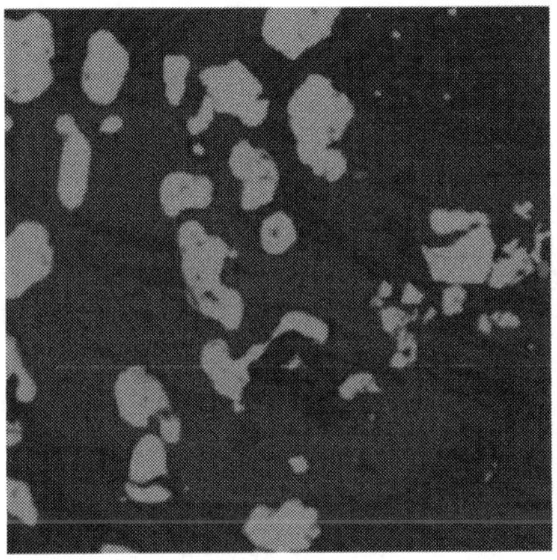

Figure 10. Pseudo coloured X-ray map of Cr (red), Nb (green) and Fe (blue). Pseudo colouring is a simple method for determining elemental associations and is also called "Primary Colour Imaging". This technique is where three elemental maps have been assigned the colours red, green and blue. Because the colours red, green and blue form specific colours when combined on a computer graphics monitor, elemental associations are clearly shown in a single "pseudo" colour image. Note, that the original X-ray maps used prior to colouring are normalized to produce good grayscale images.

8. Future improvements

The future SEM will be able to operate in many different modes, whether it is VP SEM, ESEM, low voltage and even high voltage SEM. The new Schottky emitter field emission gun microscopes are capable of 1nm resolution or better at 15 kV and can operate in VP mode and the emitters can produce greater beam currents (> 100 nA) that allow these instruments to have excellent analytical capabilities, which is of great importance for EDS, WDS, XRM and EBSD analysis. By incorporating different column designs, larger apertures, greater beam currents can be obtained, with minimal resolution loss. Obtain better and faster EBSD maps than with conventional tungsten instrument with XRM and GSR analysis completed a lot faster. With the new imaging detectors, new SDD detectors and new EBSD detectors working at closer working distances, higher resolution images and maps can be obtained no matter what mode the SEM operates in. There is currently a whole generation of detectors available that have better efficiency and can collect many different type of signals. Some of these detectors can operate under various conditions. New SEMs should be able to simultaneous collect images from all detectors BSE, SE and gaseous secondary (GSE), specimen current imaging, in-lens BSE/SE and X-rays.

Besides instruments capable of operating in a number of modes with different detectors, there are still a lot of advances for the instrument to be developed. Systems that have energy filtering, spectroscopy imaging, multi-detector systems (SDD, WDS, low energy detectors, CL) as well as correction systems on SEM columns need to be developed [48-50]. More importantly, systems have to be developed that allow for self-correcting microscopes (automated system calibration) [48-50].

9. Conclusion

There have been significant advances in the field of scanning electron microscopy, especially with electron column design, stable emitters, different SE and BSE detectors and different X-ray detectors with and without windows.

Low voltage imaging and X-ray analysis in the SEM have many challenges ahead but, there are many more opportunities and applications being developed. Low voltage microanalysis has a distinct disadvantage in that fewer X-ray lines are available and the ones that do exist quite often overlap with lines from other elements. However, if a careful approach is taken, some excellent results can be obtained on a nanometre level, even if it is only a qualitative approach.

Acknowledgements

The authors would like to thank the many that have contributed information and advice towards this paper. These include Ron Rasch (UQ), Jo Michaels (Sandia National Laboratories), Brenden Griffin (UWA), Matthew Phillips (UTS), Natasha Erdman (JEOL USA), Heiner Jaksch (Zeiss), David Bell (Harvard University), David Joy (Oak Ridge National Laboratory), Raynald Gauvin (Mc Gill), Dale Newbury (NIST), Ken Mason (Eastern Analytical), Darren Attard (Zeiss), FEI, Zeiss and JEOL. The authors wish to thank the Advanced Materials Characterisation Facility (AMC Facility) at UWS.

References

[1] Bell D C and Erdman N 2013 *Low Voltage Electron Microscopy*. (New York: Wiley)

[2] Bell D C, Mankin M, Day R W and Erdman N 2014 Successful application of Low Voltage Electron Microscopy to practical materials problems. *Ultramicroscopy* **145** 56-65

[3] Roussel L Y, Stokes D J, Gestmann I, Darus M and Young R J 2009 *Extreme high resolution scanning electron microscopy (XHR SEM) and beyond*. in: Scanning Microscopy, Proc. SPIE, vol. 7378 73780W, 1-9

[4] Young R J, Templeton T, Roussel L Y, Gestmann I, van Veen G, Dingle T and Henstra S 2008 Extreme High Resolution SEM: a paradigm shift. *Microscopy Today* **16**(4) 24-28

[5] Gauvin R, Brodusch N, Demers H and Woo P 2015 *Imaging and quantitative x-ray microanalysis at the nanoscale with a cold-field emission scanning electron microscope*. in: AMAS XIII - 13th Biennial Australian Microbeam Analysis Symposium (Hobart, Tasmania; February 2015) 32-33

[6] Griffin B J, Joy D C, Michael J R and Gauvin R 2015 *Some further aspects of secondary electron and backscattered electron imaging in the SEM and VPSEM*. in: AMAS XIII - 13th Biennial Australian Microbeam Analysis Symposium (Hobart, Tasmania; February 2015) 38-39

[7] Rasch R, Stricher A, Wood B and Truss R 2014 *Low Voltage In-Lens Backscatter SEM Imaging of Natural Fibre Polymers with Organic Surface Treatments*. in: ACMM 23 (Adelaide, South Australia)

[8] Rasch R, Stricher A and Truss R 2013 Energy Filtered Low Voltage In Lens Detector SEM and XPS of Natural Fiber Surfaces. *J. Appl. Polym. Sci.* **131** 39572

[9] Danilatos G D 1988 Foundation of environmental scanning electron microscopy. *Adv. Electronics Electr. Phys.* **71** 109-250

[10] Johnson R 1996 *Environmental Scanning Electron Microscopy-An Introduction to ESEM, Second Edition*. (Eindhoven, The Netherlands: Philips Electron Optics)

[11] Stokes D 2008 *Principles and Practice of Variable Pressure: Environmental Scanning Electron Microscopy (VP-ESEM)*. (New York: Wiley)

[12] Joy D C and Joy C S 1996 Low Voltage Scanning Electron Microscopy. *Micron* **27** 247-263

[13] Joy D C 2006 *The Hows, Whys, and Woes of the Low Voltage SEM*. in: ACMM-19 (Sydney, New South Wales)

[14] Joy D C 2006 *Charging Control*. in: ACMM-19 (Sydney, New South Wales)

[15] Joy D C 1997 Scanning electron microscopy for materials characterization. *Current Opinion Solid State Mater. Sci.* **2** 465-468
[16] Drummy L F, Yang J and Martin D C 2004 Low voltage electron microscopy of polymer and organic molecular thin films. *Ultramicroscopy* **99** 247-256
[17] Drummy L F 2014 Electron microscopy of organic–inorganic interfaces: Advantages of low voltage. *Ultramicroscopy* **145** 74-79
[18] Liu J 2000 High-Resolution and Low Voltage FE-SEM Imaging and Microanalysis in Materials Characterization. *Mater. Character.* **44** 353-363
[19] Michael J R, Joy D C and Griffin B J 2009 Challenges in achieving high resolution at low voltages in the SEM. *Microsc. Microanal.* **15** (Suppl. S2) 660-661
[20] Dusevich V M, Purk J H and Eick J D 2010 Choosing the Right Accelerating Voltage for SEM (An Introduction for Beginners). *Microscopy Today* **18** 48-52
[21] Goldstein J I, et al. 2003 *Scanning Electron Microscopy and X-ray Microanalysis.* (New York: Springer)
[22] Assa'd A M D and El Gomati M M 1998 Backscattering Coefficients for Low Energy Electrons. *Scanning Microsc.* **12** 185-192
[23] Cazaux J, Kuwano N and Sato K 2013 Backscattered electron imaging at low emerging angles: A physical approach to contrast in LVSEM. *Ultramicroscopy* **135** 43-49
[24] Bongeler R, Golla U, Kassens M, Reimer L, Schindeler B, Senkel R and Spranck M 1993 Electron-specimen interactions in low-voltage scanning electron microscopy. *Scanning* **15** 1-18
[25] Meisnara M, Lozano-Pereza S, Moodya M and Holland J 2014 Low energy EDX – A novel approach to study stress corrosion cracking in SUS304 stainless steel via scanning electron microscopy. *Micron* **66** 16-22
[26] http://www.amptek.com/products/c-series-low-energy-x-ray-windows/
[27] Wuhrer R, Guja L, Merritt D and Moran K 2014 X-Ray Mapping Investigations of Salt Migration in Seeds through use of Window and Windowless Silicon Drift Detectors. *Microsc. Microanal.* **20** (Suppl. 3) 634-635
[28] Burgess S, Li X and Holland J 2013 Improving low energy sensitivity of EDS detectors for SEM – higher spatial resolution X-ray analysis and detection of light elements including lithium. *Microsc. Analysis* **27** 8
[29] Moran L, Moran K and Wuhrer R 2014 X-ray Mapping and Chemical Phase Mapping with an Amptek SDD. *Microsc. Microanal.* **20** (Suppl. 3) 678-679
[30] Takahashi H, Murano T, Takakura M, Terauchi M, Dickinson C and Lozbin V 2015 *Soft X-ray emission spectroscopy with higher energy resolution and detectability from lithium to uranium.* in: AMAS XIII - 13th Biennial Australian Microbeam Analysis Symposium (Hobart, Tasmania; February 2015) 126-127
[31] Gauvin R 2012 What remains to be done to allow quantitative X-ray microanalysis performed with EDS to become a true characterization technique? *Microsc. Microanal.* **18** 915-940
[32] Carpenter P K 2015 *Improvements in EPMA: Spatial resolution and analytical accuracy.* in: AMAS XIII - 13th Biennial Australian Microbeam Analysis Symposium (Hobart, Tasmania; February 2015) 132-133
[33] Donovan J, Pinard P and Richter S 2015 *High speed matrix and secondary fluorescence boundary corrections based on Monte-Carlo derived k-ratio intensities.* in: AMAS XIII - 13th Biennial Australian Microbeam Analysis Symposium (Hobart, Tasmania; February 2015) 122-123
[34] Pinard P and Richter S 2015 *Methodology for automated quantitative measurements using a constant overvoltage ratio with a field emission microprobe.* in: AMAS XIII - 13th Biennial Australian Microbeam Analysis Symposium (Hobart, Tasmania; February 2015) 124-125
[35] Friel J J and Lyman C E 2006 X-ray Mapping in Electron-Beam Instruments. *Microsc. Microanal.* **12** 2-15

[36] Wuhrer R and Moran K 2006 Representation of Microstructures through Rapid X-Ray Mapping. *Materials Australia* **38**(5) 12-14

[37] Wuhrer R, Moran K and Moran L 2006 Characterisation of Materials through X-ray Mapping. *Materials Forum* **30** 63-70

[38] Moran K and Wuhrer R 2006 Quantitative Bulk and Trace Element X-Ray Mapping Using Multiple Detectors. *Mikrochim. Acta* **155** 59-66

[39] Wuhrer R and Moran K 2014 Quantitative X-ray mapping, scatter diagrams and the generation of correction maps to obtain more information about your material. *IOP Conf. Ser.: Mater. Sci. Engng.* **55** 012021

[40] Moran K and Wuhrer R 2006 X-ray Mapping and Interpretation of Scatter Diagrams. *Mikrochim. Acta* **155** 209-217

[41] Wuhrer R and Moran K 2009 Artifacts in Qualitative and Quantitative X-Ray Mapping. *Microsc. Microanal.* **15** (Suppl. 2) 30-31

[42] Wuhrer R and Moran K 2013 Post processing hyper-spectral data and generating more information from X-ray maps. *Microsc. Microanal.* **19** (Suppl. 2) 828-829

[43] Moran K and Wuhrer R 2010 Post Processing of X-Ray Maps - Theoretical BSE Mapping, Totals Mapping and Ratio Mapping. *Microsc. Microanal.* **16** (Suppl. 2) 940-941

[44] Moran K and Wuhrer R 2010 So, What Other Information can we Obtain from X-Ray Maps. *Microsc. Microanal.* **16** (Suppl. 2) 916-917

[45] Wuhrer R, Moran K and Phillips M R 2008 Multi-Detector X-Ray Mapping and Generation of Correction Factor Images for Problem Solving. *Microsc. Microanal.* **14** (Suppl. 2) 1108CD-1109CD

[46] Brodush N, Trudeau M L, Michaud P, Boselli J and Gauvin R 2012 Contribution of a New Generation Field-Emission Scanning Electron Microscope in the Understanding of a 2099 Al-Li Alloy. *Microsc. Microanal.* **18** 1393-1409

[47] Guinel M J F, Brodusch N, Verde-Gomez Y, Escobar-Morales B and Gauvin R 2013 Multi-walled carbon nanotubes decorated by platinum catalyst nanoparticles - Examination and microanalysis using scanning and transmission electron microscopies. *J. Microsc.* **252** 49-57

[48] Wuhrer R and Mason K 2012 Automated SEM and EDS System to Monitor Instrument Health and Calibration Accuracy for Forensic Science. *Microsc. Microanal.* **18** (Suppl. 2) 28-29

[49] Wuhrer R and Mason K 2012 *Using a System Called microValidator to Monitor Instrument Health and Calibration Accuracy for Forensic Science.* in: Proc. APMC 10 / ICONN 2012 / ACMM 22 (Perth, Western Australia), 1129-1

[50] Wuhrer R, Moran K and Mason K 2012 *Remote Microscopy and Intelligent Control Systems used to Monitor Instrument Health and Calibration.* in: Proc. APMC 10 / ICONN 2012 / ACMM 22 (Perth, Western Australia) 313-1

Pyrite-pyrrhotite intergrowths in calcite marble from Bistriški Vintgar, Slovenia

J Zavašnik

Jožef Stefan Institute, Center for Electron Microscopy and Microanalysis, Jamova cesta 39, 1000 Ljubljana, Slovenia

E-mail: janez.zavasnik@ijs.si

Abstract. Roman marble quarry in Bistrica gorge in southern Pohorje Mt. (north-eastern Slovenia) is situated in a 20 m thick lens of layered marble, at the contact zone between granodiorite and metamorphites. Grey and yellowish non-homogenous calcite marble is heavily included by mica, quartz, feldspars, zoisite, pyrite and amphiboles. In the present research, we have studied numerous pyrite (FeS_2) crystals associated with yellowish-bronze non-stoichiometric pyrrhotite ($Fe_{1-x}S$), not previously reported from this locality. SEM investigation revealed unusual sequence of crystallisation: primary skeletal pyrrhotite matrix is sparsely overgrown by well-crystalline pyrite, both being overgrown by smaller, well-developed hexagonal pyrrhotite crystals of the second generation. With TEM we identify the pyrrhotite as 5T-$Fe_{1-x}S$ phase, where x is about 0.1 and is equivalent to Fe_9S_{10}. The pyrite-pyrrhotite coexistence allows us a construction of f_{O_2}-pH diagram of stability fields, which reflects geochemical conditions at the time of marble re-crystallisation.

1. Introduction

Calcite marble from old Roman quarry in southern Pohorje Mt. (north-eastern Slovenia) was popular sculpturing material at the time of Trajan and Severi dynasty (96-235 AD) [1]. Due to limited resources of quality material and complicated extraction and transport methods it was only used in near vicinity; some of the tombs and statues are still preserved in archaeological parks near Celje, Šempeter and Ptuj. The original source of the marble was determined by similarity in grain size, chemical composition and characteristic assemblages of the accessory minerals [2, 3]. Marble, investigated in this research, originated from small, partially underground quarry which is still visible in Bistrica gorge (figure 1a). Pohorje metamorphic complex represent SE margin of Eastern Alps. It has complex metamorphic history with at least two stages of progressive and one stage of retrograde metamorphism [4]. Tectonic model proposes that oceanic lithosphere broke off from continental lithosphere during continental collision, explaining syn- to post-collisional magmatic and metamorphic episodes around the Periadriatic Fault and large granodiorite intrusion into central part of massif [5] (figure 1b). The majority of marble outcrops are found in E and S in medium-grade metamorphic rocks, represented by gneiss, mica schists and amphibolites. Up to now Pohorje marbles were investigated in general geological studies of Pohorje massive, where their origin and relation to metamorphic rocks were determined [4, 6, 7]. More recent researches were investigating stable isotopes characterisation focussed on the comparison between Austrian and Slovenian marbles with known ancient quarries in the Mediterranean area [2]. Furthermore the specific Roman quarry was included in extensive cathodolumniscence and petrographical study, in order to detect their local variability in texture and mineral assemblages in attempt to relate the textures to mineral reactions during the metamorphic process [8]. In latter research mineral inclusions were also identified as mica, quartz, hornblende, feldspars, zoisite, amphiboles and pyrite and from the stability of this mineral assemblage the temperature of marble re-crystallisation was estimated at < 500 °C [2].

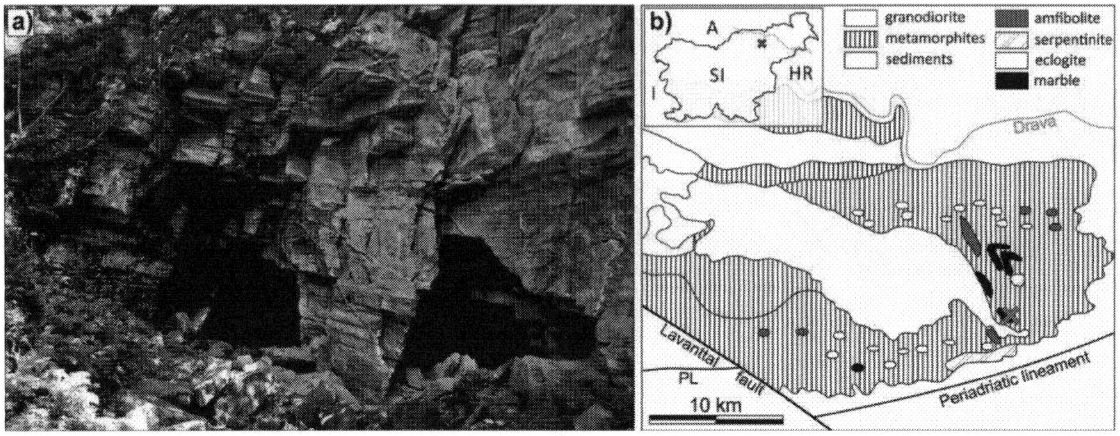

Figure 1. a) Partially underground marble quarry. Support pillar with numerous aplite veins is visible in central part of the image. Specimens for analysis were taken from bulk rock in lower part of the pillar. b) General overview of geological settings in Pohorje complex; marble outcrops are concentrated in SE of the region. Investigated outcrop is marked by red x.

During exploitation of marble from considered Roman quarry heavily included layers and pillars were left in its original position, as they were not useful for further processing. While mica, carbonate and amphibole inclusions are inert, oxidation of Fe-sulphide inclusions results in newly released sulphate, which combines with water to produce sulphuric acid, leading to rapid disintegration of the calcite marble. On the other hand preserved Fe-sulphides can be used as a geothermobarometer, which can provide us further information about conditions at the formation of the marble. In order to do so Fe-sulphides and other accessory minerals were investigated with a combination of analytical methods.

2. Materials and methods

Samples with numerous visible inclusions were taken from the central part of quarry, near main support pillar (figure 1a). Chunks of fresh, non-oxidized calcite marble were completely dissolved in 2 M HCl. Remaining insoluble inclusions were selected under stereo-microscope by hand-picking regarding their colour, transparency and crystal habit. Phase composition of the samples was analysed by powder X-ray diffractometer (XRD; Bruker AXS D4 Endeavor, Bruker AXS GmbH) using Ni-filtered Siemens KFL Cu-Kα 50 kV X-ray tube. The spectra were acquired in the range $2\theta = 5° - 75°$ with a step of 0.04° and recording time of 1 s/step. Raw data were interpreted by Philips X'Pert HighScore Software Suite using WebPDF-4+ 2014 database (The International Centre for Diffraction Data).

Chemical composition and morphology of the samples were studied by scanning electron microscope (SEM; JSM-5800, JEOL Inc.) operated at 20 kV and equipped with energy-dispersive X-ray spectrometer (EDS; Oxford Link Pentafet Mod. 6841, Oxford Instruments PLC). Prior to SEM analysis the samples were coated with a conductive carbon layer using a Balzers SCD 050 Sputter Coater (Bal-Tec AG). EDS data were compiled using Link ISIS system (Oxford Instruments) and S/Fe ratio was calculated using the accompanying SEMQuant ZAF-correction software (Z: atomic number, A: X-ray absorption, F: secondary X-ray fluorescence). Synthetic pure samples with similar S/Fe ratio (mackinawite-FeS [10] and pyrite-FeS_2 [9]) were used for calibration prior to each session; the standard with known S/Fe ratio was measured and ZAF evaluated.

Structure and chemistry of the Fe-sulphide intergrowths were studied using a 200 kV transmission electron microscope (TEM; JEM-2100, JEOL Ltd.), equipped with an energy-dispersive X-ray spectrometer (EDS; JED-2300 analyser, JEOL Inc.) suited for semi-quantitative chemical analysis. Samples for TEM analyses were mounted in 3 mm Ti-ring, thinned to 100 μm, dimpled at the disc centre down to 10 μm and further ion-milled (Bal-Tec RES 010 Rapid etching system, Bal-Tec AG) at 4 kV by Ar^+ ions until perforation.

3. Results and discussion

XRD analysis of the yellowish golden and bronze fragments (less than 2 mm in size), shows that they corresponded to a mixture of pyrite and pyrrhotite, which cannot be further mechanically separated (figure 2a). Thin brownish semi-transparent plate-like hexagonal crystals corresponded to illite (figure 2b), while white opaque, up to 4 mm big irregular chunks corresponded to a mixture of dolomite and quartz (figure 2c). Residual green up to 2 mm big elongated semi-transparent crystals corresponded to the amphibole pargasite (figure 2d). The coexistence of pyrite and pyrrhotite allows us a determination of the possible stability field and the geochemical conditions at the time of marble re-crystallisation.

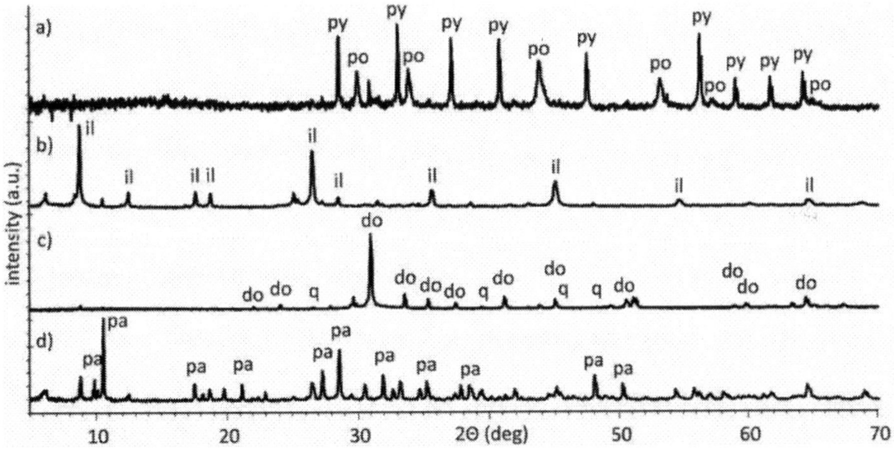

Figure 2. XRD spectra of insoluble residues. The sulphide agglomerations correspond to pyrite (py) (JCPDS card # 04-1003) and pyrrhotite (po) (# 29-0724) mixture, mica sheets to illite (il) (# 58-2017), white irregular chunks to dolomite (do) (# 75-1759) and quartz (q) (# 15-7199) mixture and d) to amphibole pargasite (pa) (# 23-1406).

Due to small size of the pyrite-pyrrhotite intergrowths their relationship was investigated in SEM. EDS analyses, collected in Table 1, show significant difference between both Fe-sulphide phases: (SEM) data were acquired in SEM on the surface of the crystals, (SEM2) data were collected on polished and Ar^+ etched cross-sections from the central part of the samples and (TEM) data were acquired in TEM on conventionally prepared sample. EDS analyses of primary pyrrhotite and pyrite surface shows slight sulphur deficiency. We attribute this inconsistency to partial surface decomposition, as thin needle-like crystals of gypsum are often found attached on and near pyrite crystals. As gypsum was not detected by XRD analysis, it was most probably formed during air exposure or due to instability of Fe-sulphides in HCl. On the other hand, EDS analyses made on the polished central part of the both Fe-sulphides constantly indicate iron deficiency. Since (at least) pyrite shows no visible alterations, we can expect near-stoichiometric composition and we attribute this Fe-depletion to the near-surface structural damage, induced during Ar^{+} etching. EDS analyses of the exact same samples, made in TEM, further confirmed our assumptions, as they showed almost stoichiometric composition for pyrite and pyrrhotite.

Table 1. EDS analyses of sulphide crystals, normalized [at%].

Pyrite	ref. [11]	SEM	SEM2	TEM	*Pyrrhotite*	ref. [12]	SEM	SEM2	TEM
S	66.7	61.3	69.1	65.5	S	52.6	41.5	54.9	49.1
Fe	33.3	38.7	30.9	34.0	Fe	47.4	58.5	45.1	50.9
Total	100.0	100.0	100.0	100.0	Total	100.0	100.0	100.0	100.0

SEM observation of Fe-sulphide intergrowths revealed unusual sequence of crystallisation: primary skeletal pyrrhotite (po1) is overgrown by well crystallized first-generation pyrite (py1), both being sparsely covered by small hexagonal pyrrhotite of second generation (po2). Almost all pyrite cubes of first generation are modified by (111) faces, while pyrite cubes of second generation (py2), modified by (110) and (111) faces, are rare and are overgrowing second generation of pyrrhotite (figure 3). According to literature [13, 14], the {100} faces attain the lowest energy at sulphur-poor and iron-rich conditions and with increasing amounts of available sulphur the {210}, {110} and {111} faces prevail [9]. Based on sequence of crystallisation and our observations in SEM, sulphide crystallisation begins with fast, irregular and skeletal pyrrhotite growth in sulphur-poor conditions. Subsequent exsolution of pyrite from pyrrhotite begins as a result of cooling, forming typical euhedral cubes rather than exsolution lamellae, in a process known as non-coherent exsolution [15]. Second generations of both Fe-sulphides are result of local variations in S- and Fe-availability or small temperature fluctuations.

To further identify possible structure variations in pyrite and pyrrhotite we investigated the Fe-sulphide intergrowths by TEM. TEM analysis of pyrrhotite was made on single pyrrhotite crystal of first generation, partially overgrown by pyrite. Pyrite showed no deviations from the normal pyrite structure and corresponds to JCPDS pattern # 04-1003. The electron diffraction (ED) pattern of pyrrhotite shows spacings of 2.04, 1.72 and 2.63 Å due to (2·0·10), (220), and (205) planes, respectively. The *d*(220) spots in ED pattern show splitting due to superstructure of $Fe_{1-x}S$. The Fe-S system has a complex phase diagram, with hexagonal Fe_7S_8 as most defined and stable phase. Our sample, based on HR-TEM images and ED pattern (figure 4), does not correspond to *hcp* lattice, but it does correspond with 5T-$Fe_{1-x}S$ phase (card # 29-0724) (Table 2). Pyrrhotite 5T, for the formula $Fe_{1-x}S$ where x is about 0.1 and is equivalent to Fe_9S_{10} with Z = 4, is characteristic for low-temperature systems [12]. The crystal system's true symmetry is orthorhombic [12] and sulphur atoms are arranged in an approximately hexagonal close-packed array, with iron atoms occupying octahedral interstices between the layers, where one in 8 of the octahedral interstices is vacant. Consequently this can be regarded as a "defect structure", with many options for disordering of the vacancies.

Table 2. Experimental and reference (JCPDS # 29-0724) d-values for pyrrhotite.

Pyrrhotite	exp.[Å]	ref.[Å] [12]
220	1.7	1.723
2·0·10	2.0	2.067
205	2.6	2.647
200	2.9	2.98

Figure 3. SEM micrographs of pyrite-pyrrhotite intergrowths. Primary pyrrhotite (po_1) is overgrown by first generation of pyrite (py_1), followed by second generation of pyrrhotite (po_2). Second generation of pyrite (py_2) crystallized last.

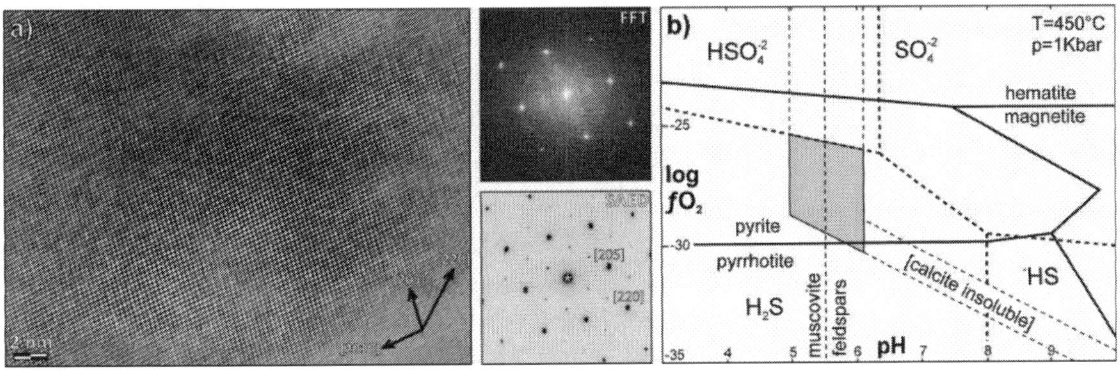

Figure 4. a) HR-TEM image of pyrrhotite ($Fe_{1-x}S$) single crystal; insets are showing FFT and experimental ED patterns. b) Log fO_2-pH diagram showing stability fields for minerals found in investigated marble [16]. Grey area represents most possible conditions of mineralising fluids.

4. Conclusions

Our research of the accessory mineral associations in unaltered marble from Bistriški Vintgar Roman quarry showed that re-crystallisation temperature was as low as < 500 °C. Newly discovered pyrrhotite improved resolution of possible stability fields at the time of marble re-crystallisation, since pyrite-pyrrhotite coexistence is possible in a very narrow window. The main conclusions can be summarized as follows:

- Non-sulphide inclusions are illite, dolomite, quartz and amphibole.
- Primary sulphides are a mixture of pyrite and pyrrhotite.
- Pyrite exsoluted from pyrrhotite as a result of cooling.
- Pyrite morphology indicate S-poor and Fe-rich conditions at the time of crystallisation.
- Investigated pyrrhotite correspont to 5T-$Fe_{1-x}S$ variety.

Acknowledgments

Help of asst. Prof. Srečo D. Škapin (IJS, K9) with XRD analyses is greatly acknowledged.

References

[1] Djurič B 2001 *Archaeologica Poetovionensis* **2** 61-70
[2] Jarc S, Maniatis Y, Dotsika E, Tambakopoulos D and Zupančič N 2010 *Archaeometry* **52** 177-190
[3] Djurič B, Hebert B, Hinker C, Hudeczek E, Karl S and Müller H 2004 *Fundberichte aus Österreich* **43** 365-431
[4] Hinterlechner-Ravnik A and Moine B 1977 *Geologija* **20** 107-140
[5] Bonin B 2004 *Lithos* **78** 1-24
[6] Hinterlechner-Ravnik A 1971 *Geologija* **14** 187-226
[7] Hinterlechner-Ravnik A, Sassi F and Visona D 1991 *Rend. Fisiche Accademia Lincei* **2** 175-190
[8] Jarc S and Zupančič N 2009 *Chemie der Erde Geochem.* **69** 75-80
[9] Zavašnik J and Rečnik A 2013 *J. Crystal Growth* **367** 18-23
[10] Zavašnik J, Stankovič N, Arshad S M and Rečnik A 2014 *J. Nanoparticle Res.* **16** 2222-2235
[11] Bither T A, Bouchard R J, Cloud W H, Donohue P C and Siemons W J 1968 *Inorg. Chem.* **7** 2208-2220
[12] Morimoto N, Gyobu A, Mukaiyama H and Izawa E 1975 *Econ. Geol.* **70** 824-833
[13] Alfonso D R 2010 *J. Phys. Chem. C* **114** 8971-8980
[14] Bernard A and Russo S 2009 *J. Mater. Chem.* **19** 3389-3394
[15] Craig J R and Vaughan D J 1994 *Ore Microscopy & Ore Petrography (2nd edition).* (New York, NY: John Wiley & Sons) 101-152
[16] Maanijou M, Rasa I and Lentz D 2013 *Econ. Geol.* **107** 683-712